Sustainable Preparation of Metal Nanoparticles
Methods and Applications

RSC Green Chemistry

Series Editors:
James H Clark, *Department of Chemistry, University of York, UK*
George A Kraus, *Department of Chemistry, Iowa State University, Ames, Iowa, USA*
Andrzej Stankiewicz, *Delft University of Technology, The Netherlands*
Peter Siedl, *Federal University of Rio de Janeiro, Brazil*
Yuan Kou, *Peking University, People's Republic of China*

Titles in the Series:

How to obtain future titles on publication:
A standing order plan is available for this series. A standing order will bring delivery of each new volume immediately on publication.

For further information please contact:
Book Sales Department, Royal Society of Chemistry, Thomas Graham House, Science Park, Milton Road, Cambridge, CB4 0WF, UK
Telephone: +44 (0)1223 420066, Fax: +44 (0)1223 420247
Email: booksales@rsc.org
Visit our website at http://www.rsc.org/books

Sustainable Preparation of Metal Nanoparticles
Methods and Applications

Edited by

Rafael Luque
Departamento de Química Orgánica, Universidad de Córdoba, Spain
Email: q62alsor@uco.es

Rajender S Varma
National Risk Management Research Laboratory, U.S. Environmental
Protection Agency, Cincinnati, USA
Email: varma.rajender@epa.gov

RSC Publishing

RSC Green Chemistry No. 19

ISBN: 978-1-84973-428-8
ISSN: 1757-7039

A catalogue record for this book is available from the British Library

Published by The Royal Society of Chemistry,
Thomas Graham House, Science Park, Milton Road,
Cambridge CB4 0WF, UK

Registered Charity Number 207890

For further information see our web site at www.rsc.org

Printed in the United Kingdom by CPI Group (UK) Ltd, Croydon, CR0 4YY, UK

Preface

Nanoscience and Nanotechnology have brought about excitement in fundamental research as well as technological advances. The word "Nano" has now become a household name. Nanomaterials can be synthesized from simple bench top methodologies all the way to advanced molecular beam epitaxy techniques. Advances made in designing new products are seen as important milestones in improving the lifestyle of developed and developing countries. Many of these products have found a niche place in the market from catalysts to consumable goods, diagnostics to drug delivery systems, and electronics to energy conversion devices. Such developments also mean that a huge production of nanoscale materials become vital to sustain the demand. An effort of this large magnitude requires changes not only in production but also in handling and transport, as well as in safety and toxicology control.

The design of semiconductor and metal nanostructures of different shapes and sizes, in particular, offers new opportunities to tailor the application of nanodevices. For example, size quantization effects in 0-D, 1-D and 3-D of semiconductors introduce unique optical and electronic properties. The use of semiconductor quantum dots in photovoltaic devices has opened up new ways to boost the efficiency of solar cells. The unique aspects such as multiple electron generation and hot electron extraction offer new opportunities to boost the efficiency of next generation of solar cells using semiconductor nanostructures. Exciton-plasmon coupling in semiconductor-metal nanostructure composites is another area of research that can aid in developing new strategies to harvest photons.

Among the large variety of nanoscale materials, metal nanoparticles are considered to be important because of the remarkable changes in their properties as compared to their bulk counterparts. Their wide range of applications is seen in diverse areas such as catalysis, biomedicine, energy conversion,

RSC Green Chemistry No. 19
Sustainable Preparation of Metal Nanoparticles: Methods and Applications
Edited by Rafael Luque and Rajender S Varma
© The Royal Society of Chemistry 2013
Published by the Royal Society of Chemistry, www.rsc.org

environmental remediation, optics or telecommunications. Such metal nano-structures with unique shapes and sizes can introduce significant enhancement in surface enhanced Raman scattering (SERS) signals, thereby enabling the detection of low level contaminants. Localized surface plasmon effects as well as quantized charging effects have been shown to improve charge separation in artificial photosynthetic and photocatalytic systems.

The production of metal nanoparticles depends on the desired applications. For example, wet chemistry methods are frequently used for biomedical applications, while gas phase deposition on solid supports is commonly employed in the preparation of catalysts and electrocatalysts. The large volume of production of such nanomaterials poses a high demand on the manufacturers to develop environmentally friendly synthetic methods. It is important not only to minimize energy consumption but also use the reactants that have negligible toxic effects. In recent years, nanosafety has become a major point of concern in manufacturing nanomaterials. The toxicity effects need to be tested for size, shape and chemical structures both during manufacture and usage by the consumers.

Researchers interested in green production and environmentally safe synthesis of metal nanoparticles will find this book highly useful. The selection of topics offers a convenient way to educate important aspects of sustainable production, safe handling, toxicology, environmental remediation and energy conversion aspects of nanomaterials.

Prof. Luis M. Liz-Marzan, University of Vigo, Spain
Prof. Prashant V. Kamat, University of Notre Dame, USA

Contents

RSC Green Chemistry No. 19
Sustainable Preparation of Metal Nanoparticles: Methods and Applications
Edited by Rafael Luque and Rajender S Varma
© The Royal Society of Chemistry 2013
Published by the Royal Society of Chemistry, www.rsc.org

CHAPTER 1
Introduction

RAFAEL LUQUE*[a] AND RAJENDER S. VARMA*[b]

[a] Departamento de Química Orgánica, Universidad de Córdoba, Campus de Rabanales, Edificio Marie Curie (C-3), Ctra Nnal IV, Km 396, Córdoba (Spain); [b] Sustainable Technology Division, National Risk Management Research Laboratory, US Environmental Protection Agency, MS 443, 26 West Martin Luther King Drive, Cincinnati, Ohio, 45268, USA
*Email: q62alsor@uco.es; varma.rajender@epa.gov

In recent years, we have experienced a "nano" revolution in which science was directly impacted with nanotechnologies forming the basis of the so-called nanoscience that are just starting nowadays to be realized as a major step forward towards future technological progress. The possibility of manipulating matter at such an ultrasmall scale (*i.e.* within the nanometer range)[1,2] has paved the way to the development of numerous nanoentities and nanosystems which currrently start to be part of our daily lives and consumer products in optics, electronic devices, sensors and even in the textile industry. The ability to directly work and control systems at the same scale as nature (*e.g.* DNA, cells) can potentially provide a very efficient approach to the production of chemicals, energy and materials (Figure 1.1).

Another important asset of nanomaterials is its inherent multidisciplinarity with a wide range of possibilities in terms of synthesis and applications that these nanoentities hold. Several subfields have been investigating nanoscale effects, properties, and applications from its infancy; every different subdiscipline is involved in modern nanoscience and technology.[2] Inputs from physicists, biologists, chemists and engineers have been a hallmark from the very early developments including the advances in nanoscience to achieve a

RSC Green Chemistry No. 19
Sustainable Preparation of Metal Nanoparticles: Methods and Applications
Edited by Rafael Luque and Rajender S Varma
© The Royal Society of Chemistry 2013
Published by the Royal Society of Chemistry, www.rsc.org

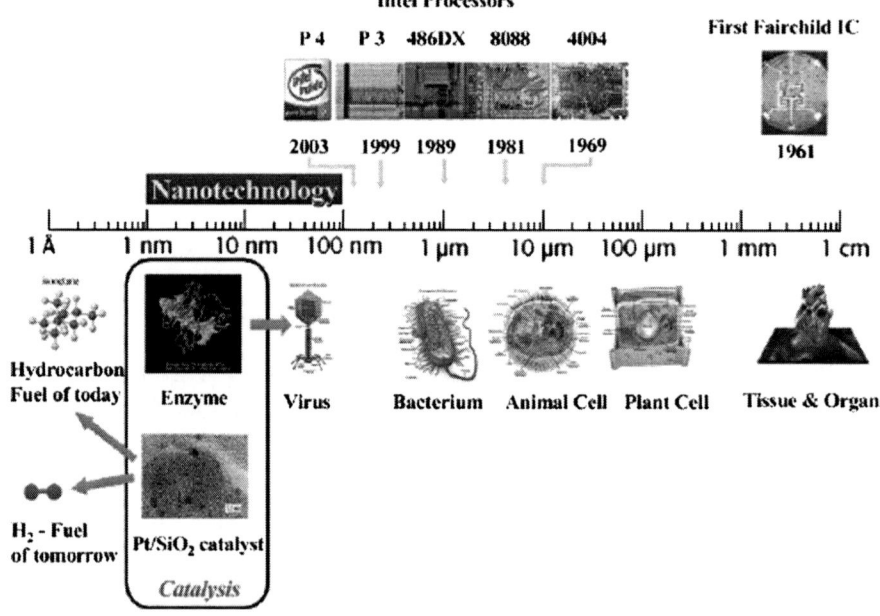

Figure 1.1   Catalysts and the nanometer regime.[2b]
                 Reproduced by permission of the Royal Society of Chemistry.

better understanding of the preparation, application and impact of these new nanotechnologies.

A *nanoparticle* can be generally defined as a particle that has a structure in which at least one of its phases has one or more dimensions in the nanometer size range (1 to 100 nm, Figure 1.1). Nanoparticles (NPs) have remarkably different properties as compared to their bulk equivalents that mainly include a degenerated density of energy states (as compared to bulk metals) and a large surface to volume ratio together with the sizes in the nanometer scale.[1–3] These nanoparticles have associated remarkable properties including a relatively high chemical activity and specificity of interaction as compared to bulk metals (*e.g.* Au). With all the aforementioned advantages and outstanding features of NPs, it is not surprising that the interest in NPs has experienced a staggering exponential increase over past years, with over 10 000 publications referring to NPs in 2010. The amplitude of research efforts is expected to continue increasing as beneficial application of the chemical properties achieved at the nanolevel become increasingly apparent.

One of the key driving forces for the rapidly developing field of nanoparticle synthesis is the contrasting physicochemical properties of nanoparticles compared to their bulk counterparts. Nanoparticles typically provide highly active centers but they are very small and not thermodynamically stable. Structures in this size regime are generally unstable due to their high surface energies and large surfaces.[1,3] To achieve stable NPs, the particle growth reaction has to be carefully controlled and minimized. This has been rendered feasible by a number of methods including the addition of organic ligands, inorganic capping materials or metal salts, colloids or soluble polymers creating core shell type particle morphologies.[4,5] These materials can be grouped in the so-called "unsupported" MNPs.

In parallel, a significant volume of research has been devoted to protocols to achieve homogeneous size dispersed nanoparticles on different supports including porous materials.[6-8] These nanoentities can consequently be grouped in the so-called "supported" nanoparticles (SNPs).

Recent advances in the design and preparation of nanomaterials have shown that a wide variety of them can be synthesized through different preparation routes and tailored to a desired size and distribution, overcoming the limitations of traditional synthetic methodologies.

In conjuction with the nanorevolution, environmental issues, growing demand for energy, political concerns and medium-term depletion of petroleum-derived products have created the need to develop sustainable technologies and low environmental impact processes not only for the production of chemicals, fuels and materials but also for the generation of nanomaterials, nanoparticles and related nanoentities. The state-of-the-art preparation techniques of many NPs attempt to follow more efficient and sustainable routes, taking special considerations to the safety and toxicity of the prepared nanoparticles. These routes include the use of alternative energy input methodologies, such as ultrasound-, microwave irradiation, and ball milling, the use of natural products and biomass (*e.g.* vitamins, fruits, agricultural residues, *etc.*) for NP preparation, and the controllable deposition and stabilization of NP using a related technology, that of nanoporous materials.

This monograph is intended to be a contribution towards the aforementioned selected methodologies for the environmentally friendly preparation of nanoparticles and their applications in various fields including energy storage, environmental remediation, biomedical applications, production of fine chemicals, and biofuels from biomass, with two additional contributions on the toxicology of designer nanoparticles and an introduction to nanosafety. Due to the rapidly expanding nature of this field, this book is hoped to provide a useful introduction to readers to this exciting research area.

Subsequent to this introductory chapter, the first part of the book commences with a chapter by Varma *et al.* that includes a description of sustainable, novel and innovative methodologies for the development of biosynthetic methods for NP preparation including the use of fungi, bacteria, algae, plants, carbohydrates and vitamins. A range of nanoparticles with different nanoparticles sizes and shapes can be achieved using these interesting methods. Chapter 3 by Özkar *et al.* then continues along the lines of sustainable ways to synthesize nanoparticles stabilized in the framework of porous materials (supported metal nanoparticles, SMNPs). This chapter reviews protocols and preparation routes of SMNPs, including physico-chemical methods, the aforementioned alternative methodologies, and detailed case studies on the utilization of various supports such as zeolites, clays, porous silica's, carbonaceous materials, MOFs and some others. This chapter also delineates some interesting catalytic applications of these materials in an array of catalytic processes including coupling and redox chemistries.

After these introductory chapters pertaining to the nanoparticle preparation and associated applications in catalysis, the second part of the book focuses on applications of nanoparticles in various research areas. Chapter 4 from Wang *et al.* deals with an interesting topic of energy conversion and storage through nanoparticles where the authors discuss the possibilities of quantum confinements in nanoparticles, preparation of quantum dots and applications in solar cells and lithium ion batteries. Following this chapter, Dionysiou *et al.* disclose the greener preparation of an assortment of nanomaterials including metal and metal oxide NPs using various methodologies for their utilization in photocatalytic applications for environmental remediation in Chapter five. The chapter includes interesting sections on the immobilization of nanoparticles and the subsequent applications to sustainable environmental systems.

Chapter 6 from Katti *et al.* deals with the uses of nanoparticles (particularly gold NPs synthesized from natural sources) for biomedical applications and treatment of tumors.

The last chapter of the applications section by Obare *et al.* comprises an overview of selected nanomaterials and nanosystems for the production of high-value added chemicals and biofuels from biomass valorization practises. This encompasses some synthetic protocols for the preparation of metallic and biometallic nanoparticles all the way to various applications in chemical processes including conversion of sugars, production of hydrocarbons, synthesis of biodiesel and the design of fuel cells, with some future perspectives in the field.

The final part of the book consists of two chapters devoted to the toxicology of designer/engineered nanoparticles by Ming *et al.* (Chapter 8) and a brief introduction to nanosafety in the lab (Chapter 9) by Balas. The later provides a novel and unique approach to issues associated to the use of nanoparticles, often missing in most nanoparticle-related books to date. Chapter 8 contains some critical information of biophysicochemical interactions at the nano/bio interface, with some important aspects on nanotoxicity. Chapter 9 wraps up the book with some fresh concepts on nanosafety, a relatively novel concept and approach. This Chapter aims to provide some discussion on the introductory issues of risks in handling nanoparticles and strategies for risk reduction together with some general guidelines on safety and prevention in a nanotechnology laboratory from control banding to techniques related to the assessment of nanoparticle emissions.

With the 21st century heralding the dawn of a new age in materials science (where scientists no longer observe the behavior of matter but with the advent of nanoparticles, materials and technology but is able to predict and manipulate matter for specific applications, with sensitivity and efficiency far surpassing previous systems), we hope this book can provide a starting point to readers in the fascinating nanoworld as well as some useful points in terms of nanosafety and nanotoxicity/environmental impact associated with nanoparticles.

Acknowledgments

The authors are grateful to Departamento de Química Orgánica, Universidad de Córdoba and the Environmental Protection Agency (EPA) in Cincinnati, respectively, for their support during the assembly and organization as well as preparation of this monograph. Rafael Luque would also like to thank Ministerio de Ciencia e Innovación, Gobierno de España, for the provision of a Ramon y Cajal (RyC) contract (ref. RYC-2009-04199) and funding under projects P10-FQM-6711 (Consejeria de Ciencia e Innovacion, Junta de Andalucia) and CTQ2011 28954-C02-02 (MICINN) as well as project IAC-2010-II granted to Rafael Luque as a "Estancia de Excelencia" at the EPA in Cincinnati from July to September 2011.

References

1. G. A. Ozin, A. C. Arsenault, L. Cademartiri, *Nanochemistry: A Chemical Approach to Nanomaterials*, Royal Society of Chemistry, Cambridge, UK, 2000.
2. (a) C. A. Mirkin, The beginning of a small revolution, *Small*, 2005, **1**, 14–16; (b) J. Grunes, J. Zhu and G. A. Somorjai, Catalysis and nanoscience, *Chem. Commun.*, 2003, 2257–2258.
3. M. Chen, Y. Cai, Z. Yan and D. W. Goodman, On the origin of unique properties of supported Au nanoparticles, *J. Am. Chem. Soc.*, 2006, **128**(19), 6341–6346.
4. (a) G. Schmid, V. Maihack, F. Lantermann and S. Peschel, Ligand-stabilized metal clusters and colloids: properties and applications, *J. Chem. Soc. Dalton Trans.*, 1996, 589–595; (b) A. M. Doyle, S. K. Shaikhutdinov, S. D. Jackson and H. J. Freund, Hydrogenation on metal surfaces: why are nanoparticles more active than single crystals?, *Angew. Chem. Int. Ed.*, 2003, **42**, 5240–5243.
5. (a) X. L. Luo, A. Morrin, A. J. Killard and M. R. Smyth, Applications of nanoparticles in electrochemical sensors and biosensors, *Electroanalysis*, 2006, **18**, 319–326; (b) Y. C. Shen, Z. Tang, M. Gui, J. Q. Cheng, X. Wang and Z. H. Lu, Nonlinear optical response of colloidal gold nanoparticles studied by hyper-Rayleigh scattering technique, *Chem. Lett.*, 2000, 1140–1141; (c) M. Harada, K. Asakura and N. Toshima, Catalytic activity and structural analysis of polymer-protected gold/palladium bimetallic clusters prepared by the successive reduction of hydrogen tetrachloroaurate(III) and palladium dichloride, *J. Phys. Chem.*, 1993, **97**, 5103–5114; (d) J. Virkutyte and R. S. Varma, Green synthesis of metal nanoparticles: biodegradable polymers and enzymes in stabilization and surface functionalization, *Chem. Sci.*, 2011, **2**, 837–846.
6. M. Boudart, Catalysis by supported metals, *Adv. Catal.*, 1969, **20**, 153–166.
7. D. Barkhuizen, I. Mabaso, E. Viljoen, C. Welker, M. Claeys, E. van Steen and J. C. Q. Fletcher, Experimental approaches to the preparation of supported metal nanoparticles, *Pure Appl. Chem.*, 2006, **78**, 1759–1769.

8. (a) M. Valden, X. Lai and D. W. Goodman, Onset of catalytic activity of gold clusters on titania with the appearance of non-metallic properties, *Science*, 1998, **281**, 1647–1650; (b) H. Sakurai and M. Haruta, Synergism in methanol synthesis from CO over gold catalysts supported on metal oxides, *Catal. Today*, 1996, **29**, 361–365; (c) J. F. Jia, K. Haraki, J. N. Kondo, K. Domen and K. Tamaru, Selective hydrogenation of acetylene over Au/Al$_2$O$_3$ catalyst, *J. Phys. Chem. B*, 2000, **104**, 11153–11156.

CHAPTER 2

Environmentally Friendly Preparation of Metal Nanoparticles

JURATE VIRKUTYTE*[a] AND RAJENDER S. VARMA*[b]

[a] Pegasus Technical Services Inc., 26 E. Hollister Street, Cincinnati, OH, 45219, USA; [b] Sustainable Technology Division, National Risk Management Research Laboratory, U.S. Environmental Protection Agency, 46 Martin Luther King's Drive, MS 443, Cincinnati, OH 45268, USA
*Email: virkutyte.jurate@epa.gov; varma.rajender@epa.gov

2.1 Introduction

Commercial and research interest in nanotechnology significantly increased in the past several years translating into more than US$9 billion in investment from public and private sources.[1]

Nanotechnology is the ability to measure, see, manipulate and manufacture things on an atomic or molecular scale, usually between one and 100 nanometers. These tiny products also have a large surface area to volume ratio, which are the most important characteristics responsible for the widespread use of nanomaterials in mechanics, optics, electronics, biotechnology, microbiology, environmental remediation, medicine, numerous engineering fields and material science.[2]

Unfortunately, high surface area often results in various drawbacks that are closely associated with the surface phenomena, e.g. outer layer atoms in the particle may have a different composition and therefore, chemistry from the

RSC Green Chemistry No. 19
Sustainable Preparation of Metal Nanoparticles: Methods and Applications
Edited by Rafael Luque and Rajender S Varma
© The Royal Society of Chemistry 2013
Published by the Royal Society of Chemistry, www.rsc.org

rest of the particle. Furthermore, nanoparticle surface will be prone to environmental changes such as redox conditions, pH, ionic strength, microorganisms, *etc.* Also, small size and large surface to volume ratio may lead to both chemical and physical differences in their properties including mechanical, biological and sterical, catalytic activity, thermal and electrical conductivity, optical absorption and melting point, compared to the bulk of the same chemical composition.[3]

Generally, the intended nanoparticle application defines its composition, *e.g.* if a nanoparticle is going to be used to interact with biological systems, functional groups will be attached to its surface to prevent aggregation and/or agglomeration.[2] Also, coatings and other surface active materials could be introduced that form transient van der Waals interactions with the surface of nanoparticles and exist in equilibrium with the free surfactant molecule. Furthermore, if the use is intended for the electronics industry, nanoparticles can be manufactured in a way that significantly enhances the strength and hardness of materials, exhibits enhanced electrical properties by controlling the arrangements within the nanoclusters, *etc.* Also, if the use is intended for environmental remediation and catalysis, increased activity could be achieved by attaching various functional side groups, doping with ions and anions and variation in size and structure. And finally, nanoparticles with nonconventional properties including superconductivity and magnetism can also be manufactured utilizing appropriate mechanisms and surface functionalization approaches.[4]

According to Christian *et al.*,[2] nanoparticle consists of three layers: i) the surface that can be functionalized, ii) a shell that may be added according to the application needs and iii) the core that can be synthesized using various methods, reaction conditions and precursors. Thus, the surface of a nanoparticle can be functionalized with various metals and metal oxides, small molecules, surfactants and/or polymers. In addition, target nanoparticle surface can be charged (*e.g.* base-catalyzed hydrolysis of tetraethyl orthosilicate, $SiO^-\ M^+$) or uncharged (citrate, sodium dodecylsulfate (SDS), polyethylene glycol (PEG), *etc.*), which highly depends on the application and the subsequent use of nanomaterials. In most cases, the shell is made of inorganic material that has a completely different structure than a core, *e.g.* iron oxide on iron nanoparticles, quantum dots (zinc sulfide on cadmium selenide) and polystyrene–polyaniline nanoparticles.[2] Importantly, the core is usually referred to as the nanoparticle itself and the physicochemical properties of nanoparticles are nearly always governed by the properties of the core. However, the environmental fate and transport most likely will be dominated by the core and shell properties rather than core alone. Also, risks associated with the occurrence of nanoparticles in the environment must be related to the surface, core and the shell.

Currently, there are two main methods to synthesize nanomaterials: "top down" and "bottom up" approaches (Figure 2.1). Briefly, the "top-down" approach suggests nanoparticle preparation by lithographic techniques, etching, grinding in a ball mill, sputtering, *etc.* However, the most acceptable and

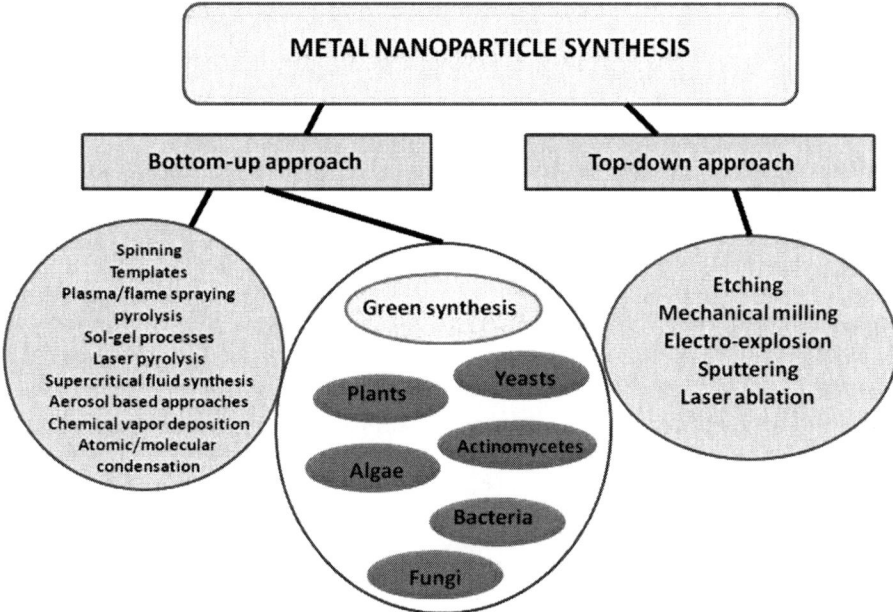

Figure 2.1   Metal and metal-oxide nanoparticle synthesis.

effective approach for nanoparticle preparation is the "bottom up" approach, where a nanoparticle is "grown" from simpler molecules – reaction precursors. In this way, it is possible to control the size and shape of the nanoparticle depending on the subsequent application through variation in precursor concentrations, reaction conditions (temperature, pH, *etc.*), functionalizing the nanoparticle surface, using templates, *etc.*

Altering of the surface properties, or, in other words, functionalization of the surface, is one of the most important aspects of nanoparticle synthesis for the desired applications. For instance, the high chemical activity of nanoparticles with a large surface is usually the main reason for undesirable and most often irreversible processes such as aggregation.[5] Aggregation significantly diminishes particle reactivity through the reduced specific surface area and the interfacial free energy. Thus, in order to avoid aggregation, nanoparticles have to be coated or functionalized with chemicals and or materials to increase their stability during storage, transportation, application and overall life cycle.

According to Stubbs and Gilman,[6] the majority of the stabilization/functionalization methods involve the addition of dispersant molecules such as surfactants or polyelectrolytes to the nanoparticle surface. These materials not only alter the chemistry of the nanoparticle surface, but also produce large amounts of waste material, because they occupy a significant (>50%) mass fraction of a nanoparticle system.[1] Thus, in order to avoid producing wastes and a subsequent contamination of the environment, there is an urge to search for environmentally benign stabilization and functionalization pathways as well as biocompatible, *i.e.* nonimmunogenic, nontoxic and hydrophilic stabilizing agents.

Research effort advancing nanoparticle functionalization and, thus, stabilization identifying new functionalization pathways and the use of benign stabilizers such as polyphenols, citric acid, vitamins (B, C, D, K), biodegradable polymers and silica has been well documented.[7–12] Therefore, regardless of the synthesis approach, synthesized nanomaterials are expected to:[13,14] i) exhibit new size-based properties (both beneficial and detrimental) that are intermediate between molecular and particulate, ii) incorporate a wide range of elemental and material compositions, including organics, inorganics, and hybrid structures, and iii) possess a high degree of surface functionality.

This chapter will summarize the "state-of-the-art" in the exploitation of various environmentally friendly synthesis approaches, reaction precursors and conditions to manufacture metal and metal-oxide nanoparticles for a vast variety of purposes.

2.2   Biogenic Nanoparticles

Metal nanoparticles have been produced using physical and chemical methods for many years now. However, the exploitation of generally accepted reducing agents such as hydrazine hydrate, sodium borohydride, *etc.* may lead to absorption of hazardous chemicals on the surface of nanoparticles and subsequently lead to undesired toxicity issues. Thus, it is vitally important to develop a reliable "green" chemistry process for the biogenic synthesis of nanomaterials. Therefore: (i) the use of organisms emerges as an ecofriendly and exciting approach that reduce waste products (ultimately leading to atomically precise molecular manufacturing with zero waste); (ii) the use of nanomaterials as catalysts for greater efficiency in current manufacturing processes by minimizing or eliminating the use of toxic materials (green chemistry principles); (iii) the use of nanomaterials and nanodevices to reduce pollution (*e.g.* water and air filters); and (iv) the use of nanomaterials for more efficient alternative energy production (*e.g.* solar and fuel cells).[15]

According to Iravani,[3] several aspects must be considered in order to produce stable and well-characterized nanoparticles using organisms:

1. *Selection of the most effective organism.* It is vital to address the important intrinsic properties of the target organisms such as enzyme activities and biochemical pathways to manufacture stable nanoparticles. For instance, certain plants have an ability to accumulate and detoxify metals, therefore they are considered perfect candidates for nanoparticle synthesis.
2. *Optimal conditions for cell growth and enzyme activity.* It is important to optimize the amount of nutrients, inoculum size, light, temperature, pH, mixing speed, buffer strength and other parameters to obtain the most effective cell growth.

3. *Optimal reaction conditions.* If the production of nanoparticles is considered on a larger scale, the yield and the production rate are the vital parameters that must be optimized through optimization of substrate concentration, biocatalyst concentration, electron donor type and its concentration, pH, exposure time, temperature, buffer strength, agitation speed and the amount of light required for the system. Furthermore, the use of additional, complementary factors such as microwave, ultrasound and visible-light irradiation should also be considered.

2.2.1 Biosynthesis of Nanoparticles

Nanoparticle synthesis involving microbial organisms (bacteria, actinomycetes, fungi, yeast, viruses, *etc.*) is a green chemistry approach that interconnects nanotechnology and microbial biotechnology.[16] Unfortunately, these biological particles are not monodispersed and the overall rate of synthesis is slow. Thus, to overcome these drawbacks, several factors such as microbial cultivation methods and the extraction techniques have been developed and optimized and the combinatorial approach such as photobiological methods are used and can be found in a very good recent review by Narayanan and Sakthivel.[17]

2.2.1.1 Fungi

Fungi have many advantages for metal nanoparticle synthesis compared with other organisms because of the presence of enzymes/proteins/reducing components on its cell surface.[18] The probable mechanism of intracellular biosynthesis includes the formation of metal nanoparticles through the reduction by enzyme (reductase) present in the cell wall or in the cytoplasmic membrane, on the inner surface of the fungal cell. For instance, Narayan and Sakthivel[18] researched the formation of Au nanoparticle in the presence of the fungus *Cylindrocladium floridanum* and found that in 7 days, the fungus accumulated fcc (111)-oriented crystalline gold nanoparticles (SPR band of UV-Vis spectrum at 540 nm) on the surface of the mycelia. These nanoparticles were effective in degrading 4-nitrophenol and the process followed a pseudofirst-order kinetic model with the reaction rate constant of $2.67 \times 10^{-2} \, m^{-1}$ with $5.07 \times 10^{-6} \, mol \, dm^{-3}$ of gold at ca. 25 nm. The authors also reported a significant increase in the reaction rates with an increase in gold nanoparticle concentration from 2.54×10^{-6} to $12.67 \times 10^{-6} \, mol \, dm^{-3}$ (ca. 25 nm) with reduced Au nanoparticle size from 53.2 to 18.9 nm, respectively.

Also, the fungus *Trichoderma viride* was used to synthesize polydispersed Ag nanoparticles with sizes from 5 to 40 nm at near room temperature (27 °C) that showed maximum absorbance at 420 nm on ultraviolet-visible spectra.[19] Antibactericidal properties were tested against four bacterial strains—namely, *Salmonella typhi* (gram-negative rods), *Escherichia coli* (gram-negative rods), *Staphylococcus aureus* (gram-positive cocci), and *Micrococcus luteus* (gram-positive cocci). An important finding of this study was that the anti-bacterial activities of ampicilin, kanamycin, erythromycin and chloramphenicol were significantly enhanced in the presence of as-prepared Ag nanoparticles. In addition, *Geotricum sp.* was found to successfully produce Ag nanoparticles with particle sizes ranging from 30 to 50 nm.[15] According to FTIR spectra, the presence of amide (I) and (II) bands of protein were identified and acted as capping and stabilizing agent on the surface of nanoparticles. The fungus *Verticillium* (from *Taxus* plant) can also be used to synthesize Ag nanoparticles with average size of 25 ± 12 nm at room temperature.[20] It is noteworthy to mention that Ag ions were not toxic to the fungal cells and the cells continued to multiply after biosynthesis of the silver nanoparticles.

Rice husk is a cheap agrobased waste material, which harbors a substantial amount of silica in the form of amorphous hydrated silica grains. Therefore, it would be an idea material to biotransform amorphous to crystalline silica nanoparticles at room temperature for numerous applications. Indeed, the fungus *Fusarium oxysporum* rapidly biotransformed amorphous plant biosilica into crystalline silica and leached out silica extracellularly at room temperature in the form of 2–6 nm quasispherical, highly crystalline silica nanoparticles capped by stabilizing proteins.[21]

2.2.1.2   Bacteria

Prokaryotic bacteria and actinomycetes have been most extensively researched for synthesis of metallic nanoparticles. One of the reasons for "bacterial preference" for nanoparticles synthesis is their relative ease of manipulation.[22] For instance, Ahmad *et al.*[23] demonstrated a novel extracellular synthesis of well-dispersed Au nanoparticles (average size of 8 nm) using the prokaryotic microorganism *Thermomonospora* sp. In addition, the bacterium *Brevibacterium casei* can be used to manufacture Ag (sizes from 10 to 50 nm) and Au (sizes from 10 to 50 nm) nanoparticles.[24] It is important to point out that FTIR data proved that the presence of proteins was responsible for the reduction and capping of nanoparticles. Coker *et al.*[25] discussed the wide applicability of Pd-ferrimagnetic nanoparticles in various industrial areas. Unfortunately, conventional synthesis methods usually result in potentially high environmental and economic costs. Thus, the use of Fe (III)-reducing bacterium *Geobacter sulfurreducens* was found to significantly reduce synthesis costs; facilitate easy recovery of the catalyst (Pd nanoparticles on biomagnetite with 10 mol% of Pd) with superior performance due to the reduced agglomeration and particle size ranging from 20 to 30 nm. Such a catalyst was highly effective in the Heck reaction coupling iodobenzene to ethyl acrylate or styrene with a complete conversion to ethyl cinnamate or stilbene within 90 and 180 min, respectively.

It was reported that TiO_2 nanoparticles (8 to 35 nm in size) can also be synthesized using microbes *Lactobacillus* sp. and *Sachharomyces cerevisae* at room temperature.[26] According to the authors, the synthesis of TiO_2 nanoparticles occurred due to pH-sensitive membrane-bound oxido-reductases and carbon-source-dependent rH2 in the culture solution. Also, the metal-reducing bacterium *Shewanella algae* can be used to manufacture gold and platinum nanoparticles. Platinum nanoparticles were prepared by reducing $PtCl_6^{2-}$ ions within 60 min at pH 7 and 25 °C.[27] It was found that platinum nanoparticles of about 5 nm were located in the periplasmic space, a preferable cell surface location for easy recovery of biogenic nanoparticles. Spherical gold nanoparticles and nanoplates can also be manufactured with *Shewanella algae* that act as reducing and stabilizing agent in the presence of H_2 as electron donor at pH 2.8 and room temperature.[28] Well-dispersed nanoparticles started to form within 1 h with a mean size of 9.6 nm. Furthermore, gold nanoplates with an edge length of 100 nm started to form after 6 h, and 60% of the total nanoparticle population was due to gold nanoplates with an edge length of 100–200 nm after 24 h. It is important to point out that the yield of gold nanoplates manufactured with *S. algae* extract was four times higher than that prepared with resting cells of *S. algae*.

2.2.1.3 Yeasts

Yeasts are eukaryotic micro-organisms in the kingdom of fungi, with more than 1500 species. There is a lack of successful studies emphasizing the efficiency of yeasts in preparation of nanomaterials, however, several studies indicate that yeasts extracts can be used to manufacture metal nanoparticles. For instance, in an elegant study, Kowshik *et al.*[29] demonstrated the extracellular formation of 2–5 nm silver nanoparticles by a silver-tolerant yeast strain MKY3. *Saccharomyces cerevisae* broth was used to synthesize gold and silver nanoparticles.[30] According to the authors, gold (size ranged from 20 to 100 nm, pH 4 to 6) and silver (from 10 to 20 nm, pH 8 to 10) nanoparticles were formed exctracellularly within 24 h and 48 h, respectively. Distinct surface plasmon peaks were observed at 540 nm and 415 nm for gold and silver nanoparticles, respectively.

In addition, Kumar *et al.*[31] showed the applicability of yeast species *Hanensula anomala* to reduce gold salt in the presence of amine-terminated polyamidoamine dendrimer as the stabilizer; cysteine could also be used as stabilizer. *Candida guilliermondii* was used to prepare silver and gold nanoparticles.[32] As-prepared Au and Ag nanoparticles exhibited a distinct surface plasmon peaks at 530 nm and 425 nm, respectively, with face-centered cubic structures. According to TEM data, near-spherical, well-dispersed gold and silver nanoparticles in the size range of 50–70 nm and 10–20 nm respectively, were formed. These nanoparticles were tested against five pathogenic bacterial strains. The highest efficiency for both gold and silver nanoparticles was observed against *Staphylococcus aureus*, which indicated the applicability of yeast-synthesized nanoparticles for environmental remediation and medical fields. Furthermore, Subramanian *et al.*[33] reported the effectiveness of marine yeasts (*Pichia capsulata*) derived from the mangrove sediments to synthesize silver nanoparticles (1.5 mM $AgNO_3$, 0.3% NaCl, pH of 6.0 and incubated at 5 °C for 24 h) that exhibited UV-Vis absorption peak at 430 nm.

Yarrowia lipolytica was reported to be an effective reducing agent to produce gold nanoparticles and nanoplates by varying concentrations of chloroauric acid at pH 4.5.[34] According to the findings, 109 cells ml^{-1} and 0.5 or 1.0 mM of the gold salt, the reaction mixtures developed a purple or golden red color, respectively, indicating the formation of gold nanoparticles; various nano-particle sizes were achieved by using 1010 cells ml^{-1} incubated with 0.5, 1.0 or 2.0 mM chloroauric acid salt. Importantly, nanoplates were synthesized in the presence of 3.0, 4.0 or 5.0 mM HAuCl$_4$ and 1011 cells ml^{-1}. It was confirmed that an increase in salt concentration and a fixed number of cells, resulted in the increase in the overall size of nanoparticles. On the other hand, an increase in cell numbers while using a constant gold salt concentration, resulted in a sig-nificant decrease in nanoparticle size. According to FTIR data, carboxyl, hydroxyl and amide groups on the cell surfaces participated and were responsible for the nanoparticle synthesis.

2.2.1.4 Algae

Similarly to yeasts, there are only a few reports of algae being used for bio-synthesis of metallic nanoparticles. Algae are diverse and a very large group of photosynthetic simple, typically autotrophic organisms, ranging from uni-cellular to multicellular forms. Therefore, the use of *Spirulina platensis* has been reported in successful preparation of silver (7–16 nm) and gold (6–10 nm) nanoparticles in 120 h at 37 °C.[35] Surface plasmon absorbance showed peaks at 424 and 530 nm for Ag and Au nanoparticles, respectively. Also, core (Au)–shell (Ag) nanostructures (17–25 nm, bimetallic 50:50 ratio) were observed when both salts were mixed during a one-pot synthesis procedure. For core–shell nanoparticles, absorption peaks were observed at 509, 486 and 464 nm at 75:25, 50:50 and 25:75 (Au:Ag) mol concentrations, respectively. FTIR data revealed that most likely proteins were responsible for the reduction and cap-ping of the metal and core–shell nanoparticles. In addition, marine microalgae *Tetraselmis suecica* can also be used to manufacture gold nanoparticles (aver-age size 79 nm) from chloroauric acid solution.[36] The UV-Vis spectrum of the aqueous medium indicated a peak at 530 nm, corresponding to the surface plasmon absorbance of gold nanoparticles.

Xie *et al.*[37] demonstrated the formation of single-crystalline triangular and hexagonal gold nanoplates by treating an aqueous solution of chloroauric acid with the extract of the unicellular green algae *Chlorella vulgaris* at room tem-perature. These findings confirmed that proteins were the primary molecules involved in the reduction and capping of gold nanoparticles. A sustained synthesis of colloidal gold (average particle size of 9 nm) using *Klebsormidium flaccidum* green algae has been demonstrated.[38] According to the findings, ca. 16 wt% of gold was cellular Au and ca. 80% metallic Au content, indicating an effective intracellular reduction process.

2.2.1.5 Actinomycetes

Actinomycetes is a member of a heterogeneous group of gram-positive, gen-erally anaerobic bacteria, notorious for a filamentous and branching growth patterns that results in an extensive colony. Due to the presence of enzymes, actinomycetes can be used to reduce metal salts and therefore manufacture nanoparticles utilizing "green" synthesis principles. For instance, Ahmad *et al.*[23] reported the benign but effective $AuCl_4^-$ reduction by *Thermomonospora sp.* (alkalothermophilic actynomycete) biomass. According to their findings, reduction of metal ions occurred extracellularly by the enzymatic process and spherical monodispersed nanoparticles with average dimensions of 8 nm at pH 9 and 50 °C were formed. However, Torres-Chavolla *et al.*[39] reported slightly larger (30 to 60 nm) though monodispersed Au nanoparticles in the presence of *Thermomonospora curvata, Thermomonospora fusca* and *Thermomonospora chromogena.* When another species of alkalotolerant actinomycete (*Rhodococcus sp.*) was used to reduce gold salt, monodisperse (5–15 nm) Au nanoparticles were formed and were mostly concentrated on both, the cytoplasmic membrane and on the cell wall. However, the predominant concentration of nanoparticles on the cytoplasmic membrane suggested that the reduction occurred by enzymes present in the membrane.[40] Importantly, the metal ions were not toxic to the cells and the cells continued to multiply after biosynthesis of the gold nanoparticles.

Spherical, crystalline and monodispersed (7 to 20 nm in size) Au nanoparticles can be synthesized (10^{-3} M $HAuCl_4$ at 45 °C in 4 h) in the presence of the marine sponge, *Acanthella elongata* that belongs to the primitive phylum Porifera.[41] It was hypothesized that water-soluble organics present in the sponge extract was responsible for the reduction of gold ions to Au nanoparticles (UV-Vis peak at 526 nm).

2.2.1.6 Plants

Over the years, plants have shown great potential in heavy-metal accumulation and detoxification.[3] Thus, they may also be used to phytosynthesize metal nanoparticles. The biosynthetic method employing plant extracts and living plants has received an increased attention during the last decade as a simple, effective and viable technique as well as a good alternative to conventional chemical and physical nanoparticle preparation methods. There are numerous plants and fruits that can be used to reduce and stabilize both, single and multi-metal nanoparticles in "one-pot" synthesis.

Several of the first reports on effective synthesis of silver and gold nanoparticles inside the living plants – alfalfa (*Medicago sativa*) were from Gardea-Torresdey and coworkers.[42–46] Interesting aspect of their studies was that alfalfa plants were grown in a target metal-salt-rich environment, thus nanoparticle synthesis was achieved within plants by silver and gold ion uptake from solid media. For instance, crystalline icosahedral (ca. 4 nm) and fcc twinned (6–10 nm) gold nanoparticles were observed when $AuCl_4^-$ was treated with alfalfa plants.[43,45,46] Also, by varying pH, crystalline gold nanoparticles with cluster radius of 6.2 Å at pH 5 and 9 Å at pH 2 were observed indicating that another layer of gold may have been deposited onto the colloid surface at lower pH.[44] Montes *et al.*[47] further researched the formation of anisotropic polyhedra and nanoplates by reducing $3 \, mM \, AuCl_4^-$ solution at room temperature with *Medicago sativa* at pH 3.5 and 3. They produced nanoparticles with sizes ranging from 30 to 60 nm with various morphologies including 30 nm dec-ahedra and 15 nm icosahedra nanoparticles in isopropanol extracts. However, when water was used as the extracting agent, triangular gold nanoplates were observed ranging from 500 nm to 4 µm in size.

Also, *Citrus sinensis* peel can be used in a simple and environmentally friendly approach to prepare Ag nanoparticles at nearly room temperature and 60 °C.[48] The size of the particles obtained was ca. 35 nm at 25 °C and ca. 10 nm for the synthesis at 60 °C. Moreover, *Citrus sinensis* reduced and stabilized nanoparticles were effective in diminishing bacterial, *e.g. E. coli, Ps. aeruginosa* (gram-negative) and *S. aureus* (gram-positive) contamination.

The extract from dried mahogany leaves (*Swietenia mahogany JACQ.*) rich in polyhydroxy limonoids can also be utilized for a rapid and benign Ag, Au and Ag/Au alloy preparation.[49] Interestingly, when target metal salts were exposed to the leaf extract, competitive reduction of Au (II) and Ag (I) occurred pro-ducing bimetallic Au/Ag alloy nanoparticles. Furthermore, *Hibiscus rosa sinensis* was used to produce stable Au and Ag nanoparticles at room tem-perature.[50] In addition to plant extracts, Indian propolis that contained hydroxyflavonoids and its chemical constituents pinocembrin and galangin in ethanol and water extracts was used to produce stable Au and Ag nanoparticles with numerous morphologies at pH 10.6.[51] Besides, Sheny *et al.*[52] demon-strated that *Anacardium occidentale* leaf extract rich in polyols can also be utilized to synthesize Au, Ag, Au-Ag alloy and Au core–Ag shell nanoparticles at room temperature without the necessity to introduce hazardous reducing and stabilizing agents. According to the XRD data, these nanoparticles were highly crystalline and predominantly in the cubic phase. Polyphenol-rich *Camellia sinensis* extract can effectively reduce gold and silver salts and produce colloidal Ag and Au nanoparticles with average sizes of 40 nm that exhibit highly efficient single photon-induced luminescence that can be manipulated by changing the concentrations of metal ions and the quantity of the reducing agent.[53]

Furthermore, Singhal *et al.*[54] demonstrated the applicability of *Ocimum sanctum* (Tulsi) leaf extract to biosynthesize extremely stable silver nanoparticles with sizes ranging from 4 to 30 nm. It was observed that *O. sanctum* leaf extract was able to reduce silver ions into silver nanoparticles within 8 min of the reaction and the resulting silver nanoparticles exhibited significant antimicrobial activity suggesting their potential application in medicinal industries.

Not only leaf extract but also stem latex of a medicinal (peptides, terpenoids, *etc.*) *Euphorbia nivulia* can be exploited in Ag and Cu nanoparticles' synthesis.[55] As demonstrated by Valodkar *et al.*,[55] 5 to 10 nm-sized nanoparticles were produced at room temperature and in the presence of microwave irradiation (30 s). These nanoparticles were exceptionally effective against gram-negative and gram-positive bacteria. In addition, Vinod *et al.*[56] reported benign synth-esis of Au, Ag and Pt nanoparticles at 45 to 75 °C in water in the presence of gum kondagogu (*Cochlospermum gossypium*). They argued that -OH groups present in the gum matrix participated in the synthesis and were responsible for the reduction of target metal salts. According to their findings, crystalline in nature with face-centered cubic geometry Au, Ag and Pt colloidal nanoparticles (7.8 ± 2.3 nm, 5.5 ± 2.5 nm and 2.4 ± 0.7 nm, respectively) were extremely stable without undergoing any subsequent oxidation.

A sunlight-induced method of nanoparticle formation may provide a new viable alternative to currently available synthesis methods. For instance, Karimi-Zarchi *et al.*[57] demonstrated a rapid sunlight-induced method using an ethanol extract of *Andrachnea chordifolia*. The smallest (3.4 nm) Ag nanoparticles were obtained when silver ion solution was irradiated for 5 min by direct sunlight radiation, and much larger nanoparticles ranging from 3 to 30 nm were formed when solution was left in the dark for 48 h. *Azadirachta indica* leaves extract containing significant amounts of terpenoids were also used to manufacture silver nanoparticles.[58] Stable nanoparticles were formed within 2 h, unfortunately agglomeration was observed after 4 h of the reaction. Nearly spherical particles (ca. 20 nm) in diameter were observed and the particle size obtained microscopically was similar to the crystallite size obtained using XRD from XRD analysis.

Coleus amboinicus Lour. (Indian Borage) is a medicinal plant containing phytochemicals such as carvacrol (monoterpenoid), caryophyllene (bicyclic sesquiterpene) and patchoulane and flavanoids (quercetin, apigenin, luteolin, salvigenin, and genkwanin). Prathna and coworkers[58] demonstrated the preparation of silver and gold nanoparticles utilizing leaf-extract-mediated green synthesis approach. In the case of silver nanoparticles, upon an increase in leaf-extract concentration, there was a significant shift in the shape of nanoparticles from anisotrophic nanostructures like triangle, decahedral and hexagonal to isotrophic spherical nanoparticles.[59] Nanoparticles ranged from 2.6 to 79.8 nm depending on the concentration of the leaf extract and the crystalline nature of fcc-structured nanoparticles was confirmed by an XRD spectrum with peaks corresponding to (1 1 1), (2 0 0), (2 2 0) and (3 1 1) planes as well as bright circular spots in the selected-area electron diffraction (SAED). However, during gold nanoparticle (sizes ranged from 4.6 to 55.1 nm) synthesis, mainly spherical, truncated triangle, hexagonal and decahedral nanoparticles were formed.[60] FTIR data confirmed the involvement of aromatic amines, amide (II) groups and secondary alcohols in reducing and capping/stabilizing of nanoparticles.

Not only the leaf extracts but also roots can be used to manufacture metal nanoparticles. For instance, Leonard *et al.*[61] used naturally occurring Korean red ginseng root to synthesize stable (excellent stability in various buffers including cysteine, histidine, saline, sodium chloride and various pHs) gold nanoparticles (size ranging from 2 to 40 nm) without the introduction of reducing or capping/stabilizing agents. The authors argued that such a wide distribution in particle sizes was related to the difference in kinetics of nanoparticle formation if compared to more conventional $NaBH_4$-induced method. For example, $NaBH_4$ is a strong reducing agent that rapidly reduces target metal ions and should likely nucleate numerous particles by depleting the gold salt before large particles are formed. On the contrary, the ginseng-roots-induced reduction process is slower, thus it might be a reason why a large variety of different sized particles are formed. An important finding was that gold nanoparticles did not aggregate, suggesting that phytochemicals acted as effective coatings providing robust shielding from aggregation. Indeed, the ginsenosides or polysaccharides, flavones and other phytochemicals within Korean red ginseng not only result in effective reduction of gold salts to nanoparticles but equally act as suitable capping agents thus preventing them from aggregation. Also, such a phytochemical coating on nanoparticle proved to be nontoxic as demonstrated through detailed cytotoxicity assays using WST-8 counting kit, performed on normal cervical cells lines.

Rubber latex extracted from *Hevea brasiliensis* provides another alternative for colloidal silver nanoparticle preparation.[62] According to their findings, UV-Vis spectra detected the characteristic surface plasmonic absorption band around 435 nm and lower silver nitrate salt concentration led to the smaller nanoparticle size. In this way synthesized silver nanoparticles were spherical with particle sizes ranging from 2 to 100 nm and had face-centered cubic crystalline structure. FTIR data implied that reduction of the silver ions was facilitated by the interaction between silver ions and amine groups from ammonia, whereas the stability of the particle resulted from cis-isoprene binding onto the surface of nanoparticles.

Macrotyloma uniflorum presents another viable alternative for noble-metal nanoparticle preparation in comparison to more conventional synthesis methods.[63,64] For instance, well-dispersed silver nanoparticles with sizes ranging from 12 to 17 nm exhibited face-centered cubic structures. According to the FTIR data, the presence of phenolic acids like *p*-hydroxy benzoic acid, 3,4-dihydroxy benzoic acid, *p*-coumaric acid, caffeic acid, ferulic acid, vanilic acid, syringic acid and sinapic acid facilitated the reduction and capping of silver nanoparticles.

In a recent study, Philip[65] demonstrated facile and rapid biosynthesis of well-dispersed silver nanoparticles using *Mangifera Indica* leaf extract. The author reported that at pH 8, the colloid consisting of well-dispersed triangular, hexagonal and nearly spherical nanoparticles having size of ~20 nm was formed. In addition, the UV-Vis spectrum of silver nanoparticles gave a characteristic surface plasmon resonance at 439 nm. According to the data, water-soluble organics present in the leaf were responsible for the reduction of silver ions and capping/stabilizing of the resulting nanoparticle.

Several studies show the applicability of ethanolic flower extract of *Nyctanthes arbortristis* (80 °C for 30 min) and water extract of *Terminalia chebula* in the preparation of gold nanoparticles with average sizes of 19.8 nm and overall sizes ranging from 6 to 60 nm, respectively.[66,67] Importantly, low reaction temperature (ca. 25 °C) favored anisotropy.[66] As confirmed by FTIR data, hydrolyzable tannins present in the extract were responsible for the reduction and capping/stabilization of gold nanoparticles.[67]

To further advance expertise in the use of plant derivatives and potential wastes in the bulk synthesis of metal nanoparticles, Njagi *et al.*[68] introduced aqueous hybrid phenolic-rich sorghum (*Sorghum spp*) in the synthesis of stable Fe and Ag nanocolloids at room temperature. Hybrid sorghums (phenolic compounds) are water-soluble, nontoxic, and biodegradable compounds, affording a green synthesis process where they act as both, the reducing and stabilizing agents. Also, most sorghum species are drought and heat tolerant, making them economical to produce. Sorghum-reduced and stabilized Ag and Fe nanoparticles were largely uniform with a narrow size distribution (the average diameter of spherical Ag and Fe nanoparticles was 10 nm and 50 nm, respectively).

2.2.1.7 Carbohydrates

Carbohydrates are the most abundant class of organic compounds found in living organisms. This huge group can be classified into simple sugars, starches, cellulose, *etc*. Due to their chemical structure and constituents, these compounds can be effectively used in fabrication of nanomaterials where they are utilized as both, the reducing and stabilizing agents. For instance, hydroxypropyl starch can be used as a reducing and stabilizing agent in the synthesis of stable Ag nanoparticles (no aggregation was observed for more than 6 months) ranging from 6 to 8 nm in diameter.[69] Gao *et al.*[70] also used biodegradable starch to manufacture stable Ag nanoparticles. They found that one of the constituents – glucose was responsible for the reduction of silver ions. According to HR-TEM findings, silver nanoparticles were covered by starch layer and formed spherical core–shell Ag/starch nanoparticles with diameters ranging from 5 to 20 nm. In addition, XRD confirmed the presence of silver nanoparticles with face-centered cubic structure.

Hydroxypropyl cellulose can also be used to manufacture silver nanoparticles and as a dopant for visible light active TiO_2.[71,72] Thus, optimum conditions for silver nanoparticles preparation were pH 12.5, 0.3% hydroxypropyl cellulose solution having molar substitution of 0.42 and carrying out the reaction at 90 °C for 90 min.[71] On the other hand, doped-TiO_2 nanoparticles were prepared using a benign "one-pot" synthesis procedure at room temperature, resulting in stable and effective nanoparticles with average size of 17 nm and BET surface area of 133.8 $m^2 g^{-1}$ that were efficient in degrading organic contaminants in the visible-light range.[72]

It is possible to manufacture Ag nanoparticles with relatively narrow size distribution and average particle size of 25 nm, varying ammonia as the reaction precursor concentration (0.2 to 0.005 mol l^{-1}), using monosaccharides (glucose and galactose) and disaccharides (maltose and lactose).[73] As-prepared silver nanoparticles (with concentrations as low as 1.69 µg ml^{-1} of Ag) showed high antimicrobial and bactericidal activity against gram-positive and gram-negative bacteria, including multiresistant strains such as methicillin-resistant *Staphylococcus aureus*. Honey can also be used to prepare Ag and Pt nanoparticles. For instance, Philip[74] demonstrated the pH-controlled synthesis of Ag nanoparticles in the presence of honey that acted as a reducing and stabilizing agent. The author discovered that by simple pH adjustment, it is possible to manufacture highly crystalline nanoparticles of various sizes at room temperature. Indeed, at pH 8.5 predominantly monodispersed and nearly spherical colloidal Ag nanoparticles of size ca. 4 nm were observed. In addition, Venu *et al.*[75] performed honey-mediated synthesis of Pt nanoparticles and nanowires and reached a conversion of platinum ions into ca. 2.2-nm sized nanoparticles at 100 °C. However, when the reaction solution was treated for extended periods of time, Pt nanowires (5 to 15 nm in length) were formed. The resulting nanoparticles were highly crystalline and the structure was face-centered cubic. According to both studies, FTIR data suggested that nanoparticles were bound to protein through the carboxylate ion group.

Furthermore, Varma and co-workers have demonstrated[76] an effective and environmentally friendly synthesis approach for Ag nanoparticles utilizing polyphenols found in tea extract and epicatechin. They also showed the benign nature of these nanoparticles when in contact with human keratinocytes (HaCaTs) used as an the *in vitro* model of exposure. Thus, during a typical synthesis, the target Ag salt was subjected to the tea extract and/or epicatechin at room temperature. As-prepared Ag nanoparticles ranged from 4 to 100 nm, depending on the ratio of water to tea extract used for the synthesis, and from 15 to 26 nm when various ratios of epicatechin were used. Nadagouda and Varma[12] synthesized Ag and Pd nanoparticles utilizing polyphenols from tea and coffee; the nanoparticles were in the size range of 20–60 nm, depending on the ratios of the reaction precursors and crystallized in face-centered cubic symmetry.

2.2.1.8   Vitamins

Due to their unique chemical structure, vitamins (B, C, D, E) can be used to manufacture highly stable and efficient metal nanoparticles. For instance, silver nanoparticles were prepared from silver nitrate in the presence of vitamin C derivative 6-palmitoyl ascorbic acid-2-glucoside (PAsAG), *via* a sonochemical route.[77] An Ag-PAsAG nanocomplex with average size of 5 nm was formed and used to protect DNA from γ-radiation–induced damage. The authors found that the presence of Ag-PAsAG complexes during irradiation inhibited the disappearance of covalently closed circular (ccc) form of plasmid pBR322 with a dose-modifying factor of 1.78. Ag-PAsAG protected cellular DNA from radiation-induced damage as evident from comet assay study on mouse spleen cells, irradiated *ex vivo*.

Chien-Jung *et al.*[78] fabricated Au nanodogbones with simple seeded mediated growth method in the presence of vitamin C. The shapes and morphology of the formed nanoparticles highly depended on the amount of added vitamin C (from 10 to 40 μL). The authors found that the aspect ratios of nanodogbones were in the range from 2.34 to 1.46, and the UV-Vis absorption measurement showed a significant blueshift on the longitudinal surface plasmon resonance band from 713 to 676 nm.

Monodispersed silver nanoparticles were synthesized by a simple "one-step" procedure in the alkaline subphase beneath vitamin E Langmuir monolayers.[79]

The authors proved that the phenolic groups in vitamin E molecules were converted to a quinone structure, and the silver ions were mainly reduced to ellipsoidal and spherical nanoparticles. According to the electron-diffraction pattern, silver nanoparticles were face-centered cubic polycrystalline.

Vitamin B$_2$ was also reported to be an effective reducing as well as capping agent due to its high water solubility, biodegradability and low toxicity in comparison to other available reducing agents.[9,80] Importantly, the addition of a solvent along with B$_2$ may alter the sizes of resulting nanoparticles. Indeed, the average particle sizes of Ag and Pd were 6.1 ± 0.1 nm and 4.1 ± 0.1 nm in ethylene glycol as well as 5.9 ± 0.1 nm and 6.1 ± 0.1 in acetic acid and *N*-methylpyrrolidinone (NMP), respectively. Interestingly, when water was used as a solvent, Ag and Pd nanoparticles self-assembled into rod-like structures. However, in iso-propanol, the nanoparticles yielded wire-like structures with a thickness in the range of 10 to 20 nm and several hundred micrometers in length. Finally, in acetone and acetonitrile, the Ag and Pd nanoparticles were self-assembled into a regular pattern making nanorod structures with thicknesses ranging from 100 to 200 nm and lengths of a few micrometers.[80]

2.3 Other Synthetic Approaches and Further Consideration

The use of environmentally friendly materials in synthesis is not limited to the preparation of metal nanoparticles. It also extends to magnetic nanoparticles, clay-supported nanoparticles, *etc*. Unfortunately, due to the vast amount of recent literature currently available, detailed discussion on preparation of other nanoparticles/nanomaterials/nanocomposites is beyond the scope of this chapter. However, some of the ideas for further consideration are presented here.

It is extremely important to recover and recycle nanoparticles inexpensively and relatively easily once their life time or function is over. Thus, manufacturing metal nanoparticles on an active or sometimes passive core, especially a magnetic one would present great advantages for numerous processes. For many years, magnetic nanoparticles have been extensively used in metal ions and dye recovery, drug delivery, enzyme immobilization, protein and cell separation, magnetic fluids, catalysis, biotechnology/biomedicine, magnetic resonance imaging, data storage and environmental remediation because magnetic separation offers unique advantages of high efficiency, cost effectiveness and rapidity in comparison to other recovery approaches.[14] Currently, magnetic nanoparticles are mostly used as heterogeneous supports in various catalytic transformations providing added benefits of easy recoverability with a simple magnet, eliminating the need of solvent swelling before or catalyst filtration after the process.[81–84] Unfortunately, one of the biggest problems is their intrinsic instability over long periods of time. As magnetic nanoparticle sizes are only <20 nm, they tend to aggregate and agglomerate, which significantly reduces the energy associated with the high surface area to volume ratio. Also, oxidation of the surface in air results in the loss of magnetism and dispersability. Therefore, these nanoparticles (bare or as active supports) must also be stabilized and/or capped with certain nontoxic and environmentally friendly species.

Moreover, to prepare monodisperse magnetic nanoparticles, various organic additives such as anionic surfactants, dispersing agents, proteins, starches, polyelectrolytes, *etc.* must be added for the subsequent use in a number of chemical, biological and physical applications. For instance, to be used as contrast agents for nuclear magnetic resonance (NMR), iron-oxide magnetic nanoparticles must be coated with a fluorescent silica shell, which would allow their visualization by optical means, and then grafted with hyperbranched polyglycerol (HPG) to attain water dispersibility and macrophage-evading properties.[85] Furthermore, bifunctional precious metal-magnetic nanoparticles are constantly gaining increased attention due to their potential applications in novel electrical, optical, magnetic, catalytic and sensing technologies. Thus, the Fe_2O_3/Ag core–shell magnetic nanocomposite (sizes ranging from 30 to 60 nm) can be efficiently prepared *via* a simple method at room temperature in the presence of sodium hydroxide (NaOH), TSA-doped polyaniline, dimethylsulfoxide (DMSO) and dimethylformamide (DMF) without calcinations.[86] In recent years, spinel ferrites[87–90] – magnetic $CoFe_2O_4$ nanoparticles in particular, became one of the most important constituents in synthetic and biological chemistry as multiferroic materials. Additionally, they can be the magnetic labels of biological systems due to their unusual properties such as doping or strain-enhanced coercivity and photoinduced magnetic effects among a few.[91] For instance, synthesis of stable spinel cobalt ferrite magnetic nanoparticles (40–50 nm) can be accomplished using a combined sonochemical and coprecipitation technique in aqueous medium without any surfactant or organic capping agent at 60 °C.[91]

Hydroxyapatites, which are the main constituent of bones and teeth and are widely used in such fields as bone repairs, bone implants, bioactive materials, purification and separation of biological molecules, can also be used to synthesize effective and inexpensive magnetic nanocomposites for various industrial applications, including remediation of soils, water and wastewater.[92] Furthermore, magnetic nanoparticles can also be functionalized with β-cyclodextrin and pluronic polymer (F127) at room temperature under a nitrogen atmosphere for successful drug-delivery applications.[93]

In addition, to avoid agglomeration, different porous solids such as clays can also be utilized as active or passive supporting media to disperse various nanoparticles of active phases. Porous materials offer high surface areas and easily accessible channels for molecules and improve their diffusion and catalytic properties.[94,95] According to Li *et al.*,[94] clay supported nanoparticles are prepared by exchanging the interlamellar alkali cations with particular metallic polyoxocations in aqueous solution. Subsequently, elevated temperatures during calcinations convert these polyoxocations to nanoparticles of the corresponding metals anchored on the silicate platelets. Essentially, the clay layers partition metal nanoparticles, preventing them from sintering, whereas a wide interlayer gallery prevents pores from being blocked by the metal-oxide particles. In this manner, clay-supported or pillared nanoparticles can be used for numerous industrial applications in view of their large surface areas, good thermal stability and special surface acidity.

For instance, Li *et al.*[94] synthesized iron, chromium, cobalt, manganese and cerium pillared clay (laponite) with the surface areas of 510 to 640 m^2 g^{-1} and total pore volume of 0.43 to 0.51 cm^3 g^{-1} in the presence of a stabilizer – acetic acid. According to the findings, metal precursors substituted sodium ions in laponite forming mesoporous solids with well dispersed metal oxides and high surface area. A significant increase in the porosity after calcination was attributed to delamination of the laponite clay caused by the reaction between the acidic metal precursor solutions and the silicate platelets.

Mesoporous alumina layers have attracted attention for their potential use in ultrafiltration of salts, as a heterogeneous catalyst support, an adsorbent in environmental cleanup, and in petroleum refinement.[96,97] To be effectively utilized for all the aforementioned applications, it is important to control the hydrolysis rate of the inorganic precursors using simple and inexpensive routes.[98–102] Thus, Wan *et al.*[99,100] introduced a novel and "green" route to synthesize mesoporous alumina thin films from inexpensive and commercially available copolymer with aluminum chloride or nitrate (salts) in an EtOH–surfactant–NH$_3 \cdot$ H$_2$O–salts (EsNs) system through the evaporation-induced self-assembly (EISA) method. They demonstrated that the binding of surfactant and NH$_3 \cdot$ H$_2$O for the formation of hydrogen bond between them in the EsNs system, controls the fast hydrolysis rate of the inorganic species and the pore structures can be changed from 2D hexagonal to cubic structures just by changing the aging time of the sol.[99] They also demonstrated a facile, effective and highly controlled synthesis of mesoporous alumina thin films with ordered 2D hexagonal or wormhole-like mesoporosity using aluminum chloride as the inorganic precursors and triblock copolymer (Pluronic F-127 or P-123) as the structure-directing agents, through the evaporation-induced self-assembly (EISA) method.[100] Jiang *et al.*[103] used a simple and facile synthesis approach and obtained *P6$_3$/mmc* [001] as well as *Fm-3m* [111] mesostructures by using triblock copolymer Pluronic P123 as the structure-directing agent and calcination at 1000 °C. These films exhibited high thermal stability up to 1000 °C, which can be utilized in the preparation of catalyst supports and separators.

Ionic liquids can also be used to synthesize mesoporous alumina. For instance, large mesoporous gamma-alumina can be synthesized through a thermal process without the postaddition of molecular or organic solvents at ambient pressure by using the dual functions of 1-hexadecyl-3-methylimidazolium chloride (C$_{16}$MimCl) as room-temperature ionic liquids (RTILs), which performs templating and cosolvent functions. Both manufactured C$_{16}$MimCl/boehmite hybrid and γ-alumina exhibited the nanostructure consisting of randomly debundled nanofibers embedded in worm-like porous networks with the length of ca. 40–60 nm and a diameter of ca. 1.5–3 nm.[104] Moreover, gamma-alumina had good thermal stability and reasonable acidic sites with the surface area of 470 m^2 g^{-1}, a pore volume of 1.46 cm^3 g^{-1} and a pore size of 9.9 nm after calcinations at 550 °C in air.

Mesoporous alumina can also be doped with various metals to increase the stability and expand the array of applications. The size compatibility between La^{3+} ions (0.115 nm) and the covalent radius of aluminum (0.118 nm) permits addition of lanthanum into the ordered mesoporous alumina (OMA) structure *via* a previously reported improvement strategy.[105] Therefore, highly thermally stable (up to 1000 °C) lanthanum-doped ordered mesoporous alumina with surface area of 305 m^2 g^{-1} after calcinations at 400 °C and 205 m^2 g^{-1} after treatment at 1000 °C has been prepared in a sol-gel system with salicylic acid and citric acid as interfacial protectors.[106] In addition, Pt can also be added to gamma-alumina structure for enhanced properties. Dandapat *et al.*[107] manufactured worm-hole-like disordered gamma-alumina mesostructures with BET surface area of 171 m^2 g^{-1}, a pore volume of 0.188 cm^3 g^{-1} and a mean pore diameter of 4.3 nm. applying single dip-coating method using boehmite (AlOOH) sols derived from aluminum tri-sec-butoxide in the presence of cetyltrimethylammonium bromide (CTAB) as structure-directing agent. To enhance the properties, $\sim$2.7 mol% of Pt were added at 500 °C, which resulted in the formation of Pt-incorporated gamma-alumina film with a BET surface area of 101 m^2 g^{-1} and a pore volume of 0.119 cm^3 g^{-1}. According to FESEM and TEM analysis, uniform distribution of Pt nanoparticles (size from 2 to 8.5 nm and diameter of 4.9 nm) was observed. Such a composite was used to reduce hexacyanoferrate III) ions by thiosulfate to ferrocyanide, and *p*-nitrophenol to *p*-aminophenol.[107]

Not only ionic liquids but also various surfactants can be used to manufacture mesoporous alumina.[108] Thus, gamma-alumina with controlled mesoporosity can be synthesized through the scaffolding of pseudoboehmite nanoparticles in the presence of a non-ionic surfactant such as the porogen (Tergitol 15-S-7 ($C_{15}H_{33}(OC_2H_4)_7OH$).[96] It was reported that calcination of the composite at 500 °C in air removed the surfactant and concomitantly converted the pseudoboehmite crystallites to gamma-alumina with pore size in the range of 3.5–15 nm and surface area of 296–321 m^2 g^{-1}, through topochemical transformation with the retention of the scaffold structure. Cetyltrimethylammonium bromide (CTAB) can also be used to prepare gamma-alumina's. CTAB addition to aluminum sulfate as a precursor, urea as precipitating agent and sodium tartrate as cotemplate were prepared using a facile on-pot synthesis.[108] According to the authors, the formation mechanism is explained whereby tartrate interacts with aluminum and CTAB simultaneously to form an intermediate as the building blocks of the final mesostructured hybrid.

In addition to mesoporous alumina, macroporous magnesium oxide, gamma-alumina and alumina zirconia solids with crystalline mesoporous walls can also be manufactured using a green chemistry and engineering approach. For instance, three-dimensionally (3D) ordered macroporous (3DOM) MgO, γ-Al_2O_3, $Ce_{0.6}Zr_{0.4}O_2$, and $Ce_{0.7}Zr_{0.3}O_2$ with polycrystalline mesoporous walls have been successfully fabricated with the triblock copolymer $EO_{106}PO_{70}EO_{106}$ (Pluronic F127) and regularly packed monodispersive polymethyl methacrylate (PMMA) microspheres as the template and magnesium, aluminum, cerium and zirconium nitrate(s), or aluminum isopropoxide as the metal source.[109]

The authors demonstrated that the introduction of surfactant F127, significantly enhanced the surface areas of the 3DOM metal oxides. With PMMA and F127 in 40% ethanol solution, it is possible to generate well-arrayed 3DOM MgO with a surface area of $243 \, m^2 \, g^{-1}$ and 3DOM $Ce_{0.6}Zr_{0.4}O_2$ with a surface area of $100 \, m^2 \, g^{-1}$; with PMMA and F127 in an ethanol $-$ HNO_3 solution, it is possible to obtain 3DOM γ-Al_2O_3 with a surface area of $145 \, m^2 \, g^{-1}$. The 3DOM MgO and 3DOM γ-Al_2O_3 can be effectively utilized for CO_2 absorption, whereas the 3DOM $Ce_{0.6}Zr_{0.4}O_2$ could be used where low-temperature reducing properties are required.

Mesoporous nanoparticles for lithium batteries can be manufactured using an electrochemical route.[110,111] Qian *et al.*[110] suggested a new electrochemical method to manufacture amorphous mesoporous $FePO_4$ with particle sizes ranging from 20 to 80 nm, BET surface area of $65.2 \, m^2 \, g^{-1}$ and a dominant pore diameter of 23.6 nm. These nanoparticles were used to prepare $LiFePO_4/C$ nanocrystals with porous structure that were deployed as $LiFePO_4/C$ cathode material with excellent cycling performances. It was demonstrated that at a $0.5 \, °C$ rate, the discharge capability was above $140.0 \, mA \, h \, g^{-1}$ and the capacity retention rate was higher than 98% after 50 cycles. Moreover, $Au/Li_4Ti_5O_{12}$ spheres based on *in situ* conversion of titanium glycolate in LiOH aqueous solution were reported using a facile approach comprising acetone and ethylene glycol (EG).[111] Titanium glycolate has several advantages over other salts that include fast and easy preparation and direct reaction with LiOH without introducing TiO_2 impurities. The as-prepared mesoporous $Au/Li_4Ti_5O_{12}$ spheres exhibited large specific surface area ($166 \, m^2 \, g^{-1}$) and good electronic conduction enhanced by Au nanoparticles when used as an anode electrode material.

Nanocomposites can be synthesized utilizing greener routes using graphene sheets and gold, zirconia, silver, platinum, SnO_2, metal alloys for batteries, sensors and other applications.[112–118] For instance, a controllable approach for electrochemical synthesis of a nanocomposite with increased electro-catalytic properties, good storage stability, reproducibility, and selectivity made from electrochemically reduced graphene oxide (ERGO) and gold nanoparticles has been developed.[112] According to SEM analysis, a homogeneous distribution of gold nanoparticles on the graphene sheets was observed and its oxidation peak current was linearly proportional to the concentration of dopamine (DA) in the range from 0.1 to 10 μM, with a detection limit of 0.04 μM (at S/N = 3). In addition, Gong *et al.*[113] reported a facile, one-step electrochemical synthesis of high-quality zirconia nanoparticles decorated graphene nanosheets (labeled as ZrO_2NPs-GNs) with thickness of about 4 nm onto a cathodic substrate. Such a nanostructured composite, combining the advantages of ZrO_2NPs (high recognition and enrichment capability for phosphoric moieties) together with GNs (large surface area and high conductivity), was highly efficient to capture organophosphate pesticides (OPs); authors reported the detection limit for methyl parathion (MP) in aqueous solutions of $0.6 \, ng \, mL^{-1}$ (S/N = 3). Liang *et al.*[114] proposed a facile, economic and effective one-step solution-based process to *in situ* synthesize

SnO$_2$/graphene (SG) nanocomposites in the presence of dimethyl sulfoxide (DMSO) and H$_2$O as both, the solvent and reactant. Furthermore, silver can also be incorporated into the graphene structure. Thus, Zhang *et al.*[115] demonstrated a green, cost-effective, one-pot route to manufacture Ag nanoparticles-graphene (AgNPs–G) nanocomposites in aqueous solution in the presence of tannic acid (TA), an environmentally friendly and water-soluble polyphenol, as a reducing agent. These nanocomposites exhibited excellent SERS activity (amperometric response time of less than 2 s) as SERS substrates and had notable catalytic performance toward the reduction of H$_2$O$_2$. Accordingly, the linear range was estimated to be from 1×10^{-4} M to 0.01 M ($r = 0.999$) and the detection limit was estimated to be 7×10^{-6} M at a signal-to-noise ratio of 3. Moreover, a glucose biosensor was further synthesized by immobilizing glucose oxidase (GOD) into chitosan–AgNPs–G nanocomposite film on the surface of a glassy carbon electrode (GCE). Such a sensor had a good response to glucose, and the linear response range was estimated to be from 2 to 10 mM ($R = 0.996$) at -0.5 V. The detection limit of 100 μM was achieved at a signal-to-noise ratio of 3.[115] A simple, fast, green and controllable approach was also developed for electrochemical synthesis of a novel nanocomposite of electrochemically reduced graphene oxide (ERGO) and gold–palladium (1:1) bimetallic nanoparticles (AuPdNPs), without the addition of any reducing reagent.[116] As demonstrated by the authors, such a nanocomposite exhibited excellent biocompatibility, enhanced electron transfer kinetics and large electroactive surface area, and was highly sensitive and stable towards oxygen reduction. The manufactured biosensor had a detection limit of 6.9 μM, a linear range up to 3.5 mM and a sensitivity of 266.6 μA mM^{-1} cm^{-2}. Also, it had acceptable reproducibility and good accuracy with negligible interferences from common oxidizable interfering species.[116] Qian *et al.*[117] proposed a novel and green approach for the synthesis of graphene nanosheets (GNS) and Pt nanoparticles–graphene nanosheets (Pt/GNS) hybrid materials, employing graphene oxide (GO) as a precursor and sodium citrate as an environmentally friendly reducing and stabilizing agent; resulting Pt/GNS exhibited excellent electrochemical activity and stability.

The reverse micelle route to synthesize metal and metal oxide nanoparticles provides excellent control over the size and shape of the nanoparticles.[119] Exchanges between the micelles in the solution result in the uniform distribution of micelles volume throughout the solution.[120–123] As reported by Heshmatpour *et al.*,[120] stable, effective and reusable (up to six times) metal nanoparticles have been prepared using a water-in-oil microemulsion system of water/dioctyl sulfosuccinate sodium salt (aerosol-OT, AOT)/isooctane at 25 °C. These nanoparticles exhibited high activity in the C–C coupling reactions of the iodobenzene with the styrene. A facile synthesis of Pd/SiO$_2$ nanobeads with tiny Pd clusters of 2 nm in diameter was reported *via* a sol–gel process for SiO$_2$ by using a water-in-oil microemulsion with Pd complexes and subsequent hydrogen reduction by heat treatment in the presence of Igepal CO-630 as a nonionic surfactant, aqueous ammonia solution, and cyclohexane.[121] These nanobeads were employed in Suzuki coupling reactions with various substrates and exhibited excellent catalytic properties. Amorphous ferric molybdate nanoparticles that can be employed as effective catalysts in many organic transformations can also be synthesized *via* microemulsion method using ferric chloride and ammonium heptamolybdate as starting materials in the presence of cationic surfactant (CTAB), isooctane as oil phase, n-butanol as cosurfactant.[122] After calcination at 450 °C, nanoparticles consisted of relatively monodispersed nanoparticles with 80–90 nm in size. He *et al.*[123] reported green synthesis of spherical carbon supported AuNi electrocatalyst (AuNi/C) with average diameter of 3–9 nm for electrochemical oxidation of borohydride using water-in-oil microemulsion of water/AOT/n-heptane.[123] They found that the molar ratios of water to surfactant AOT ($R\omega$) was less than 10, and the particle size increased with increasing $R\omega$ value, and the particle size was decreased after $R\omega > 10$ due to two-phase separation of the microemulsion.

2.4 Conclusions

Currently, there is a need to develop reliable and environmentally friendly processes to synthesize metal nanoparticles minimizing or even completely eliminating the use of hazardous chemicals. Only by adapting benign synthesis approaches will we be able to provide a background for more sustainable and "greener" nanomaterials preparative methods. It is more likely that in the nearest future manufacturing of nanoparticles will finally emerge from the laboratory scale and will be used in the synthesis of large quantities structurally well-defined, functionalized, stable and effective nanoparticles applicable for a great array of applications. To accomplish that goal, the utilization of natural resources such as biological systems becomes vital. The use of biological organisms and systems in the synthesis of nanoparticles is characterized by processes that occur close to ambient temperatures and pressures and at ~ neutral pH utilizing benign reaction precursors. It is known that of all the numerous biological systems, bacteria are relatively easy to control genetically, whereas plants and fungi are much easier to handle during downstream processing and large-scale production. Regardless the use of organisms, in order to maximize their potential in nanoparticle synthesis, it is essential to research and understand the biochemical and molecular mechanisms of nanoparticle formation.

This chapter is focused on the use of various organisms, such as plants, bacteria, fungi, yeasts, *etc.* in environmentally friendly preparation of metal nanoparticles under mild reaction conditions and without (if possible) the use of hazardous chemicals. Recent advances and state-of-the-art in terms of synthesis precursors, shape control and surface functionalization identifying potential future challenges are well described. Detailed examination of numerous syntheses examples and case studies is presented, however, suggesting that more advancement is necessary in developing large-scale continuous preparation of metal nanoparticles utilizing more wastes and "sacrificial" organisms rather than going to the field and selecting potential candidates that require land and energy to grow.

The key findings in the environmentally friendly synthesis of metal nanoparticles are as follows: developed and still underdeveloped approaches 1) eliminate the use of toxic reagents and organics solvents, 2) has very little or no hazardous byproducts formed during the synthesis process, 3) provide better control of shapes, sizes and dispersivity of metal nanoparticles and 4) reduce the need for purification of the manufactured nanoparticle, which in turn eliminates the use of extensive amounts of organic solvents that are hazardous to the environment.

Despite a good progress, there are still issues that have to be resolved in the future: 1) identification of the risks associated with biological and human exposure to nanomaterials, 2) development of even "greener" synthesis approaches, eliminating energy intensive processes and increasing the use of various wastes as reources and 3) development of analytical techniques that can perform *in-situ* routine analysis of nanoparticles for morphology, composition, structure, purity, *etc.* regardless of the experimental matrix. Therefore, despite the encouraging innovations and proposed approaches, further research is warranted to improve current methods and processes that will provide opportunities and challenges for the scientific community and general public for the years to come.

References

1. M. J. Eckelman, J. B. Zimmerman and P. T. Anastas, *J. Ind. Ecol.*, 2008, **12**, 316–328.
2. P. Christian, F. Von der Kammer, M. Baalousha and T. Hofmann, *Ecotoxicology*, 2008, **17**, 326–343.
3. S. Iravani, *Green Chem*, 2011, **13**, 2638–2650.
4. A. Huczko, *Appl. Phys. A - Mater. Sci. Process.*, 2000, **70**, 365–376.
5. A. D. Pomogailo and V. N. Kestelman, eds., *Metallopolymer Nanocomposites*, Springer Berlin Heidelberg New York, 2005.
6. D. Stubbs and P. Gilman, *Oak Ridge Center for Advanced Studies*, 2007.
7. B. Baruwati and R. S. Varma, *ChemSusChem*, 2009, **2**, 1041–1044.
8. M. N. Nadagouda, V. Polshettiwar and R. S. Varma, *J. Mater. Chem.*, 2009, **19**, 2026–2031.
9. M. N. Nadagouda and R. S. Varma, *Green Chem*, 2006, **8**, 516–518.
10. M. N. Nadagouda and R. S. Varma, *Biomacromolecules*, 2007, **8**, 2762–2767.
11. M. N. Nadagouda and R. S. Varma, *Cryst. Growth Des.*, 2007, **8**, 291–295.
12. M. N. Nadagouda and R. S. Varma, *Green Chem*, 2008, **10**, 859–862.
13. J. A. Dahl, B. L. S. Maddux and J. E. Hutchison, *Chem. Rev.*, 2007, **107**, 2228–2269.
14. J. Virkutyte and R. S. Varma, *Chem. Sci.*, 2011, **2**, 837–846.

15. A. Jebali, F. Ramezani and B. Kazemi, *J. Cluster Sci.*, 2011, **22**, 225–232.
16. D. Mandal, M. Bolander, D. Mukhopadhyay, G. Sarkar and P. Mukherjee, *Appl. Microbiol. Biotechnol.*, 2006, **69**, 485–492.
17. K. B. Narayanan and N. Sakthivel, *Adv. Coll. Interf. Sci.*, 2010, **156**, 1–13.
18. K. B. Narayanan and N. Sakthivel, *J. Haz. Mater.*, 2011, **189**, 519–525.
19. A. M. Fayaz, K. Balaji, M. Girilal, R. Yadav, P. T. Kalaichelvan and R. Venketesan, *Nanomed. Nanotechnol. Biol. Medicine*, 2010, **6**, 103–109.
20. P. Mukherjee, A. Ahmad, D. Mandal, S. Senapati, S. R. Sainkar, M. I. Khan, R. Parishcha, P. V. Ajaykumar, M. Alam, R. Kumar and M. Sastry, *Nano Lett.*, 2001, **1**, 515–519.
21. V. Bansal, A. Ahmad and M. Sastry, *J. Am. Chem. Soc.*, 2006, **128**, 14059–14066.
22. K. N. Thakkar, S. S. Mhatre and R. Y. Parikh, *Nanomed. Nanotechnol. Biol. Medicine*, 2010, **6**, 257–262.
23. A. Ahmad, S. Senapati, M. I. Khan, R. Kumar and M. Sastry, *Langmuir*, 2003, **19**, 3550–3553.
24. K. Kalishwaralal, V. Deepak, S. Ram Kumar Pandian, M. Kottaisamy, S. BarathManiKanth, B. Kartikeyan and S. Gurunathan, *Coll. Surf. B.*, 2010, **77**, 257–262.
25. V. S. Coker, J. A. Bennett, N. D. Telling, T. Henkel, J. M. Charnock, G. van der Laan, R. A. D. Pattrick, C. I. Pearce, R. S. Cutting, I. J. Shannon, J. Wood, E. Arenholz, I. C. Lyon and J. R. Lloyd, *ACS Nano*, 2010, **4**, 2577–2584.
26. A. K. Jha, K. Prasad and A. R. Kulkarni, *Coll. Surf. B.*, 2009, **71**, 226–229.
27. Y. Konishi, K. Ohno, N. Saitoh, T. Nomura, S. Nagamine, H. Hishida, Y. Takahashi and T. Uruga, *J. Biotechnol.*, 2007, **128**, 648–653.
28. T. Ogi, N. Saitoh, T. Nomura and Y. Konishi, *J. Nanopart. Res.*, 2010, **12**, 2531–2539.
29. M. Kowshik, A. Shriwas, K. Sharmin, W. Vogel, J. Urban, S. K. Kulkarni and K. M. Paknikar, *Nanotechnol.*, 2003, **14**, 95.
30. H.-A. Lim, A. Mishra and S.-I. Yun, *J. Nanosci. Nanotechnol.*, 2011, **11**, 518–522.
31. S. Kumar, A. R, P. Arumugam and S. Berchmans, *ACS Appl. Mater. Interf.*, 2011, **3**, 1418–1425.
32. A. Mishra, S. K. Tripathy and S.-I. Yun, *J. Nanosci. Nanotechnol.*, 2011, **11**, 243–248.
33. M. Subramanian, N. M. Alikunhi and K. Kandasamy, *Adv. Sci. Lett.*, 2010, **3**, 428–433.
34. P. S. Pimprikar, S. S. Joshi, A. R. Kumar, S. S. Zinjarde and S. K. Kulkarni, *Coll. Surf. B.*, 2009, **74**, 309–316.
35. K. Govindaraju, S. Basha, V. Kumar and G. Singaravelu, *J. Mater. Sci.*, 2008, **43**, 5115–5122.
36. M. Shakibaie, H. Forootanfar, K. Mollazadeh-Moghaddam, Z. Bagherzadeh, N. Nafissi-Varcheh, A. R. Shahverdi and M. A. Faramarzi, *Biotechnol. Appl. Biochem.*, 2010, **57**, 71–75.
37. J. Xie, J. Y. Lee, D. I. C. Wang and Y. P. Ting, *Small*, 2007, **3**, 672–682.

38. S. A. Dahoumane, C. Djediat, C. Yéprémian, A. Couté, F. Fiévet, T. Coradin and R. Brayner, *Biotechnol. Bioeng.*, 2012, **109**, 284–288.
39. E. Torres-Chavolla, R. J. Ranasinghe and E. C. Alocilja, *IEEE Trans. Nanotechnol.*, 2010, **9**, 533–538.
40. A. Absar, S. Satyajyoti, M. I. Khan, K. Rajiv, R. Ramani, V. Srinivas and S. Murali, *Nanotechnology*, 2003, **14**, 824.
41. D. Inbakandan, R. Venkatesan and S. Ajmal Khan, *Coll. Surf. B.*, 2010, **81**, 634–639.
42. J. L. Gardea-Torresdey, E. Gomez, J. R. Peralta-Videa, J. G. Parsons, H. Troiani and M. Jose-Yacaman, *Langmuir*, 2003, **19**, 1357–1361.
43. J. L. Gardea-Torresdey, J. G. Parsons, E. Gomez, J. Peralta-Videa, H. E. Troiani, P. Santiago and M. J. Yacaman, *Nano Lett.*, 2002, **2**, 397–401.
44. J. L. Gardea-Torresdey, K. J. Tiemann, G. Gamez, K. Dokken, I. Cano-Aguilera, L. R. Furenlid and M. W. Renner, *Environ. Sci. Technol.*, 2000, **34**, 4392–4396.
45. J. L. Gardea-Torresdey, K. J. Tiemann, G. Gamez, K. Dokken, S. Tehuacanero and M. José-Yacamán, *J. Nanopart. Res.*, 1999, **1**, 397–404.
46. J. L. Gardea-Torresdey, K. J. Tiemann, J. G. Parsons, G. Gamez and M. J. Yacaman, *Adv. Environ. Res.*, 2002, **6**, 313–323.
47. M. Montes, A. Mayoral, F. Deepak, J. Parsons, M. Jose-Yacamán, J. Peralta-Videa and J. Gardea-Torresdey, *J. Nanopart. Res.*, 2011, **13**, 3113–3121.
48. S. Kaviya, J. Santhanalakshmi, B. Viswanathan, J. Muthumary and K. Srinivasan, *Spectrochim, Acta A.*, 2011, **79**, 594–598.
49. S. Mondal, N. Roy, R. A. Laskar, I. Sk, S. Basu, D. Mandal and N. A. Begum, *Coll. Surf. B.*, 2011, **82**, 497–504.
50. D. Philip, *Physica E: Low-dimen. Syst. Nanostruct.*, 2010, **42**, 1417–1424.
51. N. Roy, S. Mondal, R. A. Laskar, S. Basu, D. Mandal and N. A. Begum, *Coll. Surf. B*, 2010, **76**, 317–325.
52. D. S. Sheny, J. Mathew and D. Philip, *Spectrochim, Acta A*, 2011, **79**, 254–262.
53. A. R. Vilchis-Nestor, V. Sánchez-Mendieta, M. A. Camacho-López, R. M. Gómez-Espinosa, M. A. Camacho-López and J. A. Arenas-Alatorre, *Mater. Lett.*, 2008, **62**, 3103–3105.
54. G. Singhal, R. Bhavesh, K. Kasariya, A. Sharma and R. Singh, *J. Nanopart. Res.*, 2011, **13**, 2981–2988.
55. M. Valodkar, P. S. Nagar, R. N. Jadeja, M. C. Thounaojam, R. V. Devkar and S. Thakore, *Coll. Surf. A.*, 2011, **384**, 337–344.
56. V. T. P. Vinod, P. Saravanan, B. Sreedhar, D. K. Devi and R. B. Sashidhar, *Coll. Surf. B.*, 2011, **83**, 291–298.
57. A. Karimi-Zarchi, N. Mokhtari, M. Arfan, T. Rehman, M. Ali, M. Amini, R. Faridi Majidi and A. Shahverdi, *Appl. Phys. A - Maters. Sci. Process.*, 2011, **103**, 349–353.
58. T. C. Prathna, N. Chandrasekaran, A. M. Raichur and A. Mukherjee, *Coll. Surf. A.*, 2011, **377**, 212–216.

59. K. B. Narayanan and N. Sakthivel, *Mater. Res. Bull.*, 2011, **46**, 1708–1713.
60. K. B. Narayanan and N. Sakthivel, *Mater. Charact.*, 2010, **61**, 1232–1238.
61. K. Leonard, B. Ahmmad, H. Okamura and J. Kurawaki, *Coll. Surf. B.*, 2011, **82**, 391–396.
62. E. J. Guidelli, A. P. Ramos, M. E. D. Zaniquelli and O. Baffa, *Spectrochim, Acta A.*, 2011, **82**, 140–145.
63. S. A. Aromal, V. K. Vidhu and D. Philip, *Spectrochim, Acta A.*, 2012, **85**, 99–104.
64. V. K. Vidhu, S. A. Aromal and D. Philip, *Spectrochim. Acta A.*, 2011, **83**, 392–397.
65. D. Philip, *Spectrochim. Acta A.*, 2011, **78**, 327–331.
66. R. Das, N. Gogoi and U. Bora, *Biopr. Biosystems Eng.*, 2011, **34**, 615–619.
67. K. Mohan Kumar, B. K. Mandal, M. Sinha and V. Krishnakumar, *Spectrochim. Acta A.*, 2012, **86**, 490–494.
68. E. C. Njagi, H. Huang, L. Stafford, H. Genuino, H. M. Galindo, J. B. Collins, G. E. Hoag and S. L. Suib, *Langmuir*, 2010, **27**, 264–271.
69. M. H. El-Rafie, M. E. El-Naggar, M. A. Ramadan, M. M. G. Fouda, S. S. Al-Deyab and A. Hebeish, *Carboh. Polymers*, 2011, **86**, 630–635.
70. X. Gao, L. Wei, H. Yan and B. Xu, *Mater. Lett.*, 2011, **65**, 2963–2965.
71. E. S. Abdel-Halim and S. S. Al-Deyab, *Carboh. Polymers*, 2011, **86**, 1615–1622.
72. J. Virkutyte, V. Jegatheesan and R. S. Varma, *Biores. Technol.*, 2012, **113**, 288–293.
73. A. Panácek, L. Kvítek, R. Prucek, M. Kolář, R. Večeřová, N. Pizúrová, V. K. Sharma, T. j. Nevěčná and R. Z. bořil, *J. Phys. Chem. B*, 2006, **110**, 16248–16253.
74. D. Philip, *Spectrochim. Acta A.*, 2010, **75**, 1078–1081.
75. R. Venu, T. S. Ramulu, S. Anandakumar, V. S. Rani and C. G. Kim, *Coll. Surf. A.*, 2011, **384**, 733–738.
76. M. C. Moulton, L. K. Braydich-Stolle, M. N. Nadagouda, S. Kunzelman, S. M. Hussain and R. S. Varma, *Nanoscale*, 2010, **2**, 763–770.
77. D. K. Chandrasekharan, P. K. Khanna, T. V. Kagiya and C. K. K. Nair, *Cancer Biother. Radiopharma.*, 2011, **26**, 249–257.
78. H. Chien-Jung, C. Pin-Hsiang, W. Yeong-Her, M. Teen-Hang and Y. Cheng-Fu, *Nanotechnol.*, 2007, **18**, 395603.
79. L. Zhang, Y. Shen, A. Xie, S. Li, B. Jin and Q. Zhang, *J. Phys. Chem. B*, 2006, **110**, 6615–6620.
80. M. N. Nadagouda and R. S. Varma, *J. Nanomater.* Doi: 10.1155/2008/782358.
81. R. Abu-Reziq, D. Wang, M. Post and H. Alper, *Chem. Mater.*, 2008, **20**, 2544–2550.
82. B. Baruwati, V. Polshettiwar and R. S. Varma, *Tetrahedron Lett.*, 2009, **50**, 1215–1218.
83. V. Polshettiwar and R. S. Varma, *Org. Biomol. Chem.*, 2009, **7**, 37–40.

84. V. Polshettiwar and R. S. Varma, *Chem.-Eur. J.*, 2009, **15**, 1582–1586.
85. L. Wang, K. G. Neoh, E.-T. Kang and B. Shuter, *Biomater.*, 2011, **32**, 2166–2173.
86. Y. Sun, G. Guo, B. Yang, X. Zhou, Y. Liu and G. Zhao, *J. Non-Cryst. Solids*, 2011, **357**, 1085–1089.
87. Ü. Özgür, Y. Alivov and H. Morkoç, *J. Mater. Sci. Mater. Electron.*, 2009, **20**, 789–834.
88. N.-H. Li, S.-L. Lo, C.-Y. Hu, C.-H. Hsieh and C.-L. Chen, *J. Haz. Mater.*, 2011, **190**, 597–603.
89. K. P. Naidek, F. Bianconi, T. C. R. da Rocha, D. Zanchet, J. A. Bonacin, M. A. Novak, M. das Graças Fialho, Vaz and H. Winnischofer, *J. Coll. Interf. Sci.*, 2011, **358**, 39–46.
90. J. D. Adam, L. E. Davis, G. F. Dionne, E. F. Schloemann and S. N. Stitzer, *Microw. Theory Technol.*, 2002, **50**, 721–737.
91. K. K. Senapati, C. Borgohain and P. Phukan, *J. Molec. Catal A - Chem.*, 2011, **339**, 24–31.
92. Z.-p. Yang, X.-y. Gong and C.-j. Zhang, *Chem. Eng. J.*, 2011, **165**, 117–121.
93. M. M. Yallapu, S. F. Othman, E. T. Curtis, B. K. Gupta, M. Jaggi and S. C. Chauhan, *Biomater.*, 2011, **32**, 1890–1905.
94. J. J. Li, Z. Mu, X. Y. Xu, H. Tian, M. H. Duan, L. D. Li, Z. P. Hao, S. Z. Qiao and G. Q. Lu, *Micropor. Mesopor. Mater.*, 2008, **114**, 214–221.
95. E. G. Garrido-Ramírez, B. K. G. Theng and M. L. Mora, *Appl. Clay Sci.*, 2011, **47**, 182–192.
96. Z. Zhang and T. J. Pinnavaia, *Langmuir*, 2010, **26**, 10063–10067.
97. Z. Wu, Q. Li, D. Feng, P. A. Webley and D. Zhao, *J. Am. Chem. Soc.*, 2010, **132**, 12042–12050.
98. C. Márquez-Alvarez, N. Žilková, J. Pérez-Pariente and J. Čejka, *Catal. Rev.*, 2008, **50**, 222–286.
99. L. Wan, H. Fu, K. Shi and X. Tian, *Micropor. Mesopor. Mater.*, 2008, **115**, 301–307.
100. L. Wan, H. Fu, K. Shi and X. Tian, *Mater. Lett.*, 2008, **62**, 1525–1527.
101. W. Cai, J. Yu, C. Anand, A. Vinu and M. Jaroniec, *Chem. Mater.*, 2011, **23**, 1147–1157.
102. L.-L. Li, W.-T. Duan, Q. Yuan, Z.-X. Li, H.-H. Duan and C.-H. Yan, *Chem. Comm.*, 2009, 6174–6176.
103. X. Jiang, H. Oveisi, Y. Nemoto, N. Suzuki, K. C. W. Wu and Y. Yamauchi, *Dalton Trans.*, 2011, **40**, 10851–10856.
104. S. H. Park, Y.-S. Yang, W. H. Jun, Hong and J. K. Kang, *Chem. Mater.*, 2007, **19**, 535–542.
105. F. Huang, Y. Zheng, G. Cai, Y. Zheng, Y. Xiao and K. Wei, *Scr. Mater.*, 2010, **63**, 339–342.
106. Q. Sun, Y. Zheng, Y. Zheng, Y. Xiao, G. Cai and K. Wei, *Scr. Mater.*, 2011, **65**, 1026–1029.
107. A. Dandapat, D. Jana and G. De, *ACS Appl. Mater. Interf.*, 2009, **1**, 833–840.

108. M. B. Yue, T. Xue, W. Q. Jiao, Y. M. Wang and M.-Y. He, *Solid State Sci.*, 2011, **13**, 409–416.
109. H. Li, L. Zhang, H. Dai and H. He, *Inorg. Chem.*, 2009, **48**, 4421–4434.
110. L. Qian, Y. Xia, W. Zhang, H. Huang, Y. Gan, H. Zeng and X. Tao, *Micropor. Mesopor. Mater.*, 2012, **152**, 128–133.
111. C. C. Li, Q. H. Li, L. B. Chen and T. H. Wang, *ACS Appl. Mater. Interf.*, 2012, **4**, 1233–1238.
112. S.-J. Li, D.-H. Deng, Q. Shi and S.-R. Liu, *Microchimica Acta*, 2012, **In press**.
113. J. Gong, X. Miao, H. Wan and D. Song, *Sens. Actuators B - Chem.*, 2012, **162**, 341–347.
114. J. Liang, W. Wei, D. Zhong, Q. Yang, L. Li and L. Guo, *ACS Appl. Mater. Interf.*, 2011, **4**, 454–459.
115. Y. Zhang, S. Liu, L. Wang, X. Qin, J. Tian, W. Lu, G. Chang and X. Sun, *RSC Advances*, 2012, **2**, 538–545.
116. J. Yang, S. Deng, J. Lei, H. Ju and S. Gunasekaran, *Biosens. Bioelectr.*, 2011, **29**, 159–166.
117. Y. Qian, C. Wang and Z.-G. Le, *Appl. Surf. Sci.*, 2011, **257**, 10758–10762.
118. S. Liu, J. Wang, J. Zeng, J. Ou, Z. Li, X. Liu and S. Yang, *J. Power Sources*, 2010, **195**, 4628–4633.
119. Y. Tamou and S.-i. Tanaka, *Nanostr. Mater.*, 1999, **12**, 123–126.
120. F. Heshmatpour, R. Abazari and S. Balalaie, *Tetrahedron*, 2012, **68**, 3001–3011.
121. M. Kim, E. Heo, A. Kim, J. Park, H. Song and K. Park, *Catal. Lett.*, 2012, **142**, 588–593.
122. M. Masteri-Farahani and M. Sadrinia, *Powder Technol.*, 2012, **217**, 554–557.
123. P. He, X. Wang, Y. Liu, L. Yi and X. Liu, *Int. J. Hydrogen Energ.*, 2012, **37**, 1254–1262.

CHAPTER 3

Preparation of Metal Nanoparticles Stabilized by the Framework of Porous Materials

MEHMET ZAHMAKIRAN[a] AND SAIM ÖZKAR*[b]

[a] Deparment of Chemistry, Yüzüncü Yıl University, 65080 Van, Turkey;
[b] Department of Chemistry, Middle East Technical University,
06800 Ankara, Turkey
*Email: sozkar@metu.edu.tr

3.1   Introduction

Metal nanoparticles are particles that have at least one dimension in the nanometer scale $(<100\,\text{nm})$[1,2] and have already found many fascinating applications in a wide variety of fields including conductors,[3] chemical sensors,[4] biosensors,[5] photovoltaic devices,[6] drug delivery,[7] fuel cells,[8] light-emitting diodes,[9] industrial lithography,[10] quantum dots,[11] quantum wires,[12] quantum devices[13] and catalysis[14] because of their unique properties, which differ from molecular or bulk species. Metal nanoparticles show typical quantum size behavior owing to the existence of discrete electronic energy levels and the loss of overlapping electronic bands.[15] In addition to the exciting electronic properties, metal nanoparticles also show different physical properties, such as small metal nanoparticles show a significantly lower melting point than the corresponding bulk metal and the other size-dependent properties that differ from bulk metal include quantum confinement in semiconductor particles, surface plasmon resonance in some metal nanoparticles and superparamagnetism in

RSC Green Chemistry No. 19
Sustainable Preparation of Metal Nanoparticles: Methods and Applications
Edited by Rafael Luque and Rajender S Varma
© The Royal Society of Chemistry 2013
Published by the Royal Society of Chemistry, www.rsc.org

magnetic nanomaterials.[1,2] Of particular importance, they have significant potential as new types of highly active and selective catalysts.[16,17] Metal nanoparticles have much higher surface-to-volume ratio, thus, a larger fraction of catalytically active atoms on the surface[18] compared to their bulk forms and these surface atoms of nanoparticles do not order themselves in the same way as those in bulk metal.[19] Moreover, the electrons in nanoparticles are confined to spaces that can be as small as a few atoms widths across, giving rise to quantum size effects.[19] In addition to these unique properties, the synthesis protocols of metal nanoparticles provide control on the size and surface composition of particles, which was not possible in the case of classical heterogeneous catalysts. Besides their superior activity and selectivity, metal nanoparticles are usually isolable, redispersible and reusable catalysts and, thus, also meet primary requirements of "green catalysis".[20]

Today, a challenging issue in the synthesis of metal nanoparticles is the achievement of the compositionally well-defined, shape- and size-controllable nanoparticles. When the reaction is structure sensitive, the catalytic activity of metal nanoparticles is drastically influenced by their size as they control the surface structure, electronic and oxidation states. Therefore, the use of well-defined metal nanoparticles as catalysts allows us to assess the nature of active sites in the catalytic reaction, which is vital for the rational design of catalysts. However, the aggregation of metal nanoparticles during their catalytic use to the bulk metal despite using the best stabilizer[21] is still the most important problem that should be overcome in their catalytic application as their agglomeration leads to a momentous decrease in their catalytic activity and lifetime (Figure 3.1).[22]

At this concern, the synthesis of metal nanoparticles in solid support materials seems to be one of the possible ways for preventing aggregation of small metal nanoparticles into bulk metal. This may additionally provide

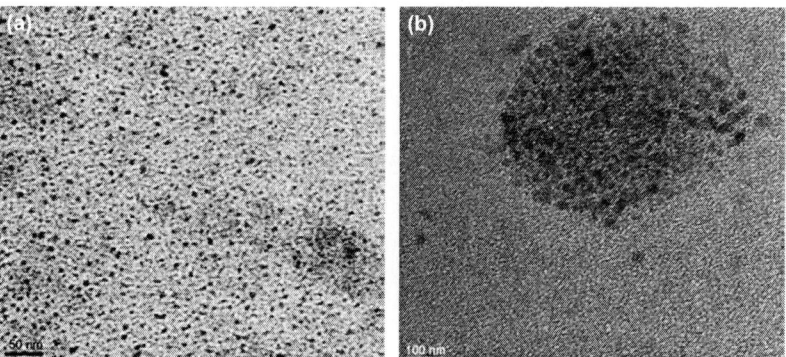

Figure 3.1 TEM images of water dispersible laurate stabilized rhodium(0) nano-particles (a) as-prepared (fresh), (b) harvested after the fifth run of cata-lytic hydrolysis of ammonia-borane, showing agglomerated rhodium(0) nanoparticles, which retain only 44% of their initial catalytic activity in the fifth run, reproduced from ref. 22.

kinetic control for the catalytic reactions,[23] which is very difficult in the case of support-free colloidal metal nanoparticles. More importantly, the recent successes in the methods to remove the surfactants/ligands from the supported metal nanoparticles surface by size control makes them more active catalysts than previously known classical supported catalysts.[24,25]

In this chapter, we will focus on recent developments in supported nanoparticles catalysts with pioneering examples after a brief survey of the main problems in the catalytic application of support-free metal nanoparticles. We will first discuss the synthesis routes of supported metal nanoparticles, then classify the supported nanoparticles according to the nature of support materials (zeolites, silica-based mesoporous materials, metal organic frameworks, hydroxyapatite, and hydrotalcite. Although carbon materials are also important supports they are not included in our discussion since they are considered as a different class of supports and need to be covered separately. Additionally, we also point out the common difficulties in the application of supported metal nanoparticles in catalysis by using selected examples plus their possible solutions and future goals.

3.2   Supported Metal Nanoparticles in Catalysis

3.2.1   Preparation Routes of Supported Metal Nanoparticles

The supported metal nanoparticles are prepared by different synthesis methods, which can be divided into three main categories, depending on the basis of the methodology: (1) chemical (wet-impregnation,[26] coprecipitation,[27] deposition–precipitation,[28] microemulsion,[29] photochemical reduction,[30] chemical vapor deposition,[31] atomic layer deposition,[32] and electrochemical reduction[33] techniques), (2) physical (sonochemical,[34] microwave-irradiation,[35] laser-ablation[36] and supercritical fluid[37] techniques) and (3) physicochemical methods (sonoelectrochemistry[38] and flame-spray-pyrolysis[39] techniques).

3.2.1.1   *Chemical Methods*

(a) Wet impregnation, also known as "impregnation", can be done in four major steps: (*i*) mixing the solid support with metal precursor in a suitable solvent for an appropriate time duration, (*ii*) removing the solvent and isolation of the solid (metal precursor@support) material, (*iii*) drying (or calcining) the resulting solid material and then (*iv*) reducing the metal ions impregnated on to the solid support by using a suitable reducing agent.[26]

Although the impregnation method involves a very simple experimental route, it usually provides a broad particle-size distribution as it is difficult to control the size of particles by this methodology. However, recent studies[40–42] have indicated that some modifications applied on the impregnation method can provide small-size metal nanoparticles with narrow size distributions. For example, Delannoy *et al.*[40] found that post-treatment of the $AuCl_3@TiO_2$, $AuCl_3@SiO_2$, or $AuCl_3@Al_2O_3$ with an aqueous ammonia solution removes most of the chlorines present in the solid samples prepared by anion adsorption and favors the formation of small gold particles during thermal activation. These supported metal nanoparticles prepared by the wet-impregnation method have been found to be highly active catalysts in various catalytic transformations such as CO oxidation,[40] ammonia synthesis,[42] C-C coupling[43] and alcohol oxidations.[44]

(b) *Coprecipitation* involves the simultaneous precipitation of metal precursor and the support material.[27] Guczi and coworkers[45] showed that gold nanoparticles on Fe_2O_3 can be prepared by the coprecipitation technique starting with $HAuCl_4$ and $Fe(NO_3)_3.9H_2O$ precursors. In a more recent study Barau *et al.*[46] reported the preparation of Pd nanoparticles on the ordered pore structure of hexagonal mesoporous silica by a sol-gel approach, which is based on a coprecipitation technique. In addition to the limited applicability of the coprecipitation technique to polymeric support, this methodology also does not provide size and shape control of the supported metal nanoparticles. Metal nanoparticle catalysts prepared by these methods have been used in different reactions such as CO oxidation,[46] oxidation and hydrogenation reactions.[47]

(c) *Deposition–precipitation* includes the complete precipitation of metal hydroxide from a pH-adjusted metal precursor solution to the support surface. Then, solid-supported metal hydroxide is calcined or treated with H_2 at elevated temperatures in order to reduce the cationic metal to the elemental state. This methodology is widely used in the preparation of supported gold nanoparticles. For instance, Sandoval *et al.*[48] have prepared a series of gold nanoparticles supported on TiO_2, CeO_2, Al_2O_3, and SiO_2, which have been tested as catalysts in the water gas shift (WGS) reaction.

(d) *A microemulsion* is a clear isotropic liquid mixture of oil, water and surfactant, and frequently in combination with a cosurfactant. The aqueous phase may contain metal salt and /or other ingredients, and the "oil" may actually be a complex mixture of different hydrocarbons and olefins. In this technique, a solid support is usually impregnated with a microemulsion containing a dissolved metal precursor in the same way as the traditional chemical impregnation method. Metal nanoparticles prepared using the microemulsion technique have been found to have a more controllable size distribution compared to the traditional impregnation, coprecipitation and deposition–precipitation techniques. As a good example, Tsang and coworkers recently reported that PtAu and Pt nanoparticles can be created inside the ceria (CeO_2) particles by a microemulsion technique.[49] Metal nanoparticles supported on different oxide materials have been used as nanocatalysts in a wide range of reactions including hydrogenation[50] and oxidation reactions.[51]

(e) *Photochemical reduction* has been used in the preparation of some supported metal nanoparticles. Choi and coworkers[52] reported the formation of well-dispersed Ag, Au, and Pd nanoparticles on the graphene oxide nanosheets under UV radiation. In another study,[53] they prepared Au nanoparticles supported on a TiO_2 *via* decomposition and photochemical deposition of a gold precursor ($HAuCl_4$) under a 125-W high-pressure mercury lamp. In a recent study, Pal and coworkers reported the synthesis of silver and gold nanoparticles supported on calcium-alginate by photochemical reduction method, which have been found to be active catalysts in the reduction of 4-nitrophenol.[54]

(f) Chemical vapor deposition (CVD)[55] *and atomic layer deposition (ALD)*[56] are gas-phase techniques and can be used to encapsulate metal nanoparticles on various types of support materials. They involve the vaporization of metal precursor and growth of the metal nanoparticles under high vacuum in the presence of an excess of stabilizing organic solvents and/or reducing agent (*e.g.* H_2, CO). However, CVD, while a gas-phase technique, does not provide adequate control over the thickness or composition of the encapsulating shell due to nonself-limiting reactions. In contrast to CVD, ALD as a more recent technique provides the possibility for atomically controlled postmodification of supported catalyst particles by applying a protective layer. Controlling the protective layer thickness can be enabled by layer-by-layer deposition feature of ALD and has expanded its application in the synthesis of both metal and metal-oxide catalytic materials.[57,58] For example, in a recent study Elam and coworkers[58] demonstrated that uniform size-controlled Pd nanoparticles (1 to 2 nm in diameter) on ZnO- and Al_2O_3-coated mesoporous silica gel can be prepared by ALD (Figure 3.2), which are active catalyst in the methanol decomposition reaction.

(g) Electrochemical reduction is usually followed for the preparation of carbon-nanotube-supported metal nanoparticles. The literature review shows that its limited to noble metals such as Pd,[59] Pt,[59] Au,[59] Ag[60] and bimetallic Pt-Ru.[61] Of particular importance is the study by He *et al.* who reported that Pt-Ru nanoparticles supported on carbon nanotube prepared by an electrochemical reduction method exhibit notable catalytic activity in the methanol oxidation.[61] As emphasized in a recent review,[62] by varying the deposition potential, substrate, and deposition time one can control the size and distribution of metal nanoparticles supported on carbonaceous materials.

3.2.1.2 Physical Methods

(a) The sonochemical method is associated with understanding the effect of sonic waves and wave properties on the chemical systems. In a sonochemical process, upon irradiation with high-intensity sound or ultrasound, acoustic cavitation usually occurs, which leads to the formation and surface growth of metal nanoparticles. Ultrasonic-wave-induced deposition and reduction of the metal nanoparticles on the solid support takes place one after the other so that the heating step normally employed in other protocols can be avoided. In addition to the synthesis of ligand- or anion-stabilized metal nanoparticles,[63] this methodology has also been applied to the preparation of supported metal nanoparticles.[64,65] In a recent study, Pan and Wai[65] reported that Pt–Rh nanoparticles can be deposited uniformly on the surface of carboxylate-functionalized multiwalled carbon nanotubes (MWNTs) using a simple one-step sonochemical method and the resulting bimetallic nanoparticles catalyst exhibits a strong synergistic effect relative to the individual Pt or Rh metal

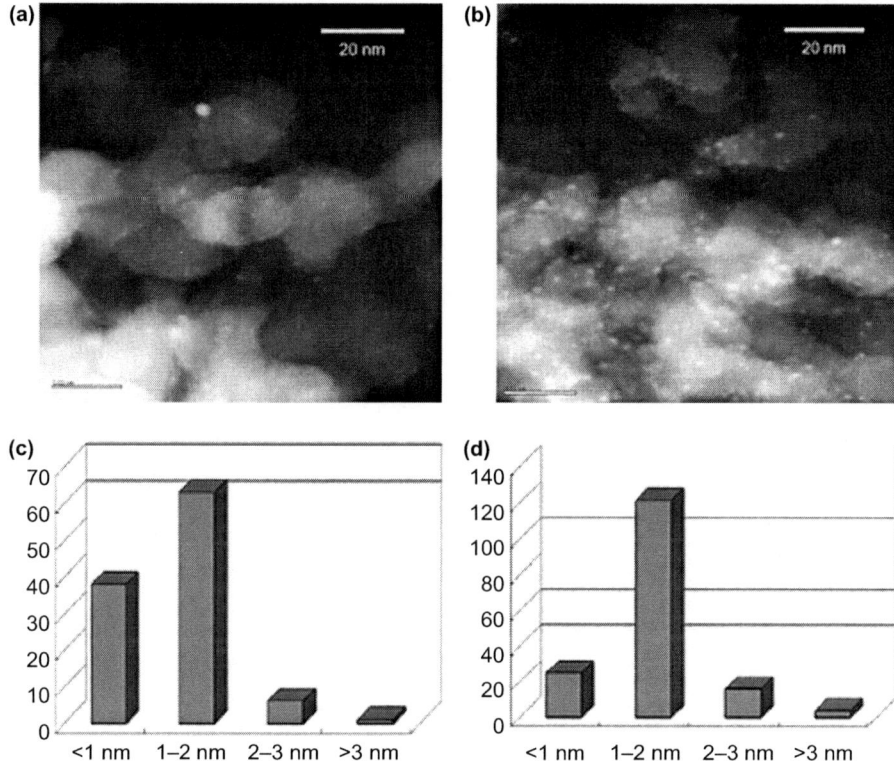

Figure 3.2 TEM images of the Pd nanoparticles supported on Al_2O_3 prepared using 100 s (a) and 400 s (b) precursor exposure times and the Pd particle-size distribution (c and d) deduced from these images, reproduced from ref. 58.

nanoparticles in hydrogenation of polycyclic aromatic hydrocarbons (PAHs), neat benzene and alkylbenzenes.

(b) Microwave irradiation is used to initiate chemical reactions. As a low-frequency energy source, the extremely rapid microwave heating is remarkably adaptable to many types of chemical reactions. In the case of preparation of metal nanoparticles, the microwave-irradiation technique has several advantages over conventional heating methods including short reaction times and the production of smaller sizes of metal nanoparticles with narrower size distributions. El-Shall and coworkers[66] comprehensively investigated the microwave irradiation for the preparation of Au and Pd nanoparticles on various metal-oxide supports such as CeO_2, CuO, ZnO. In a recent study,[67] stable ruthenium or rhodium metal nanoparticles have been supported on chemically derived graphene (CDG) surfaces with small and uniform particle sizes (Ru 2.2 ± 0.4 nm and Rh 2.8 ± 0.5 nm) by decomposition of their metal carbonyl precursors by rapid microwave irradiation in a suspension of CDG in the ionic liquid 1-butyl-3-methylimidazolium tetrafluoroborate (Figure 3.3). The resulting graphene-supported hybrid nanoparticles are active and reusable

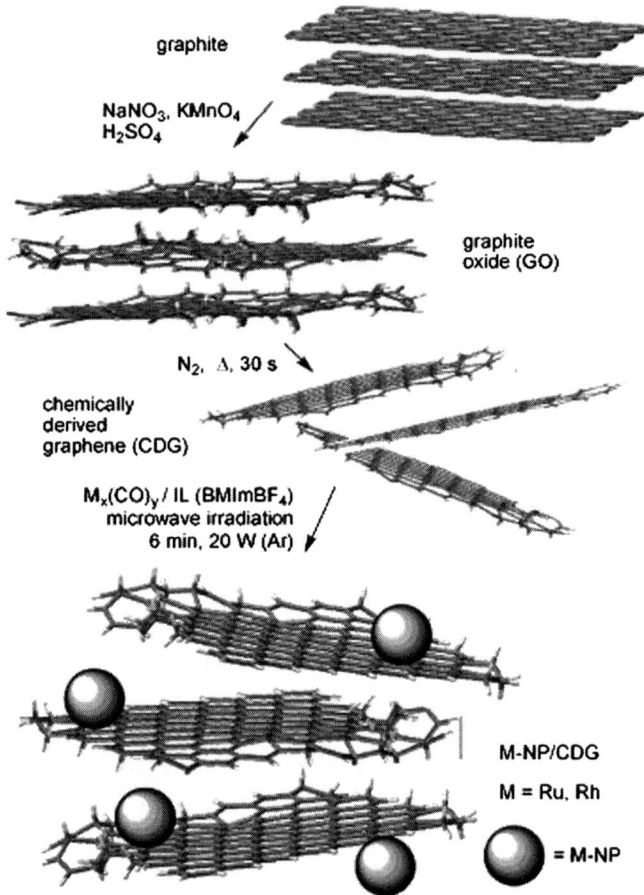

graphite

NaNO$_3$, KMnO$_4$
H$_2$SO$_4$

graphite
oxide (GO)

N$_2$, Δ, 30 s

chemically
derived
graphene (CDG)

M$_x$(CO)$_y$ / IL (BMImBF$_4$)
microwave irradiation
6 min, 20 W (Ar)

M-NP/CDG

M = Ru, Rh

= M-NP

Figure 3.3 Synthesis protocol of ruthenium and rhodium nanoparticles on chemically derived graphene (CDG), reproduced from ref. 67.

catalysts in the hydrogenation of cyclohexene and benzene under organic-solvent-free conditions.

(c) Laser ablation is the process of removing material from a solid (or occasionally liquid) surface by laser-beam irradiation. In the case of low laser flux, the material is heated by the absorbed laser energy and evaporates or sublimes. This provides significant advantages in the preparation of supported metal nanoparticles such as (*i*) it operates under well-defined conditions leading to a high control of particle geometry,[68] (*ii*) no solvents are used, so it is environmentally friendly, (*iii*) it is relatively easy to automate. Catalytic materials containing Rh, bimetallic Rh/Pt and trimetallic Rh/Pt/Au nano-particles supported on γ-Al$_2$O$_3$, CeO$_2$, TiO$_2$, SiO$_2$ and Υ-ZrO$_2$ have been tested in the partial oxidation of propylene.[68] In another case study Jimenez and coworkers[69] achieved the synthesis of well-dispersed SiO$_2$ encapsulated Ag, Au

and Au-Ag metal nanoparticles by the laser ablation of aqueous metal salt solution in the presence of SiO_2 support.

(d) A supercritical fluid involves dissolving a metal precursor in a supercritical fluid (mostly supercritical carbon dioxide, $scCO_2$) and then incorporation on a support under various conditions.[37] The supported metal precursor is then reduced to metal by either chemical reduction with a suitable reducing agent or thermal reduction. It has been demonstrated that by following aforementioned procedure Pd,[70] Ru[71] and Rh[71] nanoparticles supported on multiwalled carbon nanotubes can be prepared by using $scCO_2$ as a reaction medium and H_2 or CO as reducing agents. It has been demonstrated that Pd nanoparticles supported on multiwalled carbon nanotubes are effective catalysts in hydrogenation of olefins in carbon dioxide and the Pd nanoparticles also show a high electrocatalytic activity in oxygen reduction for potential fuel-cell application.

3.2.1.3 Physicochemical Methods

(a) Sonoelectrochemistry combines sonication (ultrasound waves) and electrochemistry. The method can be defined as a combination of ultrasonic radiation with electrode processes occurring at surfaces of electrodes immersed in a solution in an electrochemical cell. The details of this technique have already been reviewed,[72] and some case studies for the sonoelectrochemical preparation of Ag[73] and Pd[74] nanoparticles exist in the literature.

(b) Flame-spray pyrolysis is also a combination technique where the liquid containing metal precursor is fed in the center of a methane–oxygen flame *via* a syringe pump and dispersed by oxygen, forming a fine spray. The spray flame is surrounded and ignited by a small flame ring issuing from an annular gap. The resulting particles are collected on a glass fiber filter by the help of vacuum.[75] This method has been applied first for the preparation of highly stable Pd nanoparticles.[75]

3.2.2 Types of Supported Metal Nanoparticles Depending on the Nature of Support Material

In the development of supported metal nanoparticles, various types of porous and nonporous solids have been used as the host materials for the controlled preparation of guest metal nanoparticles. In this section, the majority of the most commonly used support materials and their inherent properties that provide advantages and drawbacks in the preparation of supported metal nanoparticles will be discussed. Among the wide range of support materials used in the deposition of metal nanoparticles, microporous–mesoporous materials (zeolites, silica, metal organic-frameworks, *etc.*), minerals and clay materials (hydroxyapatite, hydrotalcite, *etc.*), carbonaceous materials, and metal oxides are the main families of widely employed solid supports. Because of the existence of a great number of review articles related to the use of carbonaceous materials[76,77] metal oxides[77,78] and polymers[77,79] as catalyst supports, we will focus on the employment of zeolites, silica-based materials, metalorganic frameworks, hydroxyapatite and hydrotalcite as host materials for the guest metal nanoparticles.

3.2.2.1   Zeolites, Silica-Based Materials

The main family of microporous–mesoporous materials comprise of zeolites, which are crystalline aluminosilicates with microporous channels and/or cages in their structures. The frameworks of zeolites are formed by fully connected SiO_4 and AlO_4 tetrahedra linked by shared oxygen atoms. Typically, synthetic zeolites are crystallized in alkaline reaction mixtures containing Na^+, K^+, *etc.* cations and the charge-compensating extraframework cations are then Na^+, K^+, *etc.* cations.[80] The connectivity (topology) of the zeolite framework is characteristic of a given zeolite type, whereas the composition of the framework and the type of extraframework species can vary. Each zeolite structure type is denoted by a three-letter code.[81] As an example, Faujasite-type zeolites have the acronym FAU. Among a vast number of microporous and mesoporous materials, zeolite-Y with FAU framework is considered to be a suitable host material because of the following advantages; (*i*) it provides highly ordered large cavities with diameters of 1.3 nm for supercage (α-cage) and 0.7 nm for sodalite cage (β-cage),[82] (*ii*) this framework structure is the most open to any zeolite and is about 51% void volume, including the sodalite cages; the supercage volume represents 45% of the unit cell volume, (*iii*) in this framework, each cavity is connected to four other cavities, which in turn are themselves connected to three-dimensional cavities to form three-dimensional pore structure and it enables transferring of substrates or products throughout the framework even though one of the channels is blocked, (*iv*) the main pore structure is large enough to admit large molecules.

Commonly, the convenient procedure for generating metal nanoparticles inside the zeolite pores comprises the introduction of metal species (cations or complexes) into the zeolite by ion exchange[82] or vapor deposition[83] and then reduction by heating[84] or H_2 at temperatures higher than 300 °C and calcination up to 400–500 °C.[85] However, this high-temperature treatment may cause alteration in the zeolite framework due to the formation of an unstable acid form and may lead to the migration of a large part of the guest metal atoms out of cavities of the zeolite. For example, Ryoo and coworkers[85] have reported that Ru clusters consisting of ca. 20 Ru atoms on average can be obtained in the supercage by autoreduction of the supported ruthenium species during the evacuation of the sample under heating to 673 K, and the average number of Ru atoms per cluster gradually increased to 50 upon further heating to 823 K both in H_2 and under vacuum. A TEM image obtained at the end of their procedure indicates the formation of large-size ruthenium aggregates on the surface of zeolite-Y.

In 2005, Wang *et al.*[86] demonstrated the preparation of cobalt nanoparticles within faujasite framework by borohydride reduction of cobalt(II)-exchanged zeolite-Y. They found that the treatment of the cobalt(II)-exchanged zeolite-Y at a higher temperature before reduction leads to a lower degree of reduction, an appropriate treatment temperature is needed for obtaining cobalt nano-particles with sizes less than 5 nm. The smaller cobalt nanoparticles located inside the supercages of faujasite zeolite exhibit higher CO conversion in Fischer–Tropsch synthesis than the larger cobalt particles outside the super-cages or the cobalt on the surface of nonporous silica. This study revealed that borohydride reduction of metal-exchanged zeolite provides a lower number of agglomerated nanoparticles that can migrate outside the zeolite crystals with respect to the high temperature required for H_2 and thermal reduction tech-niques. This result motivated us to focus on the borohydride reduction of noble-metal exchanged zeolite in order to develop new catalytic materials for various catalytic transformations.[87–97]

In our first study,[87,88,90] we achieved the preparation of zeolite framework stabilized ruthenium nanoparticles (Ru(0)/NaY) by following a two-step procedure including the ion exchange of Ru^{3+} ions with the extraframework Na^+ ions of zeolite-Y, followed by the reduction of Ru^{3+} ions in the cavities of zeolite-Y with sodium borohydride in aqueous solution, all at room tempera-ture. In contrast to the results of the Ryoo study,[85] who applied high-temperature treatment, in our study the resulting ruthenium nanoparticles were found to be unagglomerated and mainly located inside the cavities of zeolite-Y. These zeolite framework-stabilized ruthenium nanoparticles were found to be highly active catalyst in the hydrolysis of sodium borohydride in water and in the hydrogenation of olefins and arenes. They provide the record catalytic activity in the hydrogen generation from the hydrolysis of sodium borohydride (Scheme 3.1) with the turnover frequency (TOF) value of 33 000 h^{-1} and total turnover number (TTO) of 103 200 at room temperature.[87,88]

Moreover, they also show exceptional activity in the hydrogenation of neat benzene even at 22 ± 0.1 °C and 3 bar initial H_2 pressure. They achieved the complete hydrogenation of benzene to cyclohexane at >99% selectivity with a TOF value of 1040 h^{-1} (Scheme 3.2). In addition to ruthenium nanoparticles we carried out the preparation of rhodium,[92,95] copper,[96] cobalt,[98,99] osmium,[94] palladium,[100,101] and gold[102] nanoparticles stabilized by zeolite framework by

$$NaBH_4 + 2H_2O \xrightarrow{\text{zeolite framework stabilized RuNPs}} NaBO_2 + 4H_2$$

Scheme 3.1   Zeolite-framework-stabilized ruthenium nanoparticles (RuNPs) cata-lyzed hydrolysis of sodium borohydride at room temperature (TOF = 33 000 h^{-1}, TTO = 103 200).

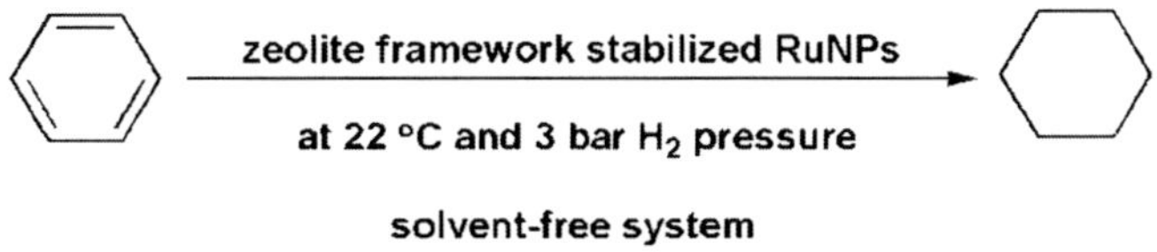

zeolite framework stabilized RuNPs

at 22 °C and 3 bar H_2 pressure

solvent-free system

Scheme 3.2   Zeolite-framework-stabilized ruthenium nanoparticles (RuNPs) cata-lyzed hydrogenation of benzene in the solvent-free system at 22 ± 0.1 °C and 3 bar initial H_2 pressure.

following the aforementioned simple procedure and tested their catalytic activity in the hydrolysis or methanolysis of ammonia-borane, Suzuki cross-coupling, and aerobic oxidation of alcohols.

In another different methodology, White and coworkers[103] reported that RuO_2 nanoparticles can be generated inside the zeolite-Y (one RuO_2 in every 2.2 supercages of FAU) by a one-step hydrothermal method, which involves addition of either $RuCl_3.3H_2O$ or $Ru(NH_3)_6Cl_3$ at the starting point of zeolite-Y synthesis into an aluminosilicate gel containing NaOH, $NaAlO_2$, SiO_2, and H_2O. By this approach they successfully incorporated RuO_2 nanoparticles (1.3 ± 0.2 nm) into the supercages of faujasite zeolite (Figure 3.4(a)). Ru K-edge X-ray absorption fine structure (EXAFS) results indicate that the RuO_2 nanoparticles anchored in the zeolite are structurally similar to highly hydrous RuO_2; that is, is a two-dimensional structure of independent chains, in which RuO_6 octahedra are connected together by two shared oxygen atoms (Figure 3.4(b)).

In their preliminary catalytic studies, they found that the RuO_2 nanoparticles exhibit extraordinarily high activity and selectivity in the aerobic oxidation of alcohols (Scheme 3.3) under mild conditions (air and ambient pressure). The physically trapped RuO_2 nanoparticles cannot diffuse out of the relatively narrow channels/pores of the zeolite during the catalytic process, making the resulting material both stable and reusable catalyst in the aerobic oxidation of alcohols.

Christensen and coworkers[104] recently reported another interesting approach to encapsulate metal nanoparticles into the silica structure. Their synthetic approach consists of three steps (Figure 3.5(a)); (1) a metal nanoparticle colloid is prepared with suitable anchor points for the generation of a silica shell, (2) the particles are encapsulated in an amorphous silica matrix, (3) the silica nanoparticle precursor is subjected to hydrothermal conditions in order for zeolite crystallization to take place. Following this approach they have prepared a material that consisted predominantly of 1–2-nm sized gold particles embedded in silicalite-1 crystals (Figure 3.5(b)).

In order to verify that the encapsulated gold nanoparticles are accessible for catalysis, they performed the aerobic oxidation of a mixture of benzaldehyde and 3,5-di-tertbutylbenzaldehyde in methanol into their corresponding methyl esters at room temperature under atmospheric pressure. They found that only benzaldehyde is oxidized in appreciable amounts using the gold–silicalite-1 catalyst. This observation was attributed to the different size of the substrates; 3,5-di-tert-butylbenzaldehyde is much more bulky than the unsubstituted benzaldehyde, and is therefore not able to diffuse into the zeolite interior where the catalytically active gold nanoparticles are located (Scheme 3.4).

Iglesia and coworkers[105] recently showed that (3-mercaptopropyl)-trimethoxysilane ligands act as good stabilizing agents for Pt, Pd, Ir, Rh, and Ag nanoparticles through their mercapto (–SH) groups which interact with cationic metal centers, while alkoxysilane moieties form covalent Si–O–Si or Si–O–Al linkages that promote zeolite nucleation around ligated metal precursors (Figure 3.6(a)). These protocols led to the successful encapsulation of clusters within the NaA zeolite, for Pt, Ir, and Rh nanoparticles with small channel apertures (0.41 nm) preclude post synthesis deposition of metal clusters

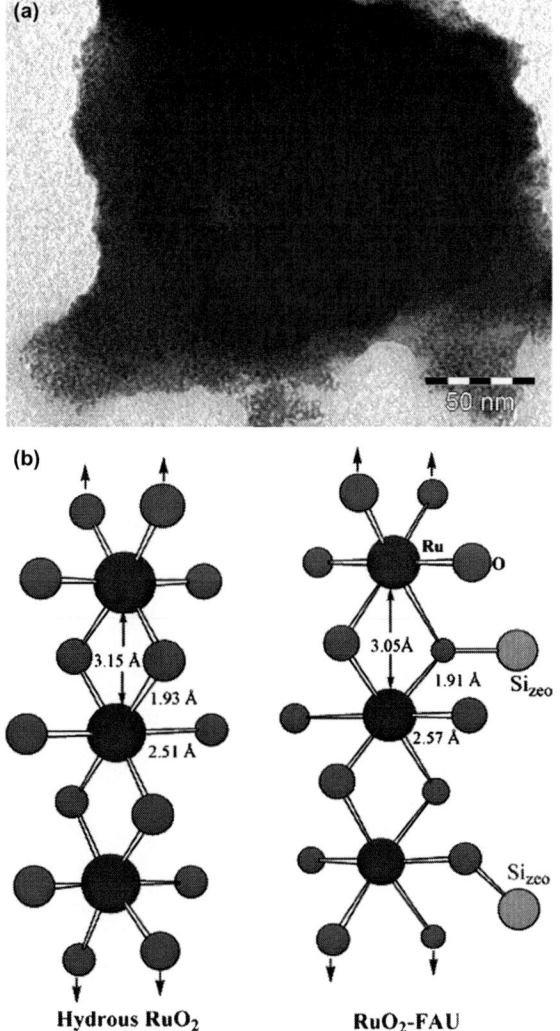

Figure 3.4   (a) High-resolution TEM image of RuO_2-FAU composite, (b) EXAFS results derived structures of hydrous RuO_2 and RuO_2-FAU, reproduced from ref. 103.

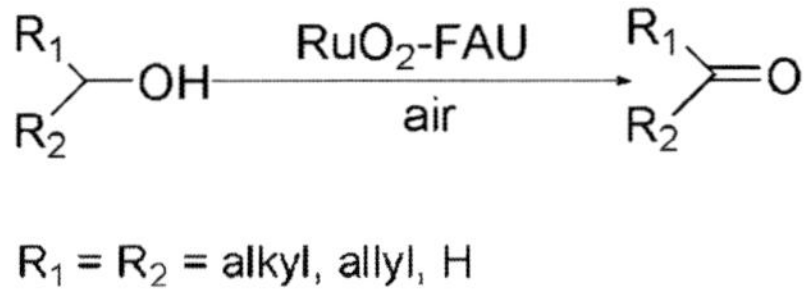

$R_1 = R_2 = $ alkyl, allyl, H

Scheme 3.3 RuO_2-FAU catalyzed aerobic oxidation of alcohols under air.

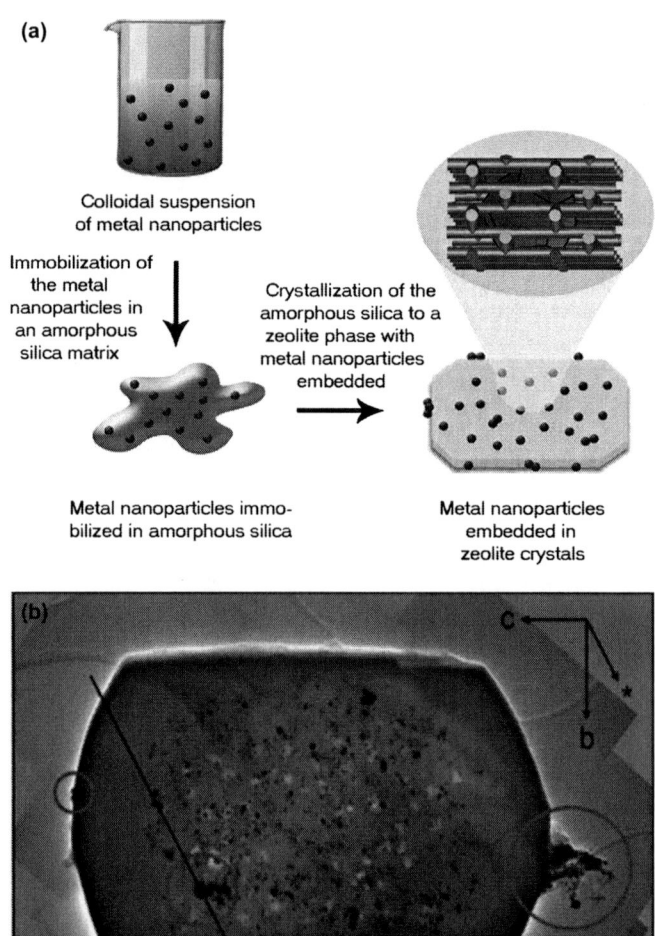

Figure 3.5   (a) Schematic illustration of the encapsulation of gold nanoparticles in zeolite crystals, (b) Mosaic composed of 10 TEM bright-field micrographs showing a representative crystallite as seen along the [1 0 0] zone axis. The crystal lattice is indexed based on the standard *Pnma* space group setting. The black areas represent high-density regions, and therefore correspond to the gold particles. Voids are observed as light regions within the zeolite crystal. Circles mark gold particles that are either obviously on the surfaces or are later shown to be on the surface of the crystallite, reproduced from ref. 104.

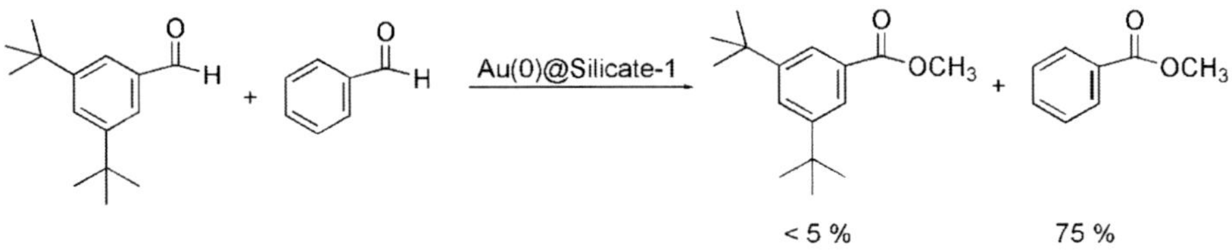

< 5 %   75 %

Scheme 3.4   Au@Silicate-1 catalyzed oxidation of 3,5-di-ter-butylbenzaldehyde and benzaldehyde mixture at room temperature and under ambient pressure.

(Figures 3.6(b–d)). Sequential treatments in O_2 and H_2 formed small (1 nm) clusters with uniform diameter. The catalytic activities of these materials were tested in both the hydrogenation of ethane/isobutene and the oxidation of methanol/isobutanol. They found that the ratio of the rates of hydrogenation of ethene and isobutene are much higher on nanoparticles encapsulated within NaA than those dispersed on SiO_2, as also found for the relative rates of methanol and isobutanol oxidation.

The grafting of metal nanoparticles *via* organosilane groups[106] onto the solid support is another common technique that is followed in the preparation of metal nanoparticles supported on microporous or mesoporous material. Chaudhari and coworkers[107] demonstrated that the preparation of palladium and platinum nanoparticles bound at high surface coverage on 3-aminopropyltrimethoxysilane (APTS)-functionalized NaY zeolites *via* borohydride reduction of aqueous solutions H_2PtCl_6 or $Pd(NO_3)_2$ in the presence of APTS-functionalized zeolite-Y. The resulting Pt-APTS-NaY and Pd-APTS-NaY were found to be highly active catalysts in the hydrogenation of styrene and Heck type arylation reaction between iodobenzene and styrene, respectively (Scheme 3.5).

In another study,[108] related to the encapsulation of silver nanoparticles within mesoporous-silica-based material, it was shown that the amino groups of APTS (aminopropyltriethoxyl silane)-modified mesoporous silica (SBA-15) can be used to anchor formaldehyde to form $NHCH_2OH$ species, on which silver precursor $Ag(NH_3)_2NO_3$ could be *in-situ* reduced (Figure 3.7(a)) and form well-dispersed silver nanoparticles within SBA-15 framework (Figures 3.7 (b) and (c)).

In their review article, Tosheva and Valtchev[109] emphasized the advantages of the use of nanozeolites in different applications, which triggered us to consider nanozeolites as host material in the stabilization of metal nanoparticles catalysts, as we thought that the diffusion of the substrate molecules through the cavities from the external surface to the cages, where the nanoclusters reside, can limit the reaction rate. Moreover, a decrease in the particle size of host material results in high external surface areas and, thus, reduces the diffusion path length compared to the large size zeolite crystals (> 1 μm). With this anticipation, we employed the nanoparticles of zeolite-Y with a narrow size distribution as the host for the stabilization of ruthenium nanoparticles in our recent study.[91] Nanozeolite-framework stabilized ruthenium nanoparticles (Figure 3.8) were prepared as a novel material by a procedure comprising of (*i*) the ion exchange of Ru^{3+} ions with the extraframework Na^+ or

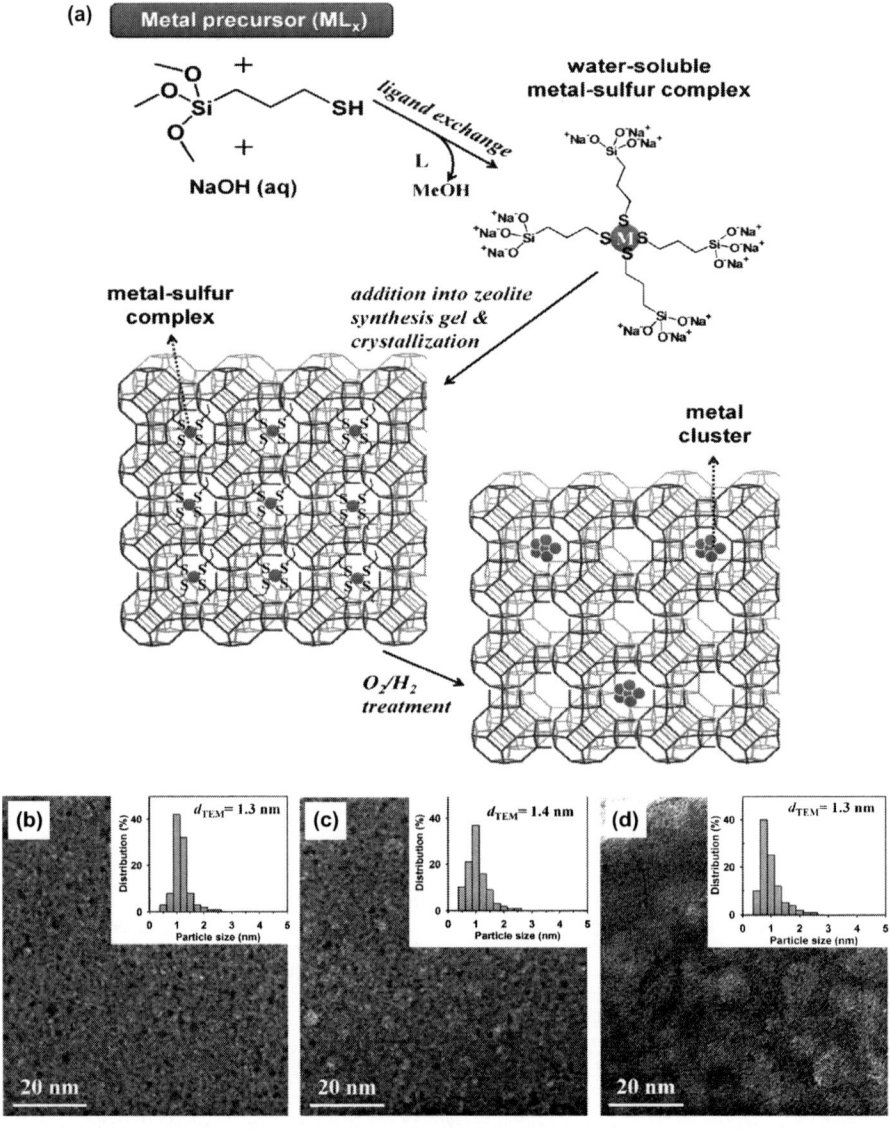

Figure 3.6   (a) Schematic representation of the process for mercaptosilane-assisted metal encapsulation during zeolite crystallization, TEM images for (a) 1.1 wt% Pt/NaA, (b) 0.98 wt% Ir/NaA, and (c) 1.0 wt% Rh/NaA samples synthesized by mercaptosilane-assisted hydrothermal crystallization, reproduced from ref. 105.

tetramethylammonium TMA$^+$ ions of nanozeolite at 50 °C for 12 h and then (*ii*) the reduction of Ru^{3+} ions within the cavities of nanozeolite with sodium borohydride in aqueous solution at room temperature, whereby the Na$^+$ ions reoccupy the cation sites left by the Ru^{3+} ions upon reduction, as tracked by far-IR spectroscopy.

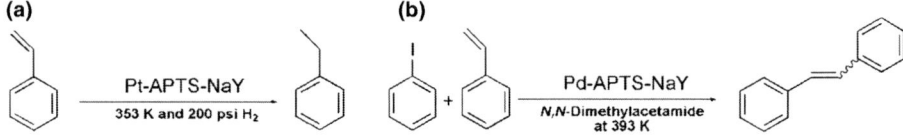

Scheme 3.5   (a) Pt-APTS-NaY catalyzed hydrogenation of styrene to ethylbenzene and (b) Pd-APTS-NaY catalyzed arylation reaction between iodobenzene and styrene.

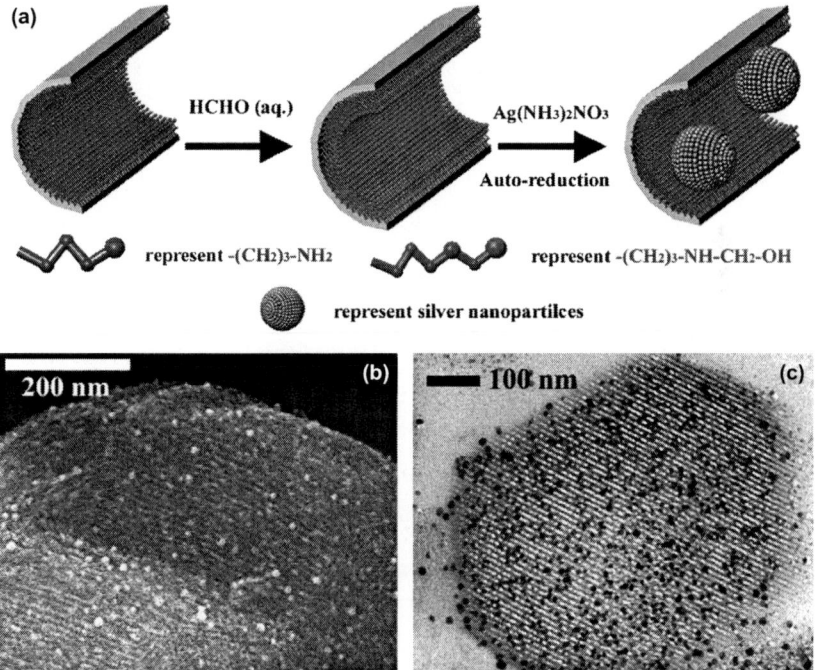

Figure 3.7   (a) Schematic representation of assemblage of highly dispersed Ag nanoparticles inside the channels of SBA-15, (b) high-resolution SEM (HRSEM) and (c) high resolution-TEM images of Ag nanoparticles formed within the framework of SBA-15., reproduced from ref. 108.

Ruthenium nanoparticles stabilized by a nanozeolite framework were found to be the most active (initial turnover frequency (TOF) = 5430 h^{-1}) and longest lifetime (total turnovers, TTO = 177 200) catalyst ever reported for the complete hydrogenation of neat benzene under mild conditions at 25 °C and 42 ± 1 psig initial H$_2$ pressure).

3.2.2.2   Metal Organic Frameworks (MOFs)

Metal organic frameworks (MOFs) represent a new class of hybrid organic–inorganic (supramolecular) materials comprised of ordered networks formed

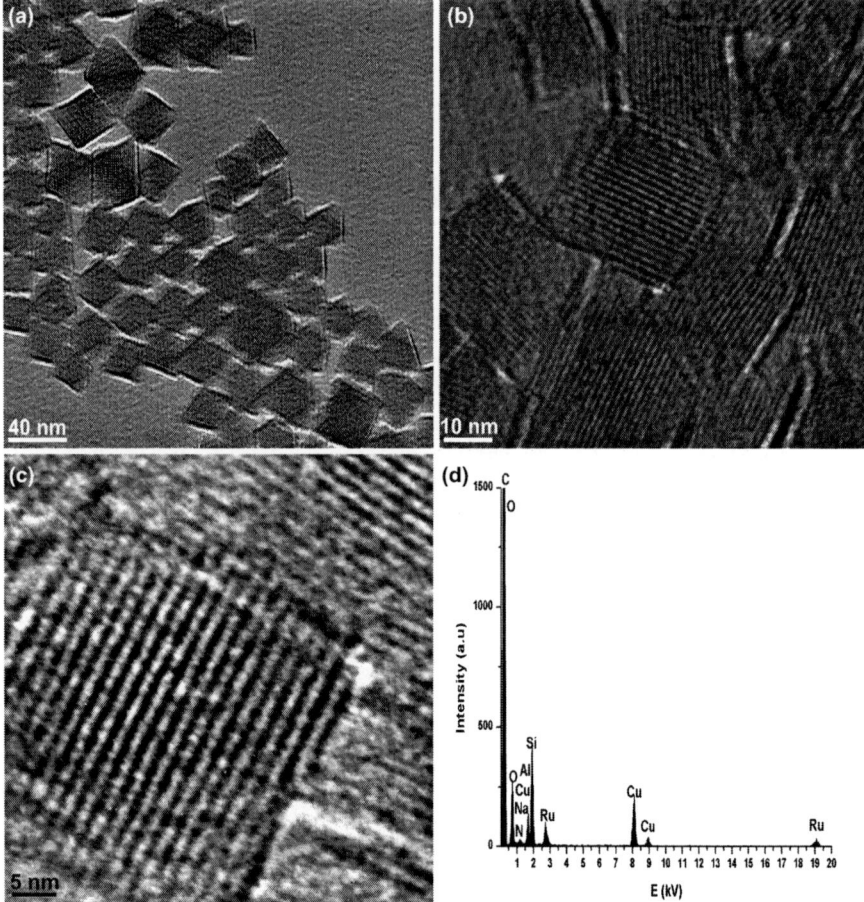

Figure 3.8   (a) and (b) are TEM images, (c) high-resolution TEM image of nano-zeolite framework stabilized ruthenium nanoparticles with different magnifications, and (d) TEM-EDX spectrum collected from the region given in (c).

from electron-donor organic linkers and metal cations. The structural porosity of MOF materials places them at the frontier between zeolites and surface metal organic catalysts.[110,111] They exhibit extremely high surface areas, as well as tunable pore size and functionality, and can act as host materials for a variety of guest molecules including metal nanoparticles.

Compared to zeolites, MOFs as hosts of metal nanoparticles can have the advantage of combining the microporous confinement for the nanoparticles and the functionality that can introduce the organic linker. The preparation of metal nanoparticles stabilized by MOFs is usually done by the incorporation of metal precursor into MOF matrix *via* (*i*) solvent-free gas-phase loading, (*ii*) solution impregnation, (*iii*) incipient wetness impregnation, (*iv*) solid grinding and followed by their reduction (thermal or chemical by using H_2 or metal hydrides).[112]

Ruthenium nanoparticles embedded within MOF-5 were prepared by chemical vapor deposition at 303 K under a static vacuum of 10^{-5} mbar for 6 days of the organometallic Ru(cod)(cot) complex (cod = 1,5-cyclooctadiene; cot = 1,3,5-cyclooctatriene) and the reduction of ruthenium precursor with H_2.[113] Ru nanoparticles stabilized by MOF-5 were used as active catalysts in the hydrogenation of benzene to cyclohexane. Moreover, the oxidation of the resulting ruthenium nanoparticles with a stream of O_2 at room temperature was found to generate RuO_2 nanoparticles inside the cavities of MOF-5 and they can catalyze the aerobic oxidation of benzyl alcohol to benzaldehyde.

A similar procedure was also applied for the preparation of Ni nanoparticles inside MesMOF-1,[114] in which highly disperse nickel nanoparticles within the framework of MesMOF-1 obtained by gas-phase loading of nickelocene into the evacuated MesMOF-1 *via* sublimation at 85 °C for 1 to 3 h and the treatment of the nickelocene-containing MesMOF-1 with H_2 at 95 °C for 5 h yielded cyclopentane molecules and Ni nanoparticles (Figure 3.9).

Fischer *et al.*,[115] reported the preparation of Cu and ZnO nanoparticles inside MOF-5 in a two-step process. First, the solvent-free gas-phase adsorption of the volatile precursors [CpCuL] ((L) = PMe₃, CN*t*Bu) and ZnEt₂ leads to the isolable inclusion. Then, they are reduced by hydrogenolysis or photo-assisted thermolysis at 200–220 °C in the case of Cu and hydrolysis or dry oxidation at 25 °C followed by annealing 250 °C in the case of ZnO. Their catalytic performance was tested in the methanol synthesis and they were found to be unstable under catalytic conditions over several hours, the metalorganic framework collapsed, and the final catalytic activities were poor.

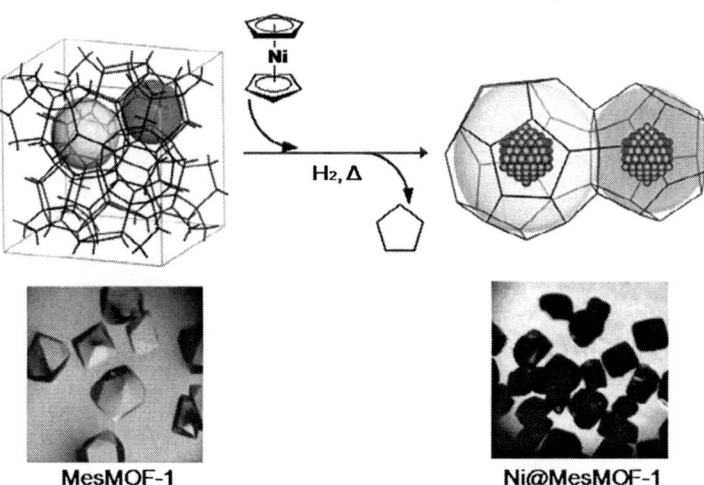

Figure 3.9   A two-step procedure for the preparation of Ni@MesMOF-1 is schematically shown with pictures for the crystals before and after Ni embedment, reproduced from ref. 114.

Shu and coworkers prepared palladium nanoparticles supported on MOF-5 by the impregnation of PdCl$_2$ into host then their reduction with hydrazine hydrate at room temperature.[116] Pd-MOF-5 was found to be a highly reactive catalyst in the ligand and copper-free Sonogashira coupling reaction between aryl iodides and terminal acetylenes (Scheme 3.6).

Pd, Cu and Pd–Cu nanoparticles of 2–3 nm were synthesized inside the pores of MIL-101 by the solution phase incorporation of palladium and copper nitrate precursors into activated MIL-101 framework followed by their hydrazine reduction under microwave irradiation (MWI) for 1–2 min (Figure 3.10).[117] As aforementioned the main advantages of MWI over other conventional heating methods are uniform and rapid heating of the reaction mixture and the simplicity of the process. Moreover, MWI can remove the coordinated water molecules around the chromium centers of MIL-101, thus creating accessible sites where nucleation of the metal nanoparticles can occur upon the reduction of the metal ions. The catalytic performance of these

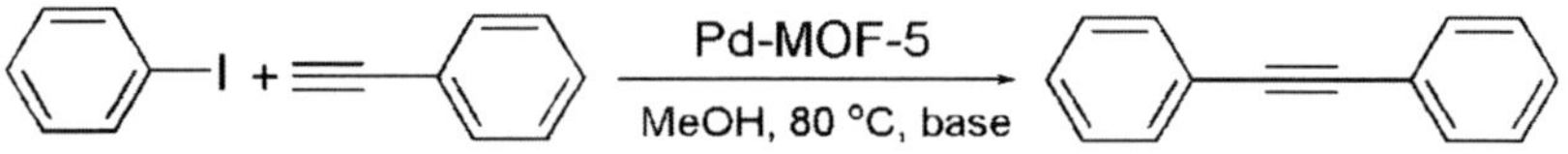

Scheme 3.6 Pd-MOF-5 catalyzed ligand and copper-free Sonogashira coupling reaction between aryl iodides and terminal acetylenes.

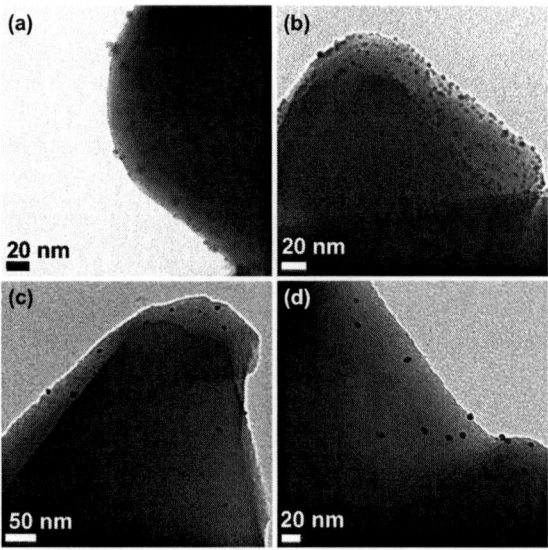

Figure 3.10 TEM images of (a) 2.9 wt% Pd and (b) 0.5 wt% Pd + 2 wt% Cu loaded on MIL prepared by direct hydrazine + MWI treatment; (c) 0.1 wt% Pd and (d) 0.3 wt% Pd loaded on MIL prepared by high-speed centrifugation followed by hydrazine + MWI treatment, reproduced from ref. 117.

resulting nanocatalysts was tested in the CO oxidation. The observed results revealed that the catalytic activities toward CO oxidation of the Pd nanocatalysts supported on the highly porous MIL-101 polymer are significantly higher than any other reported metal clusters supported on metal organic frameworks. The observed high activity was attributed to the small metal nanoparticles imbedded within the pores of the MIL-101 crystals.

In addition to the deposition of metal precursor to MOF matrix, it is also possible to impregnate preformed metal nanoparticles, *e.g.* it was reported that PVP-stabilized Au nanoparticles can be effectively deposited to the matrix of MIL-101.[118] The resulting Au/MIL-101 catalyst exhibited extremely high catalytic activities in liquid-phase aerobic oxidation of a wide range of alcohols, which could even efficiently catalyze the oxidation under ambient conditions in the absence of water or base. Moreover, the catalyst was easily recoverable and could be reused several times without leaching of metals and loss of significant activity.

Another frequently followed methodology in the preparation of MOFs-supported metal nanoparticles is solid grinding of metal precursor on host framework followed by their reduction.[119,120] Xu and coworkers achieved the formation of well-dispersed gold nanoparticles within zeolitic imidazole framework (ZIF-8) by solid grinding of $(CH_3)_2Au(acac)$ on ZIF-8 then their reduction in a stream of 10 vol% H_2 in He at 230 °C for 2.5 h.[120] The final product Au-ZIF-8 was employed as catalyst in CO and found to be highly active in this important transformation even at low Au loadings (0.5% wt). More importantly, retaining of the same activity was observed in the reusability experiments. The same author has also reported the generation of bimetallic (in both alloy and core–shell structured types) Au–Ag nanoparticles on ZIF-8 by the help of sequential deposition reduction approach under mild conditions (Figure 3.11).[121] These bimetallic Au and Ag nanoparticles supported on ZIF-8 were tested for the catalytic reduction of 4-nitrophenol (4-NPh) by $NaBH_4$ in water. It was found that core–shell structure Ag–Au nanoparticles supported on ZIF-8 provide better activity than alloy type bimetallic Ag–Au nanoparticles on ZIF-8.

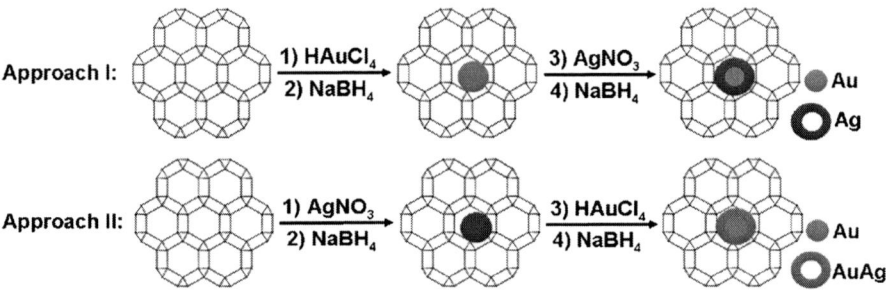

Figure 3.11   Schematic illustration of Ag@Au-ZIF-8 and Au@Ag-ZIF-8 catalysts, reproduced from ref. 121.

3.2.2.3   Hydroxyapatite (HAp)

Hydroxylapatite, also called hydroxyapatite (HAp), is a naturally occurring mineral form of calcium apatite with the formula ($Ca_{10}(PO_4)_6(OH)_2$), which possess Ca^{2+} sites surrounded by PO_4^{3-} tetrahedra parallel to the hexagonal axis (Figure 3.12),[122] which have potential as biomaterials,[123] adsorbents,[124] and ion exchangers.[125] Of particular importance, it can also be used as a support material for metal nanoparticle catalysts.[122] The employment of HAp as a catalyst support has following advantages; (*i*) active species can be immobilized on the surface of HAp, based on high ion-exchange ability and adsorption capacity of HAp, (*ii*) HAp has almost nonporous structure that can overcome problems regarding mass-transfer limitations, (*iii*) HAp structure provide weak acid–base properties that can prevent side reactions induced by the support itself.

The first well-defined supported nanoparticle catalyst by using HAp as a support was reported by Kaneda and coworkers.[126] In their work, they generated Pd nanoparticles *in-situ* during the aerobic oxidation of alcohols starting with $PdCl_2(PhCN)_2$ adsorbed-HAp. The characterization of the catalytic materials by using multiprong analyses showed that the generation of Pd(0) nanoparticles on the surface of HAp under *in-situ* conditions, which were found to be excellent catalysts in the aerobic oxidation of various alcohols in terms of activity, lifetime and reusability. Moreover, they provide a record total turnover number (TTO) up to 236 000 with a record turnover frequency (TOF) of 9800 h^{-1} for a 250-mmol-scale oxidation of 1-phenylethanol under solvent-free conditions.

In another interesting study,[127] they achieved the dispersion of magnetic Fe_2O_3 nanocrystallites in a hydroxyapatite matrix (HAp-Fe_2O_3) for the first time as a new catalyst support. Then, they utilized the cation-exchange ability of the external HAp for the equimolar substitution of Ru for Ca to form a catalytically active center (Ru-HAp-Fe_2O_3). The advanced characterization of the resulting material showed that Ru nanoparticles exist in the monomeric form in a highly dispersed manner (Figure 3.13). The Ru-HAp-Fe_2O_3 exhibited

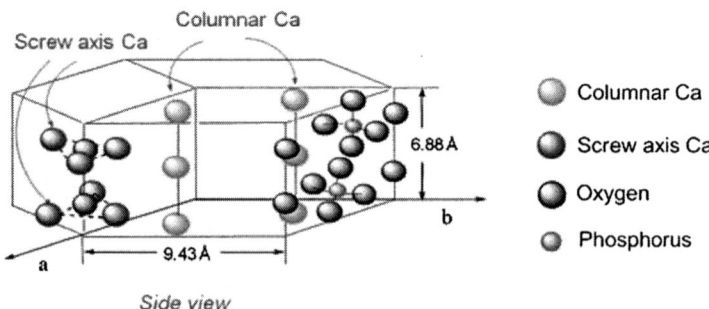

Figure 3.12   Structure of hydroxyapatite ($Ca_{10}(PO_4)_6(OH)_2$, HAp), reproduced from ref. 122.

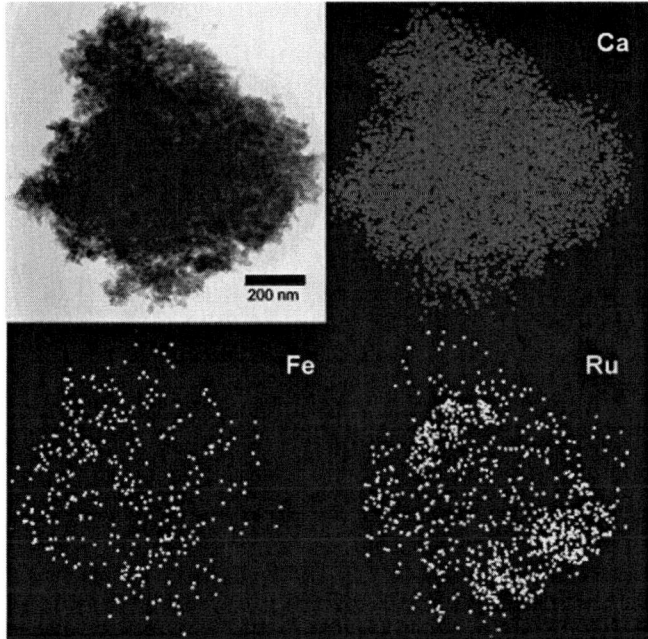

Figure 3.13 TEM and STEM elemental mapping of Ru-HAp-ϒ-Fe$_2$O$_3$, reproduced from ref. 127.

superior catalytic activity for the oxidation of various alcohols to the corresponding carbonyl compounds using molecular oxygen as a primary oxidant. More importantly, the magnetic properties of the Ru-HAp-Fe$_2$O$_3$ provided a convenient route for separation of the catalyst from the reaction mixture by application of an external permanent magnet. They did not detect Ru leaching, and the used catalyst could be recycled without appreciable loss of catalytic activity.

The same approach was also applied by the same group to prepare magnetically recoverable Pd-HAp-Fe$_2$O$_3$,[128] which were found to be highly active catalysts for the dechlorination of various organochlorides using atmospheric molecular hydrogen (H$_2$). For instance, they achieved dechlorination of chlorobenzene to benzene with an excellent turnover frequency (TOF) of 2500 h^{-1} in the presence of 1 atm of H$_2$. In addition to oxidation of alcohols and dechlorination of organochlorides, HAp-supported metal nanoparticles was also tested in the aqueous phase oxidation of phenylsilanes to silanols (Scheme 3.7) by Kaneda and coworkers,[129] in which they used Ag nanoparticles as catalytically active guest metal.[129] They prepared silver nanoparticles supported on HAp by cation exchange of Ag$^+$ with Ca^{2+} followed by reduction of surface supported Ag$^+$ ions by potassium borohydride under an argon atmosphere.

The other studies concerning the use of HAp as a support for metal nanoparticles catalysts showed the generation of active and reusable catalysts for

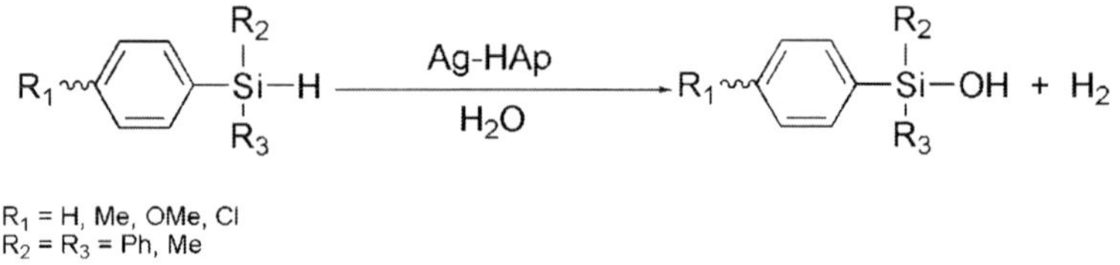

R₁ = H, Me, OMe, Cl
R₂ = R₃ = Ph, Me

Scheme 3.7　Ag-HAp-catalyzed oxidation of silanes to silanols in an aqueous phase.

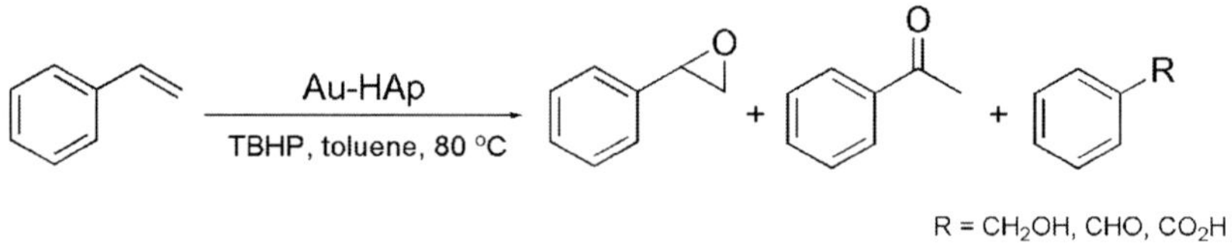

R = CH₂OH, CHO, CO₂H

Scheme 3.8　Au$_{25}$ nanoparticles supported on HAp catalyst for the selective epoxidation of styrene with *tert*-butyl hydroperoxide.

various catalytic transformations.[130–132] Che and coworkers demonstrated that the deposition of acetate-stabilized Ru nanoparticles onto a HAp surface forms highly efficient and reusable catalysts for the *cis*-dihydroxylation and oxidative cleavage of alkenes.[130] Tsukuda and coworkers[131] showed that the surface adsorption of glutathionate-stabilized Au nanoparticles on HAp followed by their calcination creates highly reactive Au$_{25}$ nanoparticles for the selective epoxidation of styrene with *tert*-butyl hydroperoxide (TBHP) (Scheme 3.8).

In our group we used HAp for the development of new supported ruthenium nanoparticles for the hydrogenation of aromatic substrates.[132] Our ruthenium nanoparticles supported on HAp were prepared by ion-exchange of Ru^{3+} ions with Ca^{2+} followed by their reduction with sodium borohydride in water at room temperature. The resulting supported ruthenium nanoparticles achieve the complete hydrogenation of various aromatic substrates (benzene, toluene, *o*-xylene, *p*-xylene, *m*-xylene and mestiylene) with high rates and selectivity under mild conditions. Moreover, they provide previously unprecedented catalytic lifetime (TTO = 192 600 turnovers over 400 h before deactivation) in the hydrogenation of neat benzene to cyclohexane. A recent study[133] revealed that the nanocrystalline HAp (80 nm in average size) can be used as a more effective supporting medium with respect to micrometer-size HAp as the reduction of the HAp matrix particle size from the microcrystalline to the nanocrystalline regime (from >1 μm to <100 nm) leads to an improved activity due to a higher external surface area, a larger number of exchange sites, and lower mass-transfer limitations. They prepared Rh nanoparticles supported on nanocrystalline HAp (Figure 3.14) by the *in-situ* H₂ reduction of Rh(III)-exchanged-nano-HAp during the hydrogenation of various arenes and olefins (cyclohexane, benzene, toluene, *o*-xylene, *p*-xylene, and *m*-xylene) in the solvent-free system under mild conditions.

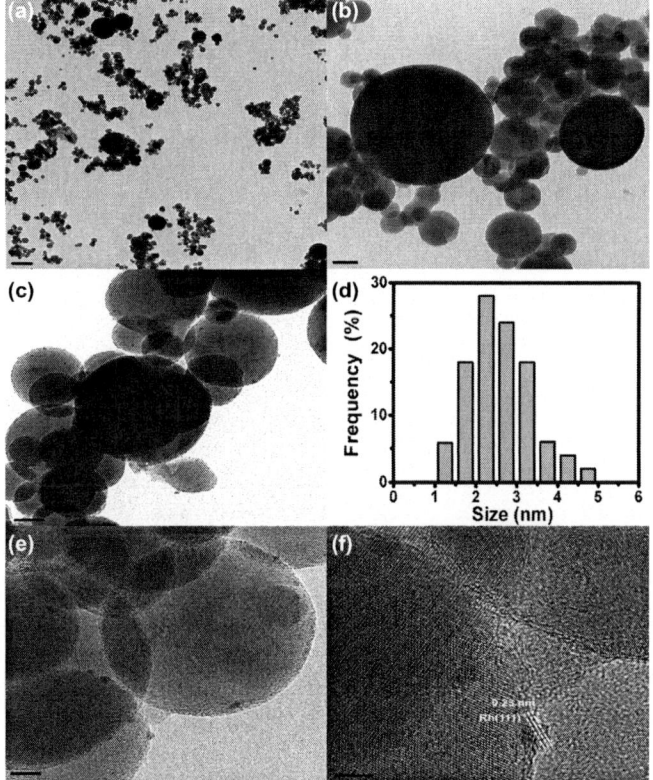

Figure 3.14   (a–c) TEM images of Rh@HAp for different magnification scale bars equal to 100, 50, and 20 nm, respectively. (d) Size histogram of Rh@HAp. (e, f) High-resolution TEM images of Rh@HAp in which scale bars correspond to 10 and 2 nm, respectively, reproduced from ref. 133.

3.2.2.4   Hydrotalcite (HT)

Hydrotalcite (HT), whose name is derived from its resemblance to talc and its high water content, is one type of inorganic material that has been studied extensively as a support material for the stabilization of metal nanoparticles catalysts.[134–140] Hydrotalcite is included in the type of anionic clay known as layered double hydroxides with general formula $(Mg_6Al_2(CO_3)(OH)_{16} \cdot 4(H_2O)$ and has anion-exchange capabilities.

   Mastalir and Kiraly,[134] incorporated Pd nanoparticles into HT framework *via* anion exchange between a dilute suspension of HT nitrate and a Pd-hydrosol stabilized by the anionic surfactant sodium dodecyl sulfate. The advanced characterization of the final material indicated the deposition of fairly monodispersed Pd particles, predominantly on the external surface of the HT layers. The Pd–HT samples proved to be efficient catalysts for the liquid-phase semihydrogenation of both terminal and internal alkynes under mild conditions. For the transformation of phenylacetylene to styrene, a bond selectivity of 100% was obtained, and the *cis* stereoselectivities for the hydrogenations of the internal alkynes 4-octyne and 1-phenyl-1-pentyne reasonably approached 100%.

Goodwin and coworkers,[135] prepared HT-supported cobalt nanoparticles by incipient-wetness impregnation of $Co(NO_3)_2$ on HT followed by thermal reduction. The catalytic performance of Co-HT sample was tested in the CO hydrogenation at different reduction temperatures (from 300 to 600 °C) used in the preparation of Co-HT. It was found that the Co-HT samples reduced at 500 °C provide the highest activity; however, CH_4 selectivity was also enhanced as the reduction temperature increased.

Hensen and coworkers investigated the promotional effect of transition-metal cations existing in the Au nanoparticles supported on HT.[137] The HT precursor was prepared by the homogeneous precipitation method using urea hydrolysis.[141] They prepared a series of transition-metal-modified M–HT ($M = Cr^{3+}$, Mn^{2+}, Fe^{3+}, Co^{2+}, Ni^{2+}, Cu^{2+}, Zn^{2+}) supports, along with a transition-metal-free sample by using the calcination–reconstruction process (also known as the "memory effect" of HT).[142] Various HT-supported gold catalysts were prepared by a modified deposition–precipitation approach with reduction by $NaBH_4$. Among these different metals promoted Au-HT catalysts they found that Au-MgCr-HT provides the best activity in the aerobic oxidation of alcohols, which was also found to be highly stable against agglomeration and leaching that makes it reusable catalyst for this reaction.

Ebitani and coworkers, recently achieved the green synthesis of 2,5-furandicarboxylic acid, one of the most important chemical building blocks from biomass, *via* oxidation of 5-hydroxymethylfurfural (Scheme 3.9) by using hydrotalcite-supported gold nanoparticle catalyst in water at 368 K under atmospheric oxygen pressure without addition of homogeneous base.[139] They prepared Au-HT catalyst by a deposition–precipitation technique using aqueous NH_3 solution followed by calcination at 473 K.

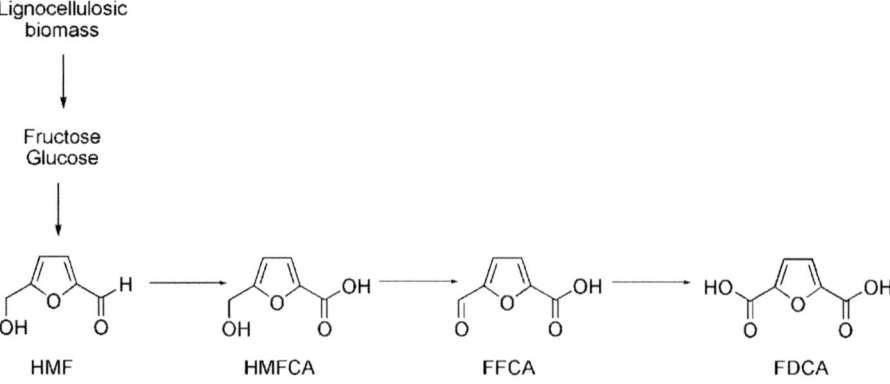

Scheme 3.9   2,5-Furandicarboxylic acid (FDCA) synthesis *via* oxidation of 5-hydroxymethylfurfural (HMF).

In a recent study,[139] Pd nanoparticles supported on HT has been synthesized in a one-step reaction using a precipitation–reduction method in which hydrolysis of hexamethylenetetramine was used to both precipitate HT by virtue of the resulting increase in pH and provide formaldehyde which leads to reduction of Pd(II) to Pd(0). The resulting Pd nanoparticles were found to be mostly tetrahedral in shape, and highly dispersed on the surface of the HT. Moreover, it was observed that the introduction of capping agent cetyl-trimethylammonium bromide (CTAB) during the synthesis changes the morphology of the Pd nanoparticles to truncated octahedral with a similar degree of dispersion (Figure 3.15). These HT-supported Pd nanoparticles were used as catalysts in the hydrogenation of acetylene, in which they provide higher activity and selectivity than conventional Pd-based impregnated catalyst.

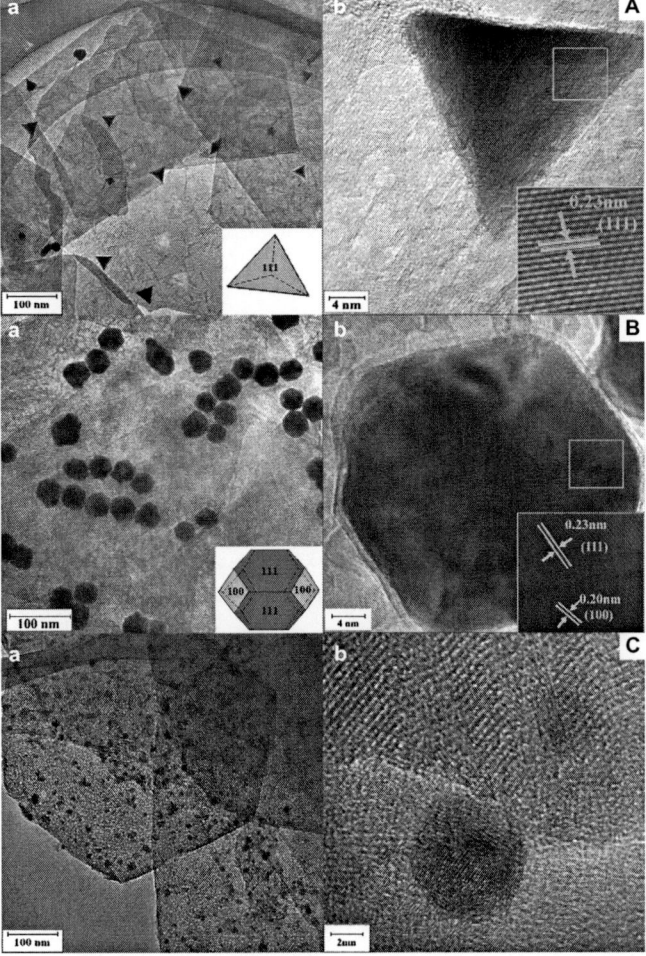

Figure 3.15 HRTEM images of Tetra-Pd/HT (A), Octa-Pd/HT (B), and Spher-Pd/HT (C), reproduced from ref. 139.

3.3 Future Goals Toward Modern Supported Metal Nanoparticles Catalysts

The preparation of supported metal nanoparticle catalysts for various types of catalytic transformations has been extensively investigated since the mid-1990s. However, there is still room for research in the improvement of the existing methodologies used in the synthesis of nanoparticles and in their catalytic employment to reach modern supported nanoparticles catalytic systems. Such research is expected to include the following:

(i) The synthesis routes of supported metal nanoparticles need to be carried out by considering the 12 green chemistry principles.[20] They should be prepared by using the least toxic metal precursor and environmentally friendly supports[143] and carried out in water or environmentally benign solvents (*e.g.* ethanol) with the lowest number of reagents, steps (*i.e.* a one-pot reaction), the minimal amount of energy as possible whilst minimizing the generation of byproducts and waste.

(ii) Although supported metal nanoparticles catalysts form a large and important group of heterogeneous catalysts, the majority of their synthesis routes do not meet the requirements of modern nanoparticles catalysis such as the control of the size, morphology and composition of nanoparticles. In this context, the development of well-designed metal precursor@support as precatalyst materials has an importance not only for controlling the size, shape, and composition but also for studying the formation kinetics and mechanism of supported nanoparticles under *in-situ* (or *operando*) conditions.[144]

(iii) In addition to the supported metal nanoparticles catalysts, it has recently been realized that metallic nanoparticles can also be greatly stabilized within inorganic layers to form yolk–shell nanoreactors and some of them provide excellent durability even under harsh reaction conditions.[145,146] Although they provide high stability throughout catalytic runs, the dense coating of core metal nanoparticles makes them less accessible for the substrate molecules. For this reason, the development of such yolk–shell structures in which active metal centers are highly accessible for the substrate molecules, whilst the shell allows the fast diffusion of reactants and products, seems to be one of the important targets for the next generation of highly durable supported nanoparticles catalysts.

(iv) The durability of supported metal nanoparticles should be tested in terms of leaching for liquid-phase reactions, reusability and recyclability performance for both gaseous and liquid-phase reactions.

References

1. J. Shwarz, C. Contescu, K. Putyera, *Encyclopedia of Nanoscience and Nanotechnology*, 2nd edn, Marcel-Dekker, New York, 2004.
2. C. A. Mirkin, *Small*, 2005, **1**, 1.

3. W. Pan, Z. R. Dai and Z. L. Wang, *Science*, 2001, **291**, 1947.
4. R. Elghanian, J. J. Storhoff, R. C. Mucic, R. L. Letsinger and C. A. Mirkin, *Science*, 1997, **277**, 1078.
5. J. N. Anker, W. P. Hall, O. Lyandres, N. C. Shah, J. Zhao and R. P. V. Duyne, *Nature Mater.*, 2008, **7**, 442.
6. M. Law, L. E. Greene, J. C. Johnson, R. Saykally and P. Yang, *Nature Mater.*, 2005, **4**, 455.
7. D. A. Lavan, T. McGuire and R. Langer, *Nature Biotechnol.*, 2003, **21**, 1184.
8. S. H. Joo, S. J. Choi, I. Oh, J. Kwak, Z. Liu, O. Terasaki and R. Ryoo, *Nature*, 2001, **412**, 169.
9. V. L. Colvin, M. C. Schlamp and A. P. Alivisatos, *Nature*, 1994, **370**, 354.
10. M. T. Reetz, M. Winter, G. Dumpich, J. Lohau and S. Friedrichowski, *J. Am. Chem. Soc.*, 1997, **119**, 4539.
11. C. B. Murray, C. R. Kagan and M. G. Bawendi, *Science*, 1995, **270**, 1335.
12. S. J. Tans, M. H. Devoret, H. J. Dai, A. Thess, R. E. Smalley, L. J. Geerligs and C. Dekker, *Nature*, 1997, **386**, 474.
13. M. Antonietti and C. Goltner, *Angew. Chem. Int. Ed. Engl.*, 1997, **36**, 910.
14. D. Astruc, F. Lu and J. R. Aranzaes, *Angew. Chem., Int. Ed.*, 2005, **44**, 7852.
15. G. Schmid and D. Fenske, *Phil. Trans. R. Soc. A*, 2010, **1905**, 1207.
16. B. Zhou, S. Han, R. Raja, G. A. Somorjai, *Nanotechnology in Catalysis*, Vol. 3, Springer, New York, 2007.
17. U. Heiz, U. Landman, *Nanocatalysis*, Springer, Heidelberg, 2007.
18. G. Schmid, V. Maihack, F. Lantermann and S. Peschel, *J. Chem. Soc., Dalton Trans.*, 1996, 589.
19. R. Pool, *Science*, 1990, **248**, 1186.
20. M. Poliakoff, J. M. Fitzpatrick, T. R. Farren and P. T. Anastas, *Science*, 2002, **297**, 807.
21. S. Özkar and R. G. Finke, *Langmuir*, 2003, **19**, 6247.
22. F. Durap, M. Zahmakiran and S. Özkar, *Appl. Catal. A: Gen.*, 2009, **369**, 53.
23. L. Tosheva and V. P. Valtchev, *Chem. Mater.*, 2005, **17**, 2494.
24. Z. Peng, J. Wu and H. Yang, *Chem. Mater.*, 2010, **22**, 1098.
25. V. Mazumder and S. H. Sun, *J. Am. Chem. Soc.*, 2009, **131**, 4588.
26. H. Choi, S. R. Al-Abed, S. Agarwal and D. D. Dionysiou, *Chem. Mater.*, 2008, **20**, 3649.
27. S. Tsubota, M. Haruta, T. Kobayashi, A. Ueda and Y. Nakahara, Preparation of Catalysts V: Scientific Bases for Preparation of Heterogeneous Catalysts. In: G. Ponecelet, P. A. Jacobs, Grange and B. Delmon, Eds. *Studies in Surface Science and Catalysis*, Elsevier Science Publ B. V., Amsterdam, 1991, **63**, 695–704.
28. M. Haruta, S. Tsubota, T. Kobayashi, T. Kageyama and M. J. Genet, *J. Catal.*, 1993, **144**, 175.
29. A. Martinez and G. Prieto, *Catal. Commun.*, 2007, **8**, 1479.
30. P. He, M. Zhang, D. Yang and J. Yang, *Surf. Rev. Lett.*, 2006, **13**, 51.

31. N. Panziera, P. Pertici, L. Barazzone, A. M. Caporusso, G. Vitulli, P. Salvadori, S. Borsacchi, M. Geppi, C. A. Veracini, G. Martra and L. Bertinetti, *J. Catal.*, 2007, **246**, 351.
32. S. M. George, *Chem. Rev.*, 2010, **110**, 111.
33. S. Dominguez-Dominguez, J. Arias-Pardilla, A. Berenguer-Murcia, E. Morallon and D. Cazorla-Amoros, *J. Appl. Electrochem.*, 2007, **38**, 259.
34. D. Nagao, Y. Shimazaki, S. Saeki, Y. Kobayashi and M. Konno, *Colloids Surf. A*, 2007, **302**, 623.
35. G. Glaspell, L. Fuoco and M. S. El-Shall, *J. Phys. Chem. B*, 2005, **109**, 17350.
36. G. Glaspell, H. M. A. Hassan, A. Elzatahry, V. Abdalsayed and M. S. El-Shall, *Top. Catal.*, 2008, **47**, 22.
37. Y. Zhang and C. Erkey, *J. Supercrit. Fluids*, 2006, **38**, 252.
38. R. G. Compton, J. C. Eklund and F. Marken, *Electroanalysis*, 1997, **9**, 509.
39. R. Strobel, S. E. Pratsinis and A. Baiker, *J. Mater. Chem.*, 2005, **15**, 605.
40. L. Delannoy, N. E. Hassan, A. Musi, N. N. Le To, J.-M. Krafft and C. Louis, *J. Phys. Chem. B.*, 2006, **110**, 22471.
41. M. Endo, Y. A. Kim, M. Ezaka, K. Osada, T. Yanagisawa, T. Hayashi, M. Terrones and M. S. Dresselhaus, *Nano Lett.*, 2003, **3**, 723.
42. A. Miyazakia, I. Balint, K.-I. Aika and Y. Nakano, *J. Catal.*, 2001, **204**, 364.
43. D. Shah and H. Kaur, *J. Mol. Catal. A: Chem.*, 2012, **359**, 69.
44. J. Zhu, K. Kailasam, A. Fischer and A. Thomas, *ACS Catal.*, 2011, **1**, 342.
45. D. Horvath, L. Toth and L. Guczi, *Catal. Lett.*, 2000, **67**, 117.
46. A. Barau, V. Budarin, A. Caragheorgheopol, R. Luque, D. J. Macquarrie, A. Prelle, V. S. Teodorescu and M. Zaharescu, *Catal. Lett.*, 2008, **124**, 204.
47. S. Schimpf, M. Lucas, C. Mohr, U. Rodemerck, A. Brückner, J. Radnik, H. Hofmeister and P. Claus, *Catal. Today*, 2002, **72**, 63.
48. A. Sandoval, A. Gomez-Cortes, R. Zanella, G. Diaz and J. M. Saniger, *J. Mol. Catal. A: Chem.*, 2007, **278**, 200.
49. C. M. Y. Yeung, K. M. K. Yu, Q. J. Fu, D. Thompsett, M. I. Petch and S. C. Tsang, *J. Am. Chem. Soc.*, 2005, **127**, 18010.
50. B. Yoon and C. M. Wai, *J. Am. Chem. Soc.*, 2005, **127**, 17174.
51. M. Turner, V. B. Golovko, O. P. H. Vaughan, P. Abdulkin, A. Berenguer-Murcia, M. S. Tikhov, B. F. G. Johnson and R. M. Lambert, *Nature*, 2008, **454**, 981.
52. G.-H. Moon, Y. Park, W. Kim and W. Choi, *Carbon*, 2011, **49**, 3454.
53. P. He, M. Zhang, D. Yang and J. Yang, *Surf. Rev. Lett.*, 2006, **13**, 51.
54. S. Saha, A. Pal, S. Kundu, S. Basu and T. Pal, *Langmuir*, 2010, **26**, 2885.
55. N. Panziera, P. Pertici, L. Barazzone, A. M. Caporusso, G. Vitulli, P. Salvadori, S. Borsacchi, M. Geppi, C. A. Veracini, G. Martra and L. Bertinetti, *J. Catal.*, 2007, **246**, 351.
56. S. M. George, *Chem. Rev.*, 2010, **110**, 111.

57. X. Jiang, H. Huang, F. B. Prinz and S. F. Bent, *Chem. Mater.*, 2008, **20**, 3897.
58. H. Feng, J. W. Elam, J. A. Libera, W. Setthapun and P. C. Stair, *Chem. Mater.*, 2010, **22**, 3133.
59. B. M. Quinn, C. Dekker and S. G. Lemay, *J. Am. Chem. Soc.*, 2005, **127**, 6146.
60. T. M. Day, P. R. Unwin, N. R. Wilson and J. V. Macpherson, *J. Am. Chem. Soc.*, 2005, **127**, 10639.
61. Z. He, J. Chen, D. Liu, H. Zhou and Y. Kuang, *Diamond Relat. Mater.*, 2004, **13**, 1764.
62. Gregory G. Wildgoose, Craig E. Banks and Richard G. Compton, *Small*, 2006, **2**, 192.
63. K. Okitsu, M. Ashokkumar and F. Grieser, *J. Phys. Chem. B*, 2005, **109**, 20673.
64. H. S. Park, J.-S. Kim, B. G. Choi, S. M. Jo, D. Y. Kim and W. H. Hong, *Carbon*, 2010, **48**, 1325.
65. H.-B. Pan and C. M. Wai, *New J. Chem.*, 2011, **35**, 1649.
66. G. Glaspell, L. Fuoco and M. S. El-Shall, *J. Phys. Chem. B*, 2005, **109**, 17350.
67. D. Marquardt, C. Vollmer, R. Thomann, P. Steurer, R. Mülhaupt, E. Redel and C. Janiak, *Carbon*, 2011, **49**, 1326.
68. S. Senkan, M. Kahn, S. Duan, A. Ly and C. Ledholm, *Catal. Today.*, 2006, **117**, 291.
69. E. Jiménez, K. Abderrafi, R. Abargues, J. L. Valdés and J. P. Martínez-Pastor, *Langmuir*, 2010, **26**, 7458.
70. X. R. Ye, Y. Lin and C. M. Wai, *Chem. Commun.*, 2003, 642.
71. X. R. Ye, Y. Lin, C. Wang, M. Engelhard, Y. Wang and C. M. Wai, *J. Mater. Chem.*, 2004, **14**, 908.
72. R. G. Compton, J. C. Eklund and F. Marken, *Electroanalysis*, 1997, **9**, 509.
73. L.-P. Jiang, A.-N. Wang, Y. Zhao, J.-R. Zhang and J.-J. Zhu, *Inorg. Chem. Commun.*, 2004, **7**, 506.
74. X.-F. Qiu, J.-Z. Xu, J.-M. Zhu, J.-J. Zhu, S. Xu and H.-Y. Chen, *J. Mater. Res.*, 2003, **18**, 1399.
75. L. Madler, W. J. Stark and S. E. Pratsinis, *J. Mater. Res.*, 2003, **18**, 115.
76. G. G. Wildgoose, C. E. Banks and R. G. Compton, *Small*, 2006, **2**, 182.
77. R. J. White, R. Luque, V. L. Budarin, J. H. Clark and D. J. Macquarrie, *Chem. Soc. Rev.*, 2009, **38**, 481.
78. D. Astruc, F. Lu and J. R. Aranzaes, *Angew. Chem., Int. Ed.*, 2005, **44**, 7852.
79. M. Králik and A. Biffis, *J. Mol. Catal. A:Chem.*, 2001, **177**, 113.
80. S. Bhatia, *Zeolite Catalysis: Principles and Applications*, CRC Press, Florida, 1989.
81. C. H. Baerlocher, W. M. Meier, D. H. Olson, *Atlas of Zeolite Framework Types*, 5th edn, Elsevier, Amsterdam, 2001.
82. D. W. Breck, *Zeolite Molecular Sieves*, Wiley, New York, 1984.

83. S. Özkar, G. A. Ozin, K. Moller and T. Bein, *J. Am. Chem. Soc.*, 1990, **112**, 9575.
84. G. Riahi, D. Guillemot, M. Polisset-Thfoin, A. A. Khodadadi and J. Fraissard, *Catal. Today*, 2002, **72**, 115.
85. S. J. Cho, S. M. Jug, Y. G. Shalt and R. Ryoo, *J. Phys. Chem.*, 1992, **96**, 9922.
86. Y. Wang, H. Wu, Q. Zhang and Q. Tang, *Mic. Mes. Mater.*, 2005, **86**, 38.
87. M. Zahmakıran and S. Özkar, *Langmuir*, 2008, **24**, 7065.
88. M. Zahmakıran and S. Özkar, *Langmuir*, 2009, **25**, 2667.
89. M. Zahmakıran and S. Özkar, *J. Mater. Chem.*, 2009, **19**, 7112.
90. M. Zahmakıran and S. Özkar, *Appl. Catal. B:Env.*, 2009, **89**, 104.
91. M. Zahmakıran, Y. Tonbul and S. Özkar, *J. Am. Chem. Soc.*, 2010, **132**, 6541.
92. M. Zahmakıran, F. Durap and S. Özkar, *Int. J. Hyd. Energ.*, 2010, **35**, 187.
93. E. Bayram, M. Zahmakıran, S. Özkar and R. G. Finke, *Langmuir*, 2010, **46**, 12455.
94. M. Zahmakıran, S. Akbayrak, T. Kodaira and S. Özkar, *Dalton Trans.*, 2010, **39**, 7521.
95. S. Çalışkan, M. Zahmakıran and S. Özkar, *Appl. Catal. B: Env.*, 2010, **93**, 387.
96. M. Zahmakıran, T. Kodaira and S. Özkar, *Appl. Catal. B: Env.*, 2010, **96**, 533.
97. M. Zahmakıran, T. Ayvalı, S. Akbayrak, S. Çalışkan, D. Çelik and S. Özkar, *Catal. Today.*, 2011, **170**, 76.
98. M Rakap and S. Özkar, *Appl. Catal. B: Env.*, 2009, **91**, 21.
99. M Rakap and S. Özkar, *Int. J. Hyd. Energ.*, 2010, **35**, 3341.
100. F. Durap, M. Rakap, M. Aydemir and S. Özkar, *Appl. Catal. A: Gen.*, 2010, **382**, 339.
101. M. Rakap and S. Özkar, *Int. J. Hyd. Energ.*, 2010, **35**, 1305.
102. M. Zahmakıran and S. Özkar, *Mater. Chem. Phys.*, 2010, **121**, 359.
103. B-Z. Zhan, M. A. White, T.-K. Sham, J. A. Pincock, R. J. Doucet, K. V. R. Rao, K. N. Robertson and T. S. Cameron, *J. Am. Chem. Soc.*, 2003, **125**, 2196.
104. A. B. Laursen, K. T. Højholt, L. F. Lundegaard, S. B. Simonsen, S. Helveg, F. Schüth, M. Paul, J.-D. Grunwaldt, S. Kegnæs, C. H. Christensen and K. Egeblad, *Angew. Chem. Int. Ed.*, 2010, **122**, 3582.
105. M. Choi, Z. Wu and E. Iglesia, *J. Am. Chem. Soc.*, 2010, **132**, 9129.
106. A. Cauve, D. Brune, F. D. Renzo, P. Moreau and F. Fajula, *Stud. Surf. Sci. Catal.*, 1995, **94**, 286.
107. S. Mandal, D. Roy, R. V. Chaudhari and M. Sastry, *Chem. Mater.*, 2004, **16**, 3714.
108. J. Sun, D. Ma, H. Zhang, X. Liu, X. Han, X. Bao, G. Weinberg, N. Pfander and D. Su, *J. Am. Chem. Soc.*, 2006, **128**, 15756.
109. L. Tosheva and V. P. Valtchev, *Chem. Mater.*, 2005, **17**, 2494.

110. J.Y. Lee, O. K. Farha, J. Roberts, K. A. Scheidt, S. T. Nguyen and J. T. Hupp, *Chem. Soc. Rev.*, 2009, **38**, 1450.
111. D. Farrusseng, S. Aguado and C. Pinel, *Angew. Chem. Int. Ed.*, 2009, **48**, 7502.
112. A. Corma, H. Garcia and F. X. L. Xamena, *Chem. Rev.*, 2010, **110**, 4606.
113. F. Schröder, D. Esken, M. Cokoja, M. W. E. Berg, O. I. Lebedev, G. V. Tendeloo, B. Walaszek, G. Buntkowsky, H.-H. Limbach, B. Chaudret and R. A. Fischer, *J. Am. Chem. Soc.*, 2008, **130**, 6119.
114. Y. K. Park, S. B. Choi, H. J. Nam, D.-Y. Jung, H. C. Ahn, K. Choi, H. Furukawa and J. Kim, *Chem. Commun.*, 2010, **46**, 3086.
115. M. Müller, S. Hermes, K. Kahler, M. W. E. Berg, M. Muhler and R. A. Fischer, *Chem. Mater.*, 2008, **20**, 4576.
116. S. Gao, N. Zhao, M. Shu and S. Che, *Appl. Catal. A: Gen.*, 2010, **388**, 196.
117. M. S. El-Shall, V. Abdelsayed, A. R. S. Khder, H. M. A. Hassan, H. M. El-Kaderi and T. E. Reich, *J. Mater. Chem.*, 2009, **19**, 7625.
118. H. Liu, Y. Liu, Y. Li, Z. Tang and H. Jiang, *J. Phys. Chem. C*, 2010, **114**, 13362.
119. T. Ishida, M. Nagaoka, T. Akita and M. Haruta, *Chem. Eur. J.*, 2008, **14**, 8456.
120. H.-L. Jiang, B. Liu, T. Akita, M. Haruta, H. Sakurai and Q. Xu, *J. Am. Chem. Soc.*, 2009, **131**, 11302.
121. H.-L. Jiang, T. Akita, T. Ishida, M. Haruta and Q. Xu, *J. Am. Chem. Soc.*, 2010, **136**, 19332.
122. K. Kaneda and T. Mizugaki, *Energy Environ. Sci.*, 2009, **2**, 655.
123. L. C. Palmer, C. J. Newcomb, S. R. Kaltz, E. D. Spoerke and S. I. Stupp, *Chem. Rev.*, 2008, **108**, 4754.
124. M. J. Mura-Galelli, J. C. Voegel, S. Behr, E. F. Bres and P. Schaaf, *Proc. Natl. Acad. Sci.*, 1991, **88**, 5557.
125. J. Reichert and J. G. P. Binner, *J. Mater. Sci.*, 1996, **31**, 1231.
126. K. Mori, T. Hara, T. Mizugaki, K. Ebitani and K. Kaneda, *J. Am. Chem. Soc.*, 2004, **126**, 10657.
127. K. Mori, S. Kanai, T. Hara, T. Mizugaki, K. Ebitani, K. Jitsukawa and K. Kaneda, *Chem. Mater.*, 2007, **19**, 1249.
128. T. Hara, T. Kaneta, K. Mori, T. Mitsudome, T. Mizugaki, K. Ebitani and K. Kaneda, *Green Chem.*, 2007, **9**, 1246.
129. T. Mitsudome, S. Arita, H. Mori, T. Mizugaki, K. Jitsukawa and K. Kaneda, *Angew. Chem. Int. Ed.*, 2008, **47**, 7938.
130. C.-M. Ho, W.-Y. Yu and C.-M. Che, *Angew. Chem., Int. Ed.*, 2004, **43**, 3303.
131. Y. Liu, H. Tsunoyama, T. Akita and T. Tsukuda, *Chem. Commun.*, 2010, **46**, 550.
132. M. Zahmakıran, Y. Tonbul and S. Özkar, *Chem. Commun.*, 2010, **46**, 4788.
133. M. Zahmakıran, Y. Roman-Leshkov and Y. Zhang, *Langmuir*, 2012, **28**, 60.

134. A. Mastalir and Z. Király, *J. Catal.*, 2003, **220**, 372.
135. Y.-T. Tsai, X. Mo, A. Campos, J. G. Goodwin and J. J. Spivey, *Appl. Catal. A: Gen.*, 2011, **396**, 91.
136. A. Tsuji, K. Tirumala, V. Rao, S. Nishimura, A. Takagaki and K. Ebitani, *ChemSusChem*, 2011, **4**, 542.
137. P. Liu, Y. Guan, R. A. V. Santen, C. Li and E. J. M. Hensen, *Chem. Commun.*, 2011, **47**, 11540.
138. W. Fang, Q. Zhang, J. Chen, W. Deng and Y. Wang, *Chem. Commun.*, 2010, **46**, 1547.
139. N. K. Gupta, S. Nishimura, A. Takagaki and K. Ebitani, *Green Chem.*, 2011, **13**, 824.
140. J. T. Feng, X. Y. Ma, Y. F. He, D. G. Evans and D. Q. Li, *Appl. Catal. A:Gen.*, 2012, **413**, 10.
141. M. Ogawa and H. Kaiho, *Langmuir*, 2002, **18**, 4240.
142. S. K. Jana, Y. Kubota and T. Tatsumi, *J. Catal.*, 2007, **247**, 214.
143. V. Budarin, J. H. Clark, R. Luque, D. J. Macquarrie and R. J. White, *Green Chem.*, 2008, **10**, 382.
144. J. E. Mondloch, Q. Wang, A. I. Frenkel and R. G. Finke, *J. Am. Chem. Soc.*, 2010, **132**, 9701.
145. J. Y. Kim, S. B. Yoon and J.-S. Yu, *Chem. Commun.*, 2003, 790.
146. M. Kim, K. Sohn, H. B. Na and T. Hyeon, *Nano Lett.*, 2002, **2**, 1383.

CHAPTER 4

Energy Conversion and Storage through Nanoparticles

SHENQIANG REN*[a] AND YAN WANG*[b]

[a] University of Kansas, Department of Chemistry, 1251 Wescoe Hall Drive, Lawrence, KS 66045; [b] Worcester Polytechnic Institute, Department of Mechanical Engineering, 100 Institute Road, Worcester, MA 01609
*Email: shenqiang@ku.edu; yanwang@wpi.edu

4.1 Introduction

Nanoparticles (NPs) are small particles less than 100 nm in diameter.[1] They have properties that are different from the bulk materials.[2,3] There have been enormous research activities in developing strategies for the colloidal synthesis of semiconductor and metal NPs. The composition, size, shape, and surface protection of NPs can be controlled to an exceptionally high degree. Colloidal NPs are both the subject of fundamental studies in terms of their optoelectronic properties, and promising candidates for applications in light-harvesting photovoltaics, nanoplasmonics, and nanocatalysis by design.[4-6] One particular interesting emerging class of colloidal nanostructures are semiconductor NPs, also termed "quantum dots" (QDs, the size less than 25 nm),[7,8] such as CdS, CdSe, CdTe, InP, GaAs, GaP, GaN, GaInP, and others. QDs are composed of an inorganic core, made up of between a few hundred and a few thousand atoms, surrounded by an outer layer of organic surfactant molecules (ligands).[9-13] At this length scale, the dimensions of the confinement become less than the de Broglie wavelength of charge carriers, giving rise to quantization effects. For example, the separation between valence and conduction bands becomes dependent on the

RSC Green Chemistry No. 19
Sustainable Preparation of Metal Nanoparticles: Methods and Applications
Edited by Rafael Luque and Rajender S Varma
Published by the Royal Society of Chemistry, www.rsc.org

particle size; decreasing size of the semiconductor nanoparticle yields an increase in the bandgap width. As a consequence, it is possible to tune the electrical properties of semiconductors by preparing quantum dots of different size, thus mimicking the photoelectrical behavior of a wide range of semiconductors using only a single material. This effect is accompanied by an exaltation of the Coulomb interaction between the charge carriers. For most semiconductors, this observation normally occurs when the particle size is reduced to a few nanometers. These effects significantly modify the processes of intraband and interband relaxation. The small size of QDs implies a smaller number of atoms contributing to molecular orbitals in the formation of energy bands (Figure 4.1). As a consequence, the bandgap and energy-level position are a function of the QD size, leading to size-tunable absorption and emission characteristics. Furthermore, in very small QDs, the conduction bands and valence bands are discontinuous and electrons occupy discrete, quantized states, analogous to the particle in a box model.[14] This is shown in Figure 4.1. The size-dependent optical properties of QDs have been the focus of significant research over the past two decades. QDs have attracted intensive interest for optoelectronic applications since their optical and electronic properties are tunable by changing their size and/or the shape. For example, cadmium selenide (CdSe) QDs vary in luminescence color when irradiated with ultraviolet (UV) light, as shown in Figure 4.2.[15,16] This color change is caused by the quantum confinement, which will be explained in the following section. The nanometric QD size also results in a very high surface-to-volume ratio. The coordination sphere of this high population of surface atoms partially occurs *via* complex formation with the stabilizing ligands. Nevertheless, a significant fraction of these organically passivated QDs typically exhibit surface-related trap states acting as fast nonradiative de-excitation channels for photogenerated charge carriers, thereby reducing the fluorescence quantum yield (QY).

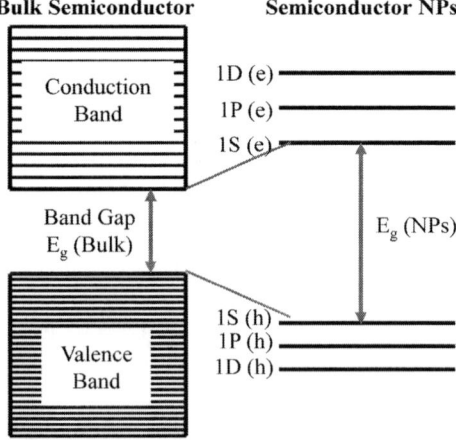

Figure 4.1   Schematic illustration of bandgap variation of semiconducting bulk and QDs, which show the associated quantization of energy states for QDs.

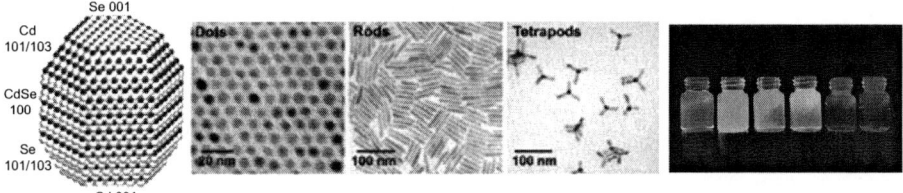

Figure 4.2 The CdSe QD fluorescence dictated by their size, can be emitted from a single material with different size.
Figure credit: Adapted with permission from refs. 17 and 18.

4.1.1 Quantum Confinement of Nanoparticles

The electron–hole pair (exciton) in a semiconductor is bound within a characteristic length, known as the Bohr radius, which varies depending on the material. When the size of a semiconductor QD is smaller than the Bohr diameter, the charge carriers in the QD are spatially confined. In that case, the energy of the charge carriers will be raised and the properties change from the bulk regime to the quantum-confinement regime (Figure 4.3).[17] In the quantum-confinement regime, the optical and electrical properties depend on the QD size. A simple model for considering the confined electronic states of a spherical QD is the particle-in-a-box model, where the particles are confined in an infinite spherical well potential. The radius of the potential well is a, and the potential V experienced by the particle is $V = 0 (r \leq a)$ or $V = \infty$ $(r > a)$.[18] The wave function of the particle is zero outside the well. By solving Schrödinger's equation inside the well, the confinement energy of the particle is given by

$$E_{n,l} = \frac{h^2}{2ma^2} \beta_{n,l}^2$$

where m is the effective mass of the particle, $\beta_{n,l}$ is the nth zero value of a spherical Bessel function of order l. The most qualitative information from this expression is that the confinement energy of the particle is inversely proportional to a^2, which explains the increased bandgap with decreasing the size.[19,20] For example, the bandgap of CdSe can be increased from 1.8 eV (its bulk value) to 3 eV.

The particle-in-a-box model considers a free particle in an empty box. In reality, we should take the effects of the periodic crystal lattice and the electron–hole Coulomb energy into consideration. The most popular and successful theory for CdSe QD electronic structure that incorporates these facts is the multiband effective mass theory. In this model, the confined electron and hole levels are considered independently, which is a reasonable assumption considering the wide bandgap of CdSe. The interband transitions are only allowed between states with the same angular momentum, *i.e.* S–S and P–P.[21] The room-temperature absorption spectrum of 4.6 nm diameter CdSe QD is shown in Figure 4.4, with the first few transitions labeled.

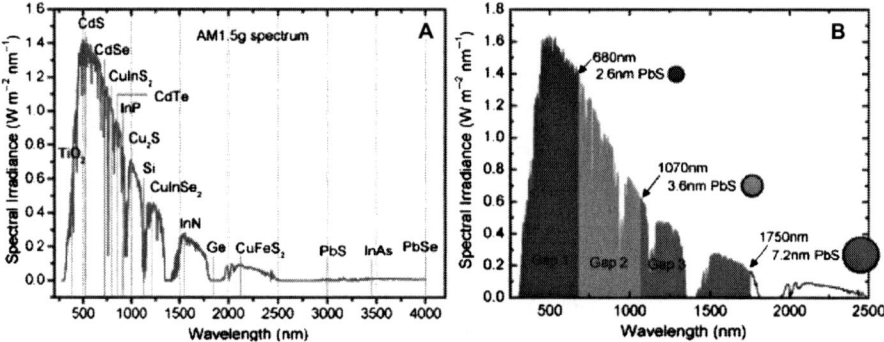

Figure 4.3   The Sun irradiation spectrum and corresponding semiconductor bandgap
of (A) Bulk and (B) QDs.
Figure credit: reference 19.

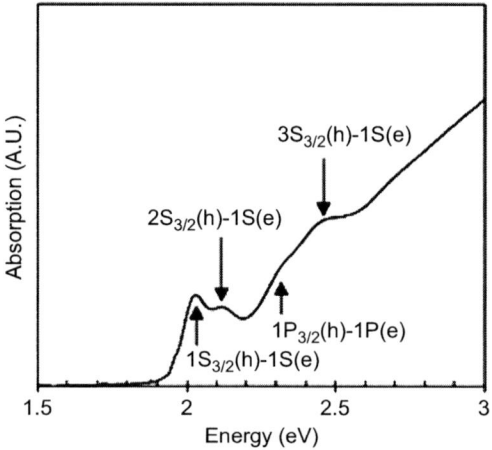

Figure 4.4   Absorption spectrum of CdSe QDs with 4.6 nm diameter. The first few
resolvable transitions are labeled, where S and P refer to angular
momentum.
Figure credit: ref. 23.

4.1.2   Synthesis of Quantum Dots

Quantum dots are employed in solar cells in the following way: QD are brought
into close contact with a semiconductor having a wide bandgap (TiO_2, ZrO_2)
which serves as an electron acceptor. At the same time, the quantum dot
interacts with a redox system that functions as a hole acceptor. Now, we
consider examples of how a solar-cell system can be prepared.

Two groups of synthesis methods for QD production exist: chemical
synthesis and QD growth *via* vapor-phase deposition. In the chemical route,
QD are formed as colloidal particles stabilized by the ligands, whereas in

vapor-phase deposition nanoparticles grow on a semiconductor support. The QD synthesis from the chemical approach involves the hot-injection method. This approach was first developed by Louis Brus in the late 1970s,[22] and followed by Alivisatos and coworkers[23] and Bawendi *et al.*[24] The QD nanoarchitecture is getting more sophisticated, for example, onion-like core–shell structures formed by coating one material with another. These structures can be used to shield a chemically or electrically sensitive QD from their environment. The QDs have been applied in biological imaging applications, and magnetic metal NPs show promise as the recording media. As an example of the former method, consider the formation of CdSe nanoparticles from a "green" precursor – CdO.[25]

In CdSe preparation, CdO and two nanoparticle stabilizers – trioctylphosphine oxide (TOPO) and hexylphosphonic acid (HPA) – are loaded into a three-neck flask. At about 300 °C the powder of CdO is dissolved yielding a homogeneous solution. Then Se stock solution (Se dissolved in TOPO) is introduced to the mixture resulting in formation of a CdSe colloidal solution. To obtain QDs in a form easy to store, methanol can be added to flocculate colloidal particles. Then, the flocculate can be separated from the solution and completely redispersed in a hexane/butanol/TOPO mixture to produce optically clear colloidal solution. The flocculation/redispersion procedure is repeated several times to obtain pure powder of CdSe nanoparticles capped with TOPO. Good-quality nanoparticles synthesized in this method and their monodispersity does not require additional fractionation. It is important to note that TOPO capping can be exchanged for other types of capping agents, such as thiols, furan, pyridines, polymers, *etc.* Subsequently QD powders can be redissolved to form transparent colloidal solutions.

Solution-processed solar cells involving colloidal quantum dots (CQDs) are prospective candidates for production of large-area, low-cost photovoltaics with improved efficiencies. The intensive ongoing research of CQDs is aimed at finding an optimal solar-cell configuration that will provide maximum photovoltaic efficiency. For the potential industrial application, not only is efficiency important, but also stability and lifetime must be taken into account. In many reports, however, these devices have been found to suffer from extreme air sensitivity. At least two mechanisms deleterious to device performance are expected to be at work. First, oxidation produces increased p-type doping, reducing the spatial extent of the depletion region and negatively affecting photocurrent. Secondly, oxidation of metal chalcogenide surfaces has been shown to produce deep traps for electrons. These may act as recombination centers and also impede electron transport. To prevent QD surface oxidation, Debnath *et al.*[26] proposed a novel CQD synthesis technique involving strongly bound ligands.

As mentioned above, organic ligands such as trioctylphosphine/trioctylphosphine oxide are employed in the synthesis of highly monodispersed nanoparticles. Because these long ligands hinder efficient charge transport within CQD solids, exchange of the capping ligands to short, weakly bound molecules is often practiced. The authors hypothesized that the use of weakly bound ligands, such as butylamines and pyridine, is responsible for the ready access of oxygen and water to the unprotected surface of CQDs. Therefore, they chose a strongly bound ligand – N-2,4,6-trimethylphenyl- N-methyl-dithiocarbamate (TMPMDTC) for stabilization of PbS quantum dots. They showed that the ligand exchange did not negatively affect the charge transfer between QDs and electrodes; the corresponding solar-cell device possessed a promising efficiency of 3.6%. To measure aging effects in QDs exposed to air, they used the fact that deep traps on QD surfaces emerge in the presence of oxygen, and their number alters the cell capacitance. The capacitance was measured with respect to the control; the control exhibited a large increase in low-frequency capacitance consistent with the emergence of deep traps. In contrast, the TMPMDTC device capacitance did not change significantly across the same frequency regime.

Once the QD colloidal solution is prepared, the next step is to attach QDs to the electron acceptor (*e.g.*, TiO_2). The system consisting of QDs bound to mesoporous metal oxides can be prepared using two approaches:[27] (1) colloidal QDs capped with surface ligands are attached to metal-oxide surfaces through linker molecules or other attractive forces; (2) QDs are grown directly onto TiO_2 electrodes in chemical bath deposition (CBD) processes. In the CBD approach, dissolved cationic and anionic precursors are reacted slowly in two separate beakers and the bare electrode dipped alternatively into each, to grow the target QDs. This process has been called "successive ionic layer adsorption and reaction" (SILAR). The SILAR process is regarded as the best way to allow deposition of well-defined QD layers onto mesoporous metal oxides in the solution process, and very precisely controlled multilayers can be deposited over QD cores by alternating injection of cationic and anionic precursors. However, the SILAR process has a disadvantage – it is not effective for preparation of metal selenides or tellurides because of the difficulties in preparing stable Se^{2-} and Te^{2-} precursors.

Lee *et al.*[28,47] reported the development of a new procedure for preparing selenide and telluride ions, which enables the use of the SILAR process for preparation of selenides and tellurides. In the process, the corresponding dioxide precursors are reduced in ethanol allowing SILAR growth of CdSe and CdTe QDs over mesoporous TiO_2 films. The innovative approach was generation of selenide ions *in situ* by reducing SeO_2 with $NaBH_4$ in ethanol. When cation- and anion-containing solutions were prepared, the SILAR process was performed following the usual procedure; the optimized TiO_2 film/FTO electrode was successively immersed into two different solutions for about 30 s each, one consisting of $Cd(NO_3)_2$ dissolved in ethanol and the other containing the *in situ* generated Se^{2-} in ethanol. This immersion cycle was repeated several times. The resulting homogeneous and well-defined QDs, when employed as sensitizers in photoelectrochemical cells, provided a promising overall efficiency of 4.2%.

In the vapor-phase deposition method of QD synthesis, one of the approaches for producing ordered QD systems is based on seeded self-ordering, in which physical and/or chemical templates provide nucleation sites where the QDs self-form during the epitaxial growth step. Nucleation control not only produces ordered arrays of QDs at desired locations, but also fixes the growth duration, thus resulting in more uniform size distributions. These self-assembled QDs have been used to fabricate lasers. In 2002, Toshiba demonstrated fabrication of InP QDs on a p-type (hole) semiconductor layer then applied an n-type (electron) layer on top.[29] The result was a device capable of generating single photons. A similar approach was applied to create a single-photon detector. QD fluorescence at very precise frequencies, can be stimulated either electrically or by a wide variety of wavelengths, so that they absorb light of one frequency but continue to emit at the specific frequency dictated by their size. The particles also reflect and absorb light in ways that can be affected by an applied voltage, offering potential in photochromic and electrochromic applications (ones that change color with the application of light or electricity, respectively), as well as in light-harvesting solar cells. Mohan *et al.*[30] developed a technique in which QD formation is based on the chemical and physical patterning of a (111) GaAs substrate for controlling the local density of source atoms, the nucleation site, and the dot shape and size. To fabricate pyramidal recesses onto the (111)B GaAs substrate, they first formed openings in a SiO_2 hard mask by using electron-beam lithography in a poly(methyl methacrylate) (PMMA, molecular weight 950k) resist layer. Subsequent wet etching in a 1% bromine/methanol solution led to the formation of tetrahedral pyramidal recesses exposing {111}A facets due to the anisotropic etching of the GaAs substrate. Low-pressure (20 mbar) organometallic chemical vapor deposition with trimethylgallium, trimethylindium, and arsine as precursors and nitrogen as ambient atmosphere, was employed for epitaxial growth on the patterned substrates. The grown structures consisted of a 1.3-nm thick GaAs buffer layer grown at 590 °C, followed by 5 nm GaAs (inferior cladding), 0.5 nm $In_{0.3}Ga_{0.7}As$ QD layer, and 20 nm GaAs layers (superior cladding), all grown at 590 °C.[30]

4.1.3 The Basic Working Principles of Nanostructured Solar Cells

As solution-processed material, QDs can be readily applied to large-scale, flexible roll-to-roll processing. In addition to their low-cost manufacturing promise, third-generation solar cells based on the QDs have been studied to improve the power-conversion efficiency. Efficiency is typically reported in terms of a standard AM1.5 solar illumination of 100 mW/cm^2 (one-sun) power intensity. This spectrum is analogous to light impingent on the earth after it has traveled through the earth's atmosphere a distance of 1.5 atmosphere thickness. Solar cells can also be described by several photovoltaic parameters (Figure 4.5), such as the open circuit voltage (V_{oc}), short circuit current (I_{sc}) and fill factor (FF).[31] While under illumination, the open-circuit voltage is the voltage across the cell when the cell current is zero and the-short circuit current is the current across the cell when the cell voltage is zero. The fill factor is defined as: Fill Factor $= FF = \frac{V_m I_m}{V_{oc} I_{sc}}$, where V_m is the maximum power point

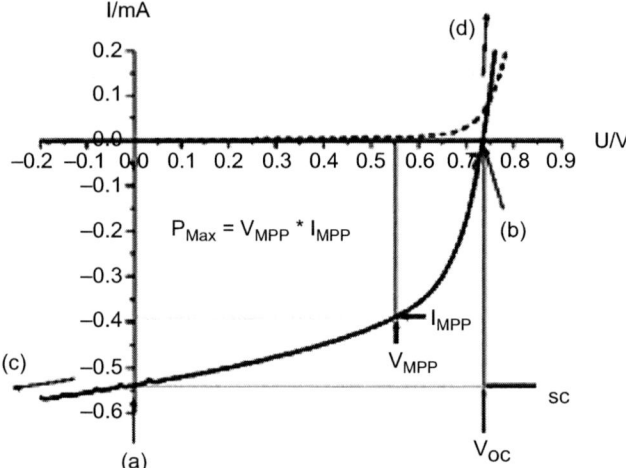

Figure 4.5   Schematic illustration of *J–V* characteristics measured in the dark (dashed line) and under illumination (solid line) for a typical solar cell. The characteristic intersections with the abscissa and the ordinate are the open-circuit voltage (V_{OC}) and the short-circuit current (I_{SC}), respectively. The largest power output (P_{Max}) is determined by the point where the product of voltage and current is maximized. Division of P_{Max} by the product of I_{SC} and V_{OC} yields the filling factor FF.
Figure credit: ref. 33.

voltage and I_m is the maximum power point current. The FF is a measure of the closeness to an ideal solar cell, which would have a rectangular shape in the fourth quadrant of an *I–V* output graph for a given light exposure. The power-conversion efficiency may then be determined as:

$$\eta_{PCE}\,(\%) = \frac{FF * V_{OC}I_{SC}}{P_{light\text{-}source}}$$

where $P_{light\text{-}source}$ is the incident light-source power. The external quantum efficiency (EQE) value is also a critical photovoltaic parameter, as it measures the number of carriers collected per number of photons directed to the solar cell at a given wavelength. This value is frequently referred to as the incident-photon-to-carrier-efficiency (IPCE), and may be found as:

$$IPCE = \eta_{QE} = \frac{hcI}{e\lambda P} = \frac{1240 * I_{SC\,(\lambda)}}{\lambda * P(\lambda)}$$

where $I_{SC}(\lambda)$ is the wavelength-dependent short-circuit photocurrent density in mA/cm^2, λ is the exposure light wavelength in nm and $P(\lambda)$ is the wavelength-dependent light intensity in mW/cm^2.

Early applications of QDs into solar cells were unsuccessful due to their high level of carrier confinement. For example, CdSe QD film can be deposited on a transparent, indium tin oxide (ITO) electrode and integrated into a simple liquid-junction solar cell. Under light exposure, an electron is promoted to a higher state within the QD; however, since the exciton Bohr radius is larger than the radius of the particle, a strong attractive Coulombic force prevent the exciton from dissociating into an electron and a hole. There is a very minimal driving force to separate the exciton and therefore no photocurrent is generated. Indeed, it is far more probable for the exciton to undergo direct recombination and emit a photon through photoluminescence. One of the major breakthroughs for the third-generation photovoltaic was the coupling of QDs with both organic and large-bandgap inorganic semiconductor materials. In this manner, the excited state of the QD is very quickly deactivated by a fast electron or hole transfer to the adjacent material, as shown in Figure 4.6. Gerischer and Liibke[32] were the first to demonstrate TiO_2 sensitized by colloidal CdS QDs; however, Grätzel and coworkers[33] showed that dye-sensitized nanostructured TiO_2 films can be further facilitated by the application of QDs as the dye.

Efforts are now underway to further improve the efficiency of the QD–semiconductor photoactive composites. Early work in this area, even prior to the Grätzel cell, focused on understanding the fast electron transfer from the sensitizer QD to the wide-bandgap colloidal semiconductor. While the individual components of this system are very promising, the earliest power-conversion efficiencies for liquid-junction solar cells based on these materials were relatively low. The low efficiencies have been attributed to both band mismatch and, more importantly, high levels of carrier recombination due to the large number of interfaces. The interfaces generally have a higher number of deep trap states and therefore act as efficient recombination centers. As the consequence, QD solar cells pertain to the need for novel device architectures that can reduce interfacial recombination and drive the solar-cell efficiencies in comparison with the current technology. In order to surpass first-generation silicon technology, such architectures must also allow for either hot-carrier extraction or multiple exciton generation, as will be discussed throughout this review.

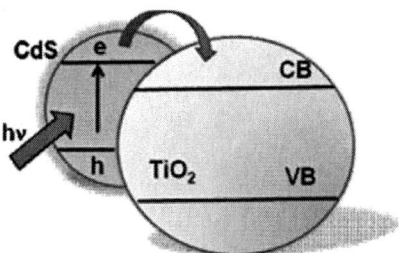

Figure 4.6 Energy-band diagram of CdS/TiO_2 coupled semiconductor system to facilitate the charge separation in the type-II heterojunction solar cells. Figure credit: ref. 23.

4.2   Quantum Dot Solar Cells

Early work in QD solar cells can be separated into three general structural configurations, as detailed by Nozik.[34,35] In the past three years, an additional fourth configuration has also been identified, which shows higher power-conversion efficiencies and the light absorption in the infrared (IR) region. Four different solar-cell configurations will be examined below that show promise in competing with current solar-cell technologies. The extended hot-carrier lifetime and multiple exciton generation effects will also be briefly discussed in terms of the experimental results that have been achieved to date.

(1)   First Structure Configuration: Vacuum-Deposited Quantum Dot Solar Cells

The first configuration utilized a vacuum-deposition technique to form QDs on the solid crystal film, primarily act as interband dopants (Figure 4.7).[36] It has been suggested that the QD array placed in the intrinsic region of a p-i-n structure potentially make use of hot-carrier and multiple exciton generation effects. These solar cells are typically based on expensive, epitaxially grown III-V semiconductors and are studied for high efficiency and outer space applications that do not pay much attention to the cost of photovoltaics.[37] Although the introduction of solar concentrators to decrease costs further complicates their use, the optics used in concentrators will be costly. The economics and vacuum deposition of such systems are beyond the scope of this review. Therefore, this category of QD solar cells will not be discussed in detail throughout this chapter – the reader is encouraged to look at review articles for further information on these topics.[38]

(2)   Second Structure Configuration: Liquid-Junction Quantum Dot Solar Cells

The second configuration is QDs coupled to wide-bandgap inorganic oxide semiconductors within a liquid junction cell. In this configuration, the

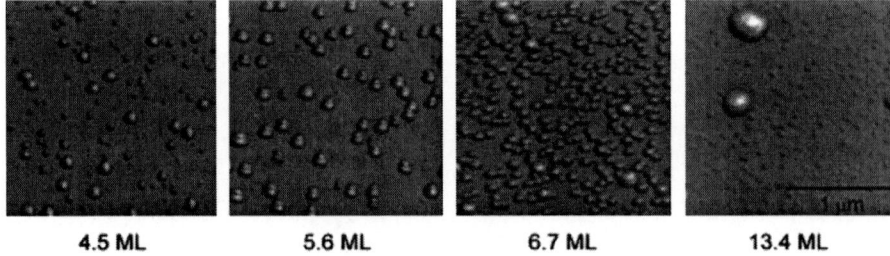

| 4.5 ML | 5.6 ML | 6.7 ML | 13.4 ML |

Figure 4.7   Evolution of Stranski – Krastanow InP islands grown on (100) AlGaAs at 620 °C by MOCVD for increasing amounts of deposited InP [expressed as monolayers (ML)]. The scale of each scan is 2 μm × 2 μm.
Figure credit: ref. 39.

QD-sensitized film is deposited from solution onto a transparent electrode defined as the working electrode. Opposite to the working electrode is a high surface area counterelectrode. Excitons are generated within the QD and then separated into holes and electrons at the QD/oxide interface. Electrons are then collected on the working electrode while holes are scavenged by the electrolyte solution and transported to the counterelectrode to complete the photovoltaic circuit. The experimental schematic image is shown in Figure 4.8.

A significant amount of QD-sensitized solar-cell research is based on the novel architectures in order to improve the power-conversion efficiency of liquid junction cells, such as Kamat[39] Carbon-based 1D (carbon nanotube) and 2D (graphene) nanostructures have been examined to facilitate charge transport.[40] However, power-conversion efficiency has remained relatively low due to the high level of interfacial carrier backtransfer and charge recombination. The presence of a liquid species in the solar cell greatly complicates manufacturing and later packaging efforts. Leakage-free encapsulation and increased toxicity due to the use of a liquid electrolyte will become a serious concern that will drive up the cost of solar cells. In addition, the stability and corrosion-resistant properties of electrode surfaces is of great concern for solar-cell lifetime. The use of chemically inert counter electrodes, such as Pt electrodes, partially addresses this concern, but would greatly increase the cost of solar cells. The QD-coated working electrode would inevitably suffer from degradation as the electrolyte dissolves the photoactive layer over time. Solar heating of the electrolyte during operation will also cause variation of solar performance, the effects of which have not yet been fully studied. This type of configuration of solar cells will not become competitive with current technologies in the near future. We will not discuss this in detail.

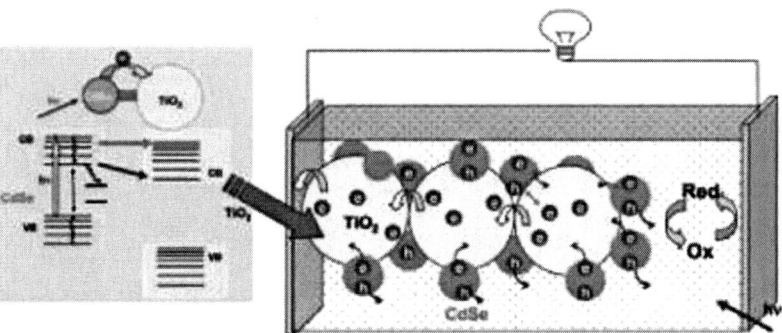

Figure 4.8 Principle operation of QD sensitized solar cells (QDSSC). Charge injection from excited CdSe QDs into TiO_2 nanoparticles is followed by collection of charges at the electrode interface. The redox electrolyte scavenges the holes and thus regenerates the CdSe.
Figure credit: ref. 42.

(3) Third Structure Configuration: Quantum Dot Hybrid Solar Cells

The third structure configuration of QD solar cells, called as the organic–inorganic hybrid solar cells, is by coupling QDs with organic semiconducting polymers. By offsetting the energy levels of the QD and the organic semiconductors, it is feasible to break apart photogenerated excitons in the QD through the fast hole transfer from the QD to the neighboring conjugated polymer matrix, where the hole will be diffused and collected at the anode electrode. The remaining electron diffuses along the QD phase to the opposite cathode electrode, as dictated by the hopping mechanism or the multiple trapping model typically used in the QD films. Alternatively, a photogenerated exciton in the organic phase may undergo fast electron transfer to the QD with the same effect. The conduction of holes in the organic phase and electrons in the QD phase is preferable, as most polymeric semiconducting materials have higher hole mobility than the electron mobility.

In 1996, Alivisatos and coworkers[41] developed the proof-of-concept hybrid solar cells based on CdS and CdSe QDs coupled to MEH-PPV conjugated polymer, following the results that showed efficient electron transfer from CdS QDs to a polyvinylcarbazole polymer. The CdS QDs were used to sensitize the film and it was observed that only hole transport occurred throughout the composite film. Alivisatos and coworkers[42] increased the relative amount of CdS/CdSe QDs in the nanocomposite above the percolation threshold in an attempt to form a percolated QD phase within the conjugated polymer matrix. In this manner, it was shown that efficient bipolar carrier collection could occur, with electron transport occurring in the QD phase and hole transport occurring in the polymer phase. This concept is analogous to the organic bulk heterojunction solar cell, instead of the high electron affinity buckminsterfullerene that is replaced by the QD phase. In this study, both trioctylphosphineoxide (TOPO)-coated CdSe QDs and "naked" QDs were mixed with the polymer material prior to spun-cast film. Both systems exhibited relatively poor photovoltaic performance, with peak quantum efficiencies of 12% and 0.005% for the naked and TOPO-coated systems, respectively. A power-conversion efficiency of 0.1% was achieved for the naked CdSe solar-cell device under an AM1.5, one-sun illumination. The efficiency difference of naked and TOPO-coated QDs highlights one of the most critical findings in this report: charge transfer may only occur at the QD/polymer interface. It is therefore necessary to either exchange or remove the QD ligand (which is generally present as a consequence of the synthesis method detailed by Bawendi and coworkers[43]) prior to integration in any photovoltaic system. The authors attribute the overall poor solar device behavior to electrons becoming trapped at "dead ends" within the QD phase. A possible explanation is that the majority of photogenerated carriers recombined at the deep trap states present at the numerous QD/polymer interfaces.

Later, Konstantatos and Sargent[44] continued this work with relative success using PbS QDs coupled with MEH-PPV polymer in a bulk heterojunction

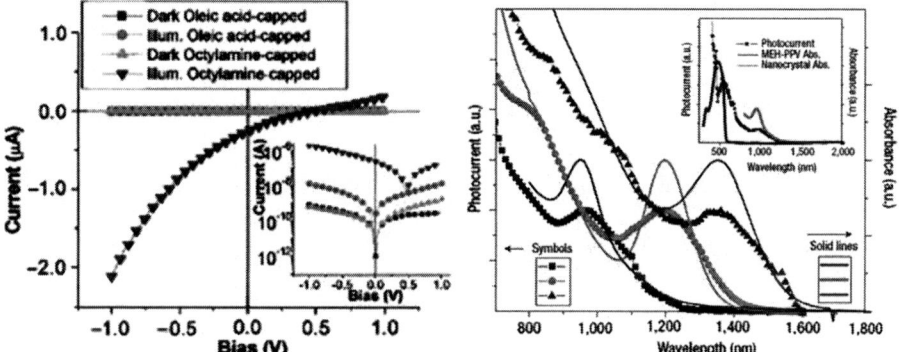

Figure 4.9 Dark and illuminated *J–V* curves of devices made with oleic acid-capped (squares and circles) and octylamine capped (triangles) PbS nanocrystals. Inset shows the same data in a semilog plot (left), and photocurrent spectral responses and absorption spectra (right).
Figure credit: ref. 47.

(Figure 4.9). Qi *et al.*[45] simultaneously developed PbSe QDs coupled with MEH-PPV for the photodetector applications. MEH-PPV has a relatively low ionization potential (~ 4.9 eV to 5.1 eV), which matches well with the ionization potential of PbS QDs (~ 4.95 eV) allowing for the favorable hole transport to the organic material. In this area, Sargent and coworkers[46] used PbS QDs capped with octylamine ligands, which are shown to have superior performance over the as-produced PbS QDs capped with oleic acid ligands. This variation is shown in Figure 4.9, and follows as a consequence of the relative carbon chain sizes of the ligands (8 carbon atoms for octylamine and 18 carbon atoms for oleic acid). As with the naked *vs.* capped CdSe data above, the longer ligand separates the QD from the polymer and therefore hinders exciton dissociation. Naked PbS QDs are not feasible in this case as a consequence of their poor uncapped stability.

The performance of organic-PbS/PbSe hybrid solar cells was so poor that they only showed peak external quantum efficiencies of 0.0064%, on the same level as the capped CdSe QD samples. Annealing improved the external quantum efficiency to 0.15% and allowed for a measurable power-conversion efficiency of 0.001% at relatively low power intensity. Within all samples, increasing the power density of the light source decreased the quantum efficiency, which is a clear sign of bimolecular recombination. In general, bimolecular recombination becomes more prevalent at high light intensities when there are a larger number of photogenerated free charge carriers. Sargent and coworkers[47] were, however, successful in sensitizing the MEH-PPV polymer to absorb in the IR region, as shown in Figure 4.9, where the spectral response matches well with the absorbance of the PbS QDs.

Günes *et al.*[48] had slightly more success by coupling PbS QDs with P3HT in a bilayer heterojunction structure. They were able to achieve I_{sc}, V_{oc}, fill factor, power-conversion efficiency values under one-sun, AM1.5 light illumination of

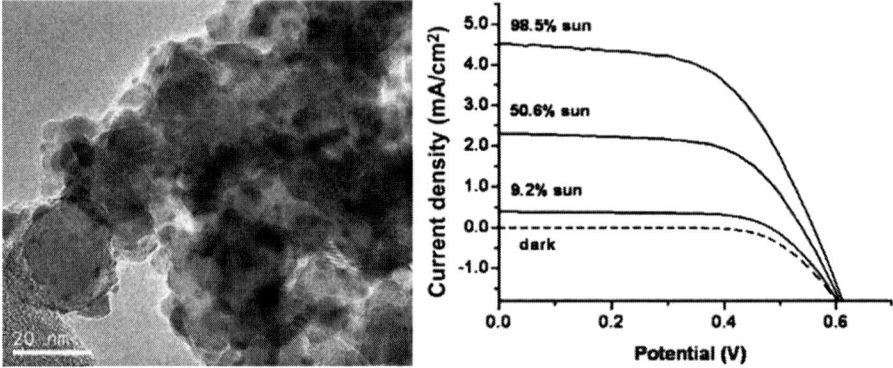

Figure 4.10   HRTEM image of PbS/TiO$_2$ and J–V curves obtained from PbS QDs sensitized solid-state cell using spiro-OMeTAD as a hole conductor. Figure credit: ref. 51.

0.3 mA/cm^2, 0.35 V, 0.35 and 0.04%, respectively. In more recent work, Grätzel and coworkers[49] have achieved a much higher power-conversion efficiency of 1.46% for a significantly modified version of the PbS QD hybrid cell (Figure 4.10). In this work, the PbS QDs are formed directly on a TiO$_2$ NP layer through a process known as successive ionic adsorption and reaction, where a substrate is successively and repeatedly dipped into Pb^{2+} and S^{2-} solutions. The advantage of this approach is that PbS QDs can be stable without the ligand capping layer. They completed the cell by forming a junction with the organic hole-transport material spiro-OMeTAD, to present the final device structure of: ITO/TiO$_2$ NP film/PbS QD film/spiro-OMeTAD/gold electrode. The I–V photoresponse and external quantum efficiency characteristics for this structure are shown in Figure 4.10. Both Larramona and coworkers and Grätzel and coworkers have successfully applied this structure to CdS-sensitized hybrid solar cells.[50]

The shape anisotropy of QDs can also vary the photovoltaic performance in the hybrid solar cells. In 2002, Alivisatos and coworkers[51] published results showing that CdSe nanorods instead of CdS QDs significantly enhance the photovoltaic properties of hybrid solar cells (Figure 4.11). The improvement was attributed to three major effects: directed charge-carrier motion through the 1D structure, enhanced percolation network of the nanorods by minimizing "dead ends" (shown in Figure 4.11) and electron hopping, which can now transport continuously along the long axis of the nanorods. It used a 90 wt% of 7 nm diameter by 60 nm length CdSe nanorods in P3HT as the active layer in the bulk heterojunction solar cell. Their final solar-cell structure of ITO/PEDOT:PSS/ P3HT:CdSe nanorods/Al achieved V_{oc}, FF, power conversion values under one-sun, AM1.5 solar illumination of 0.7 V, 0.4 and 1.7%, respectively. While this solar cell is a significant improvement over the 0.1% power-conversion efficiency in Alivisatos' previous work in 1996, these values are still far below the levels required to be competitive with alternative technologies.[52]

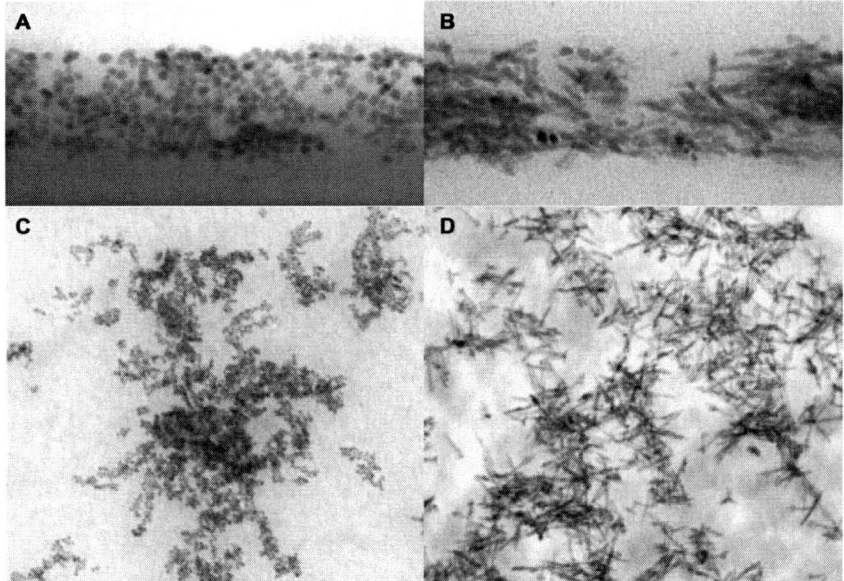

Figure 4.11 In (A), TEM cross-sectional image of a film with 60% by weight 10 × 10 nm CdSe QDs in P3HT reveals the distribution and organization of QDs across the 110 nm thick film. In (B), TEM cross-sectional image of a 100-nm thick film consisting of 40 % by weight 7 × 60 nm CdSe nanorods in P3HT reveals that most nanorods are partially aligned perpendicular to the substrate plane. Thin films of 20% by weight (A) 7 nm by 7 nm (B) and 7 nm by 60 nm CdSe nanocrystals in P3HT were studied *via* transmission electron microscopy.
Figure credit: ref. 53.

Since this initial work on CdSe nanorods, research on QD hybrid solar cells has been rigorously shifted toward the development of new geometrical shapes to enhance percolation and thereby improve electron transport. The following work on CdTe nanorod, branched CdSe and hyperbranched nanocrystal hybrid solar cells has made a further step to push forward the power-conversion efficiency. From the initial discussion on the characteristics of QDs and their relevance in photovoltaics, the desire to alter the geometric shape of the nanocrystalline materials seems counterintuitive. The exciton Bohr radii of CdSe and CdTe are 5.7 nm and 7.3 nm, respectively. In the nanorods and branched structures, at least one dimension is no longer smaller than the exciton Bohr radius and, as a consequence, carriers are no longer confined in three dimensions. It is therefore unlikely that such structures will exhibit the extended hot-carrier lifetimes and multiple exciton-generation capabilities that are unique to third-generation photovoltaics. These unique properties were desired at the beginning of this research effort to allow QD solar cells to achieve higher efficiencies than single-crystal silicon solar cells.

It should be emphasized that the study of novel self-assembled organic/inorganic QD hybrid nanostructured systems is now playing an important role in determining the power-conversion efficiency. It provides fully functional solar cells with high levels of efficiency, and also suggests a number of important data that will be critical in future QD hybrid solar-cell research efforts. For example, Ren *et al.*[53] recently demonstrated that the self-assembled CdS QDs can be grated onto conjugated polymer P3HT nanowires to reach a power-conversion efficiency of 4.06% under AM1.5 solar illumination (Figure 4.12). The self-assembled coaxial nanowire structures address the inorganic QD vertical phase-separation issue due to the polarity difference between inorganic and organic phases. The self-assembled CdS QDs formed a percolation network to enhance the charge dissociation and collection efficiency. Given that the baseline levels of efficiency have been established, it is now appropriate to begin examining the more complex nanostructured elements to enhance the properties of the simplistic systems.

It is worthwhile to consider the role of QDs in new-generation photovoltaics, as well as competing technologies. As a third-generation technology, QD solar cells appeal to both increased efficiency and lowered manufacturing costs. When used solely as sensitizers or simple absorbing species, QD solar cells may instead be classified as second-generation solar cells

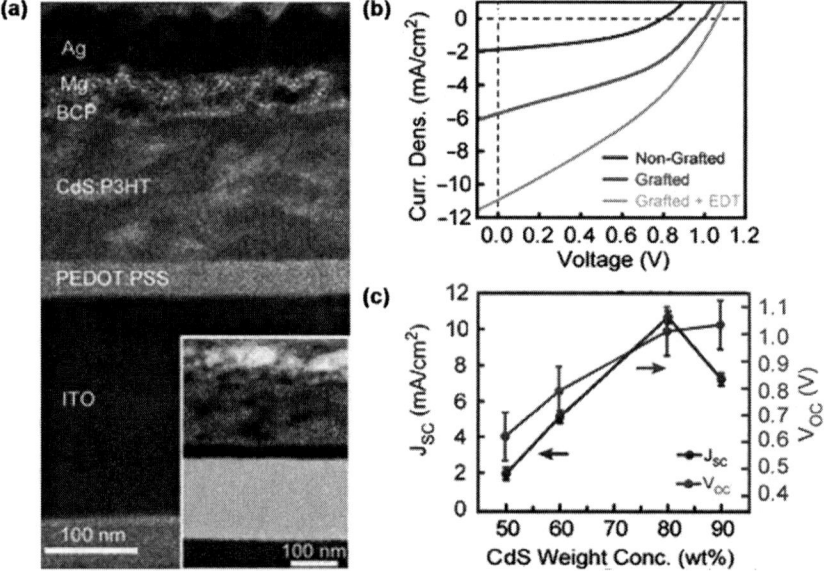

Figure 4.12   (a) Cross-sectional TEM image of P3HT/CdS hybrid solar cells. Inset image shows the EDS elemental mapping. (b) Current–voltage characteristics of P3HT/CdS hybrid solar cells from nongrafting, grafting and the subsequent ligand exchange. (c) The photovoltaic performance summary (J_{SC} and V_{OC}) of P3HT/CdS after the chemical grafting and ligand exchange, as related to the CdS weight concentration. Figure credit: ref. 55.

and are therefore competing with all thin-film photovoltaics. For simplicity, let us consider one such competitor, organic solar cells, which are at a similar level of research and development. While QD hybrid solar cells have reached just under 6% power-conversion efficiency, purely organic solar cells have, in a similar period of time, managed to reach 9.2% power-conversion efficiency with the likely capacity to reach 10% power-conversion efficiency by the end of 2012. Furthermore, organic solar cells have maintained their relative ease of fabrication and do not require processing of toxic cadmium and lead, as is the case with QD hybrid solar cells. In order to make a real impact in the realm of hybrid solar cells, research efforts would be better spent addressing the initial goals of third-generation photovoltaics: developing materials and structures to achieve hot-carrier extraction or efficient multiple exciton generation in an economically viable manner.

(4) Fourth Structure Configuration: Schottky and Depleted Quantum Dot Solar Cells

In both the second and third structure configurations of QD solar cells, a substantial amount of research has been dedicated to the development of more complex nanostructures and systems. These systems generally have more interfacial regions in which carrier trapping and recombination can occur. Furthermore, by increasing the complexity of device fabrication, it is difficult to foresee the development of feasible large-scale manufacturing efforts. In contrast, the fourth configuration greatly simplifies the solar-cell architecture and yet still manages to achieve power-conversion efficiencies far superior to the QD solar cells examined previously. This configuration was first developed by Sargent and coworkers[54] in 2007 and initially relied on a simple ITO/PbS QDs/Al structure. The QDs form a Schottky contact with the Al electrode to create a depletion layer to aid exciton dissociation.

It has been previously established that nanocrystals generally do not exhibit significant band bending. In liquid-junction cells, this is further emphasized by the fact that the electrolyte scavenges excess charges to hinder space-charge-layer formation (Figure 4.13).[55] This scavenging effect may be expected to occur to some degree with the organic species in organic–QD hybrid solar cells. In these studies, a single QD is treated as individual system with negligible electronic wave function overlap with neighboring QDs. However, for very densely packed QD thin film, such as a spun-cast film of short-ligand PbS QDs, the film can be assumed as a homogeneous medium to show macroscopic electrical properties. Therefore, PbS QD film may be regarded instead as a rather poor quality semiconducting film. In contrast to the studies for isolated QDs, a densely packed PbS QD layer coupled to an appropriate metal layer should exhibit band bending, form an associated space-charge layer and display simple Schottky contact properties. This stipulation holds true for ITO/PbS QD film/Al solar-cell devices, with initial studies showing device rectification, very good fit to the simple diode model and a measurable depletion width of

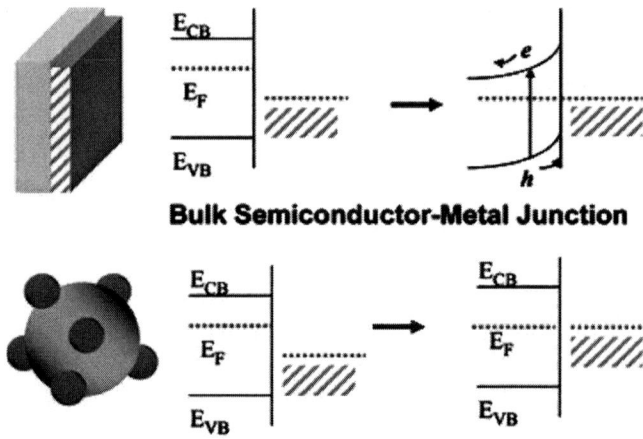

Bulk Semiconductor-Metal Junction

Semiconductor-Metal Nanocomposite

Figure 4.13   Fermi-level equilibration at bulk metal–nanocrystal (top) and metal nanoparticle–semiconductor nanoparticle junction (bottom). Figure credit: ref. 54.

90–150 nm. As a point of note, similar films made from the related PbSe QDs have been shown to have mobilities on the order of 0.1–1 cm^2/(V s) after chemical treatment, which supports the possibility of fast and efficient carrier extraction. Surprisingly, it has also been shown that the QDs in these Schottky-barrier structures retain their individual quantum-confinement effects, despite strong interparticle carrier transport. This property is important for the future possibility of hot-carrier extraction and multiple exciton generation.

Initial work on the planar QD Schottky solar cells showed power-conversion efficiency of 1.8% under a one-sun, AM1.5 solar illumination, with an IR power-conversion efficiency of 4.2%, which is a 3-times improvement in IR efficiency over previous hybrid devices. The final structure for the optimized device was: ITO/PbS QD film/LiF/Al/Ag. LiF was used to improve the metal junction – a common procedure in top cathodes used in organic solar cells. Densely packed films were feasible because the ∼2.5 nm oleate ligands on the QDs were exchanged with ∼0.6 nm 4-carbon n-butylamine ligands. The band structure and output characteristics for this system are shown in Figure 4.14.

From the band diagram in Figure 4.14, it is clear that the diffusion transport of carriers across the quasineutral region is a limiting factor in carrier extraction.[56] In PbS QD thin films, the diffusion lengths have been estimated to be 0.4 μm and 0.1 μm for holes and electrons, respectively. Therefore, for active layer films on the order of 200–300 nm, carrier recombination due to slow diffusion should be reasonably small. Unfortunately, the use of thin films of QDs hinders device performance due to the greater number of photons transmitted through the film. Thicker active layers allow for the absorption of a larger number of impingent photons. Nozik and coworkers[35] observed this

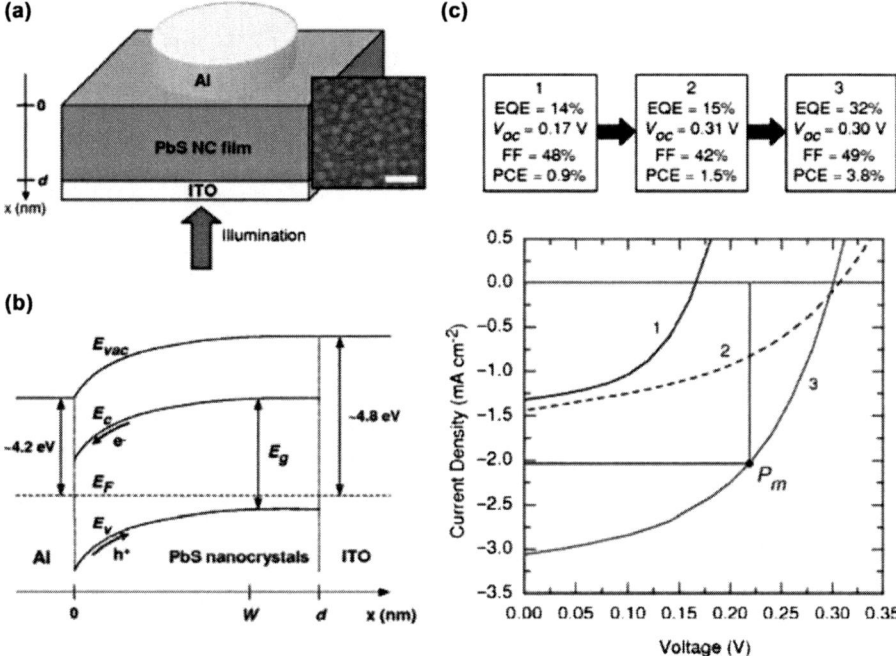

Figure 4.14 (a) Depiction of the architecture of a typical CQD-Schottky device. The inset shows an SEM image of the nanocrystal film (scale bar is 20 nm). (b) The energy band model. (c) The J–V curve and photovoltaic performance (under 975 nm, $12 \, \text{mW cm}^{-2}$ illumination) for the baseline device (1), device processed using the fast cooling PbS CQDs (2) and device further employed butylamine ligand exchange (3). Figure credit: ref. 58.

problem and found that 250-nm to 400-nm PbSe QD films exhibited lower quantum efficiencies for high-energy photons. This reduction in efficiency is because high-energy photons are absorbed near the surface of the cell and in the quasineutral region. These absorbed photons do not receive any benefit for either exciton dissociation or subsequent carrier transport from the internal electric field provided by the Schottky contact. In addition to decreasing the quantum efficiency, thicker films have also been shown to lower the V_{oc} and increase the series resistance of the solar cell.

Sargent and coworkers[57] examined the possibility of using a rough, microporous ITO electrode as a matrix for the PbS QDs in order to decrease the total length of the quasineutral region and allow for a thicker QD film. Through careful annealing, the microporous ITO/PbS/Mg cells were able to attain an IR power-conversion efficiency of 2%, a slight improvement over the work described above. They did not report a standard one-sun, AM1.5 power-conversion efficiency. While such a device structure may offer a shorter path for free carriers, it should be noted that a rough, microporous ITO electrode also introduces a number of potentially deep-trap interfaces. The added presence of these deep trap states may serve to hinder instead of help device performance by further increasing carrier recombination. In addition, the conductivity across the porous ITO is likely to be worse than the bulk film and may contribute to significant increases in the solar cell series resistance.

Following the success of PbS-based QD Schottky solar cells, Nozik and coworkers as well as Sargent and coworkers investigated the use PbSe QDs in the Schottky solar cell architecture. Sargent *et al.* used multiple spin-coating approach combined with a benzenedithiol crosslinking ligand that allowed for the formation of insoluble thin films of tightly packed PbSe QDs. In this work, they used an ITO/PbSe QD film/Mg/Ag structure to achieve the power-conversion efficiency of 1.1% under AM 1.5 solar illumination. PbSe-based Schottky-QD solar cells appeared to rely strongly on the diffusion transport of carriers when compared to PbS-based cells due to their smaller depletion layer width and larger quasineutral region. Hanna and Nozik [58] avoided the complicated ligand exchange process by using a simpler and more effective layer-by-layer deposition method (Figure 4.15). In this manner, it formed a thin layer of QDs on a substrate through dip coating and subsequently removed the oleate ligands by dipping the substrates with 1,2-ethanedithiol (EDT). This process was continuously repeated until a suitably thick film was formed. In their ITO/PbSe QD film/Ca/Al device, they were able to achieve the power-conversion efficiency of 2.1% under AM 1.5 solar illumination. Alivisatos and coworkers [59] later adopted Nozik's approach for solar-cell fabrication with ternary QDs composed of $PbS_{0.7}Se_{0.3}$, allowing for a much higher power-conversion

Treatment	d(nm)	ε_s	n_{ave}	μ (cm^{2}V^{-1}s^{-1})
Oleic Acid	1.8	2	1.57	–
Aniline	0.8	2	2.2	–
Butlyamine	0.4	5.4	2.46	7.4
ethylenediamine	0.4	16	2.62	47.0
Hydrazine	0.25	52	2.69	29.4
NaOH	0.1	1	2.4	35.0

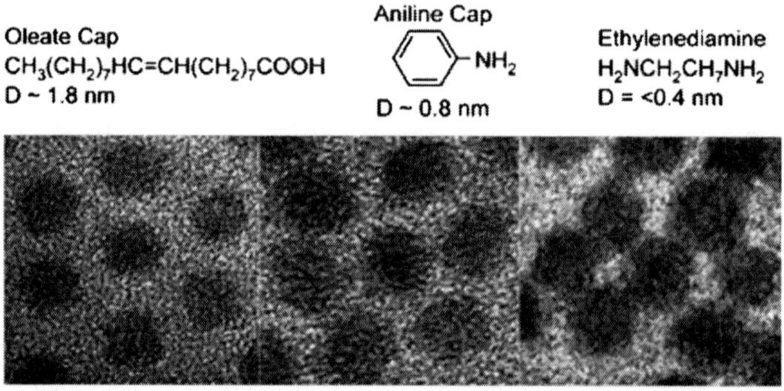

Oleate Cap
$CH_3(CH_2)_7HC=CH(CH_2)_7COOH$
D ~ 1.8 nm

Aniline Cap
$-NH_2$
D ~ 0.8 nm

Ethylenediamine
$H_2NCH_2CH_7NH_2$
D = <0.4 nm

Figure 4.15   Effect of different chemical treatments of PbSe QD films on interdot distance and carrier mobility as measured by THz spectroscopy. Figure credit: ref. 60.

efficiency of 3.3% under AM 1.5 solar illumination. They attribute this improvement to two major factors: (1) Increased J_{sc} due to a larger exciton Bohr radius (as a consequence of the PbSe contribution to the ternary structure), which provides better electronic coupling among neighboring QDs (2) Increased V_{oc} due to a redistribution of trap states, following from the different surface energies of a ternary compared to a binary QD.

Due to Fermi-level pinning, QD Schottky solar cells are inherently limited to a maximum V_{oc} of half of the bandgap of the QD. Bandgap widening due to quantization effects allows for reasonable open circuit voltages for PbS- and PbSe-based systems. However, in order to further improve device efficiency, it is desired to achieve higher open-circuit voltages. For this reason, research has recently shifted to the formation of planar heterojunctions with wide-bandgap films adjacent to QD films. Leschkies *et al.*[60] recently examined ZnO-PbSe QD planar heterojunctions for this reason and successfully achieved open-circuit voltages on the order of 0.45 V. The full solar-cell structure along with a cross-sectional scanning electron microscope (SEM) image and the associated energy-band diagram are shown in Figure 4.16. By separating excitons into free carriers at the PbSe/ZnO interface, this structure also reduces both geminate and bimolecular recombination. This is due to the fact that electrons are isolated to the ZnO layer, whereas the holes are isolated to the PbSe layer. The ZnO layer was formed by sputtering and the PbSe QD layer was formed through the LBL deposition method.

The solar cell shown in Figure 4.16 makes use of a 15-nm small-molecule hole-transport layer that was thermally evaporated prior to the deposition of the gold electrode. This layer is used as an electron-blocking layer to prevent the transfer of electrons to gold that would oppose the short-circuit current.[62] In the best devices, they were able to achieve J_{sc}, V_{oc} and power-conversion efficiency values under AM1.5 solar illumination of 15.7 mA/cm^2, 0.39 V and 1.6%, respectively. Interestingly, the authors also observed a linear increase in the open circuit voltage of the solar cell with increasing PbSe bandgap (by decreasing the size of the PbSe QDs). Such behavior has been witnessed for organic solar cells and supports the donor–acceptor, also known as the "excitonic," solar-cell model, where excitons must diffuse to the heterojunction to undergo dissociation. Sargent and coworkers also examined a QD planar heterojunction structure based on a TiO$_2$ film coupled with a PbS QD film formed by a layer-by-layer method (Figure 4.17).[61] In contrast to the above study, a gold electrode was used directly without an electron blocking layer. The optimized device for this structure was able to achieve the most impressive data for QD solar cells to date, with J_{sc}, V_{oc}, FF and power-conversion efficiency values under AM1.5 solar illumination of 16.2 mA/cm^2, 0.51 V, 0.58 and 5.1%, respectively. Toward the goal of low-cost photovoltaics, Sargent also showed that this system could be made by using the LiF/Ni electrode instead of a gold electrode, while still achieving a power-conversion efficiency value of 3.5% under AM1.5 solar illumination.

In contrast to the simple donor–acceptor model offered by Leschkies *et al.*,[62] Sargent and coworkers provided convincing evidence that their TiO$_2$-PbS

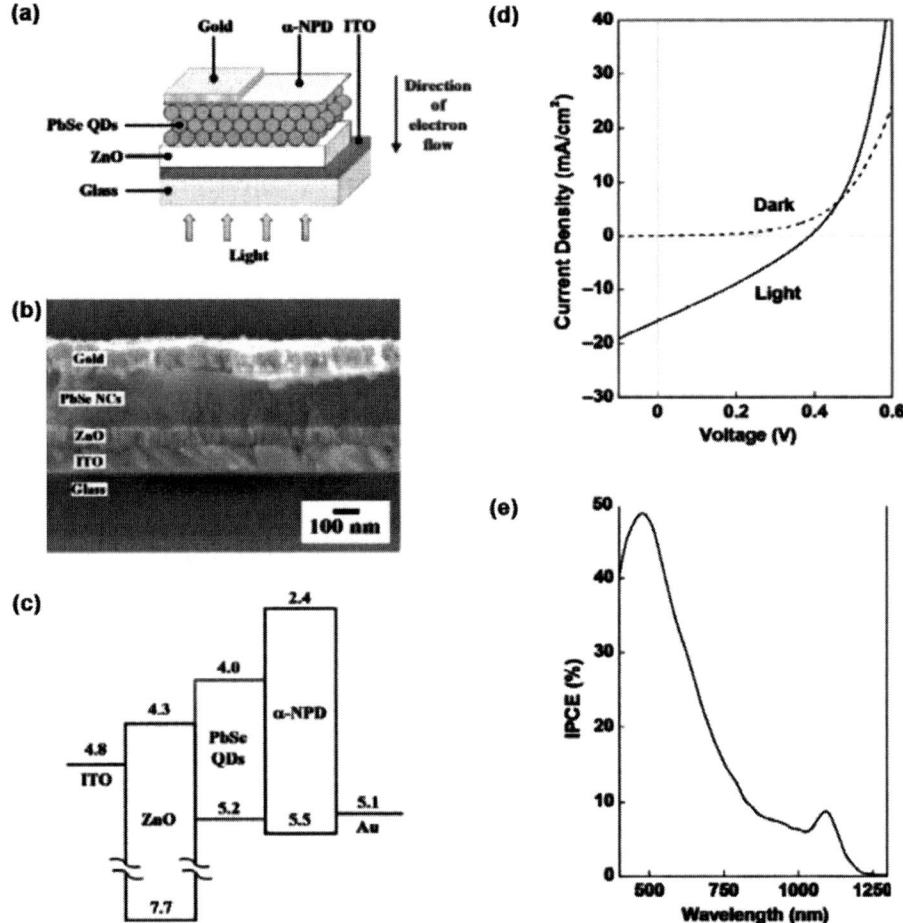

Figure 4.16 (a) Schematic and (b) cross-sectional SEM image and (c) energy band diagram of the ITO/ZnO/PbSe CQDs/ α -NPD/gold heterojunction solar cell. (d) *I–V* characteristics of the cell recorded in the dark and under simulated AM1.5G illumination and (e) its corresponding EQE spectrum. Figure credit: ref. 62.

system follows a depleted heterojunction model that exhibits an internal electric field.[63] They found that larger PbS QDs with no conduction-band offset to the TiO_2 layer provided a solar cell that exhibited nearly the same photocurrent as smaller PbS QDs with significant offset. Therefore, an additional driving force for exciton separation beyond a simple heterojunction must be present, such as an internal electric field. Furthermore, with an excitonic solar-cell model and primarily Förster energy transfer (a valid assumption given that Dexter electron transfer only occurs over very short ranges), excitons would have to "hop" an unreasonably large number of times to reach the heterojunction. Such diffusion time scales are much greater than the exciton lifetime. As a point

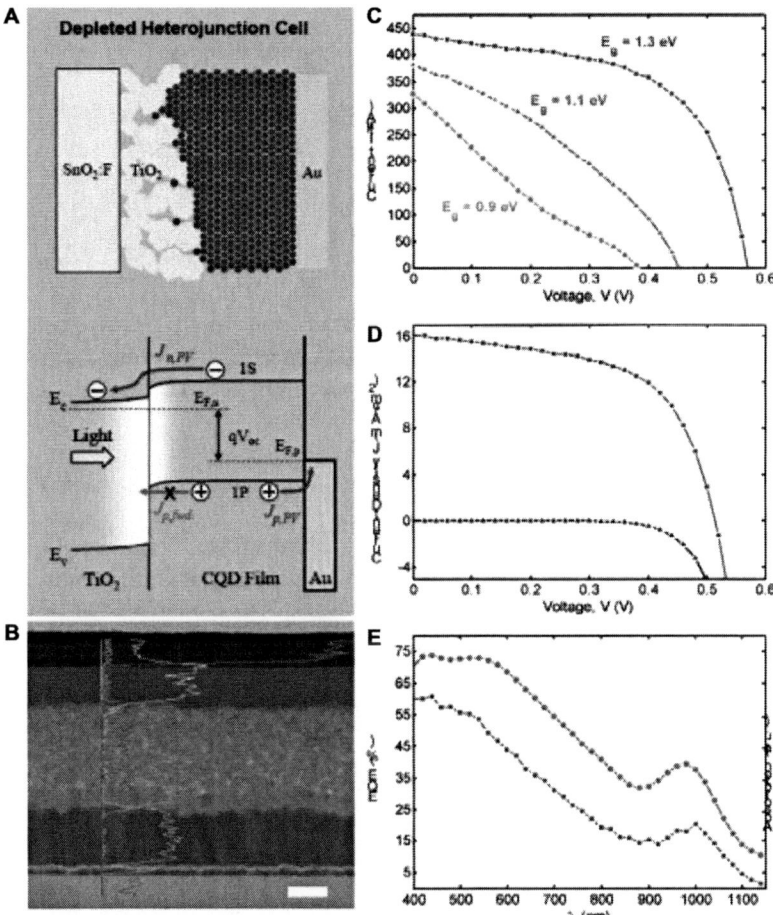

Figure 4.17 Device architecture and performance of TiO$_2$/PbS heterojunction solar cells. (A) Schematic demonstration of FTO/TiO$_2$ /PbS/Au and its band diagram; (B) Cross-sectional TEM as well as elemental distribution plot of a photovoltaic device. The scale bar is 200 nm; (C) I–V response of devices using three different size colloidal QDs; (D) J–V curves for the champion device recorded with and without one sun illumination; (E) EQE and absorption spectra of the champion device.
Figure credit: ref. 65.

of note, depletion widths for both p-n junctions and Schottky contacts are strongly dependent on built-in voltage and free-carrier density. Extending this knowledge to the current system, the variation in solar-device behavior between the ZnO-PbSe heterojunction and the TiO$_2$-PbS heterojunction could be due to material quality impacting free-carrier density or due to the relative band offsets of the constituent materials. It is likely that the ZnO-PbSe heterojunction exhibits a thin interfacial depletion layer that does not fully deplete the device active layer.[64] In order to better understand the discrepancy observed in these results, additional data regarding the electron affinities of the different layer materials in addition to depletion layer widths and positions in the active layers would be required. The former may be acquired by photoelectron spectroscopy, whereas the latter can be obtained through simple capacitance–voltage measurements and calculations.

While this fourth structure configuration of QD solar cells allows for entirely inorganic structures, all studies have shown that the solar cells exhibit fast device degradation in oxygen and ambient environments. This instability has been attributed to both the surface oxidation of the QDs, resulting in the formation of deep trap states, and to the oxidation of the low work function top electrode. Sargent and coworkers[65] examined a ligand-exchange encapsulation method to address the prior issue. This work allowed for stable Schottky-QD solar cells with a 3.6% power-conversion efficiency for 0.5 h in air under AM 1.5 solar illumination. In order to address the latter issue, the authors made use of a LiF passivation layer prior to the deposition of the Al top electrode, granting stability comparable with organic solar cells. X-ray photoelectron spectroscopy (XPS) studies indicated that the LiF layer slowed the oxidation of the Al electrode and reduced diffusion of oxygen into the QD film. However, as is the case with organic solar cells, it is necessary to further improve the solar-cell stability in order to become competitive with current technologies.

4.3   Hot Carriers and Multiple Exciton Generation Effects

When electrons are excited by photons with energy E_{hv} in excess of a semiconductor bandgap, they tend to rapidly lose their excess energy in the form of heat; in this context, the maximum solar to electrical energy conversion efficiency for an optimal single bandgap (E_g) semiconductor absorber is limited to about 31%.[66]

To exceed this efficiency value, one possible mechanism currently under active investigation is to convert the excess energy of incident photons with $E_{hv} \geq 2E_g$ (E_g – energy gap) into additional free carriers in the material. In the bulk material this process is called impact ionization and can be described as reverse Auger recombination. Although the impact ionization has been already investigated for nearly a half of a century, it was not employed in conventional solar cells due to its low efficiency. In the bulk material, formation of the secondary excitons becomes apparent only if the energy of incident photos exceeds the bandgap significantly. In practice the impact ionization is apparent only for ultraviolet radiation, which constitutes 1% of the solar spectrum.

It was shown that impact ionization can occur at much lower energies if the semiconductor particles are of nanometer size because of their electronic structure associated with carrier confinement in three dimensions. In QDs, this process is referred to as multiple exciton generation (MEG). Signatures of MEG has been identified by different research groups spectroscopically, and in 2010 Sambur *et al.*[67] were the first who obtained evidence of MEG from investigating photocurrents. They used PbS quantum dots stabilized by bifunctional passivating ligand – 3-mercaptopropionic acid – and attached to TiO$_2$ crystals *via* thiolate and carboxylic acid moieties as a model system for investigating MEG. The use of MPA as a ligand resulted in an effective system configuration in which a single layer of QDs was covalently bound to the atomically flat single-crystal substrates with no three-dimensional QD clusters. Their choice of PbS as a QD material was made due to its narrow bandgap, which remains small for PbS nanoparticles despite the effect of quantum confinement. PbS QDs can be readily synthesized with bandgap energies ranging from 0.5 to 2.0 eV, making it possible to measure sensitized photocurrents associated with MEG by using photons sufficiently low in energy to avoid direct excitation of the TiO$_2$ bandgap (3.0 eV for rutile and 3.2 eV for anatase). TiO$_2$ was chosen in a form of anatase due to its greater bandgap. The authors synthesized four QD samples with particle diameters of 9.9 nm, 4.5 nm, 3.1 nm, and 2.5 nm with corresponding bandgap energies of 0.85 eV, 0.96 eV, 1.27 eV, and 1.39 eV, respectively. The semiconductor electrode was a nearly atomically flat anatase (001) surface, uniformly covered with MPA-capped PbS quantum dots, as confirmed by AFM imaging. Photocurrent spectroscopy was employed to resolve the sensitized photocurrents as a function of incident photon energy for each QD size. The light power was measured to calculate the incident photon-to-current efficiency (IPCE) spectra. To quantify MEG effects, accurate measurement of the absorbed photon-to-current efficiency (APCE) is needed. However, doped anatase used in photocurrent studies is opaque, and thus the task of measuring photon absorption becomes difficult. Sambur *et al.* found the solution and performed absorbance measurements on undoped semitransparent rutile (001) TiO$_2$ single crystals with both sides by using the photospectrometer. Multiple regions of the rutile and anatase crystal surfaces were imaged with AFM in order to assure that the densities of QDs on all the surfaces were nearly identical. Upon analysis of APCE data, no increase in the quantum yields, indicative of MEG, was observed despite crossing the threshold of illumination with photon energies of twice the bandgap for the 4.5-nm PbS QDs (0.96 × 2 = 1.92 eV). However, at 2.8 and 3.1 eV illumination, the QDs with $E_g = 0.96$ eV (corresponding to photon energies of 2.9 and 3.2 times the bandgap) exhibited APCE values that exceeded unity, indicating that the MEG effect takes place.

While there have been reports on PbS, PbSe, PbTe, Si and InAs QDs exhibiting multiple exciton generation effects,[68] many authors have been critical of the inability of these researchers to present these effects in experimental solar cells.[69] Indeed, most of this preliminary research is based on pump probe laser experiments with lasers exhibiting powers much greater than realistic AM1.5 conditions. Even more recent work on the promising Schottky-QD solar cells have exhibited internal quantum efficiencies that fall short of 100%. Similarly, while there has been some recent success in experimentally showing hot-carrier transfer, it is uncertain if structures can be designed to efficiently harvest these carriers under realistic conditions. Given the above considerations, there has yet to be any direct proof that MEG or hot carriers can provide additional

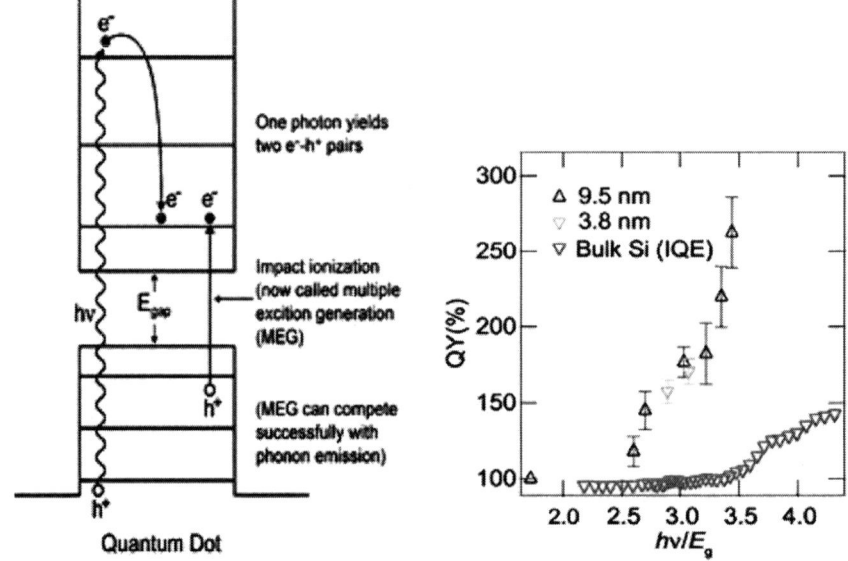

Figure 4.18   Compilation of all MEG QYs for the 9.5 and 3.8 nm Si QD samples, and
impact ionization quantum yields for bulk Si.
Figure credit: ref. 71.

current or cell potential that would improve the efficiency of third-generation
solar cells under realistic conditions (Figure 4.18). However, given that QD
solar cells have only begun to see significant improvements in their power-
conversion efficiencies in the past couple of years, it is too early to dismiss this
technology. With the development of depleted heterojunction solar cells, significant improvements in QD solar-cell characteristics are expected in the very
near future.[70]

4.4   Nanoparticle-Based Li Ion Battery

4.4.1   Introduction

Lithium (Li) ion batteries have received more attention than any other battery
technology since Sony made the first commercial cell in 1991. Two decades
later, they are still the most widely used; they account for about 60% of
worldwide sales in the portable batteries segment. Li ion batteries are currently
the system of choice, offering high energy density, flexible and lightweight
design, and long lifespan in comparison with competing battery technologies.[71,72] Figure 4.19 shows the comparison among different battery technologies. Li ion battery technologies are clearly superior to other types of batteries
with respect to energy density, which is a critical parameter for portable electronics, as well as hybrid and electric vehicles.

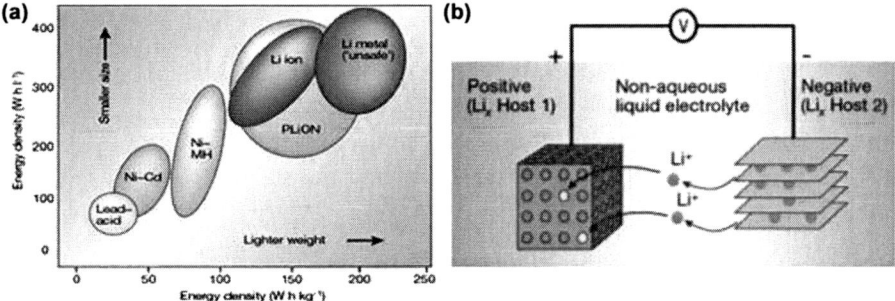

Figure 4.19 Comparison among different battery technologies.[74]

The working principle of Li ion batteries is the following (Figure 4.19). When charging, Li ions transfer from cathode materials to anode materials through the electrolyte, and electrons are transferred from the cathode to the anode *via* the external circuit. When discharging, Li ions will transfer from anode materials to cathode materials through the electrolyte, and electrons are transferred from the anode to the cathode through the external circuit. With every cycle of charging and discharging, Li ions traverse back and forth between anode and cathode. It is for this reason that a Li ion battery is also called a rocking chair battery. The typical anode material is graphite, and cathode materials include $LiCoO_2$, $LiNi_{1/3}Mn_{1/3}Co_{1/3}O_2$, $LiMn_2O_4$ and $LiFePO_4$, *etc.* The electrolyte is flammable organic solvent with a dissolved salt such as $LiPF_6$. The working voltage is more than 3 V, which is one of the most important advantages compared to batteries with aqueous solution. The charge and discharge reactions are shown in eqns (4.1)–(4.3).

$$\text{Anode reaction}: x\text{Li}^+ + \text{C} + \text{e}^- \underset{\text{Charge}}{\overset{\text{Discharge}}{\longleftrightarrow}} \text{Li}_x\text{C} \tag{4.1}$$

$$\text{Cathode reaction}: \text{LiCoO}_2 \underset{\text{Charge}}{\overset{\text{Discharge}}{\longleftrightarrow}} \text{Li}_{1-x}\text{CoO}_2 + x\text{Li}^+ + x\text{e}^- \tag{4.2}$$

$$\text{Overall cell reaction}: \text{C} + \text{LiCoO}_2 \underset{\text{Charge}}{\overset{\text{Discharge}}{\longleftrightarrow}} \text{Li}_x\text{C} + \text{Li}_{1-x}\text{CoO}_2 \tag{4.3}$$

With the development of personnel electronics, hybrid and electric vehicles, high power and energy density batteries are required. However, Li ion diffusion and electronic conductivity in the solid phase can be the rate-limiting step for the conventional active materials with microsize particles. In order to reduce the diffusion length for Li ions and electrons, adopting the nanosize particles is a direct and effective way. Nanomaterials have very small feature size in the range of 1–100 nm and they are not simply another step in miniaturization. At the nanomaterial level, some material properties are affected by the laws of atomic physics, rather than behaving as traditional bulk materials do.

Therefore, nanomaterials can offer the potential to create a unique Li ion battery with both high energy and power density. We will mainly discuss the current developments and applications of nanomaterials for anode and cathode in Li ion batteries.

4.4.2 Cathode

Currently, four major cathode materials, including $LiCoO_2$, $LiMn_2O_4$, $LiFePO_4$, and $LiNi_{1/3}Mn_{1/3}Co_{1/3}O_2$, are being used in commercial Li ion batteries, although many other compounds are in the R&D stage.

$LiCoO_2$: $LiCoO_2$ is layered structure and its theoretical specific capacity is 137 mAh/g. Microsize $LiCoO_2$ is normally used in the applications of portable electronics, which does not require high power. For $LiCoO_2$, Li ion diffusion in bulk materials is important during the charging and discharging process and can be the limiting step for the electrochemical reactions.

The Li ion diffusion coefficient for $LiCoO_2$ is in the range of 10^{-13} and 10^{-8} cm^2 s^{-1} based on different literature values.[73–83] Normally, a high number is based on the theoretical calculation. The defects of the materials and testing methods can lower the Li diffusion coefficient. The Li ion diffusion length at different rates and Li ion diffusion coefficients can be calculated based on eqn (4.1) when Li ion diffusion in the bulk material is the limiting step for the reaction. The results are shown in Table 4.1. With accurate Li ion diffusion coefficients, we can determine the ideal particle size based on the discharge rate.

$$l = 2\sqrt{Dt} \tag{4.4}$$

where l is the diffusion length, D is diffusion coefficient, t is time.

From Table 4.1, we can see that Li ion diffusion can be the rate-limiting step at high rate. For example, when the Li ion diffusion coefficient is 10^{-11} cm^2 s^{-1}, the Li ion diffusion length is only 3.8 μm at 1 Charging (C) rate. If the radius of $LiCoO_2$ powder is more than 3.8 μm, only part of the particle can be electrochemically charged/discharged even if other steps are fast enough. Nanosize $LiCoO_2$ powder has been successfully synthesized in a few groups,[84,85] which can offer high rate performance than the microsize powder. Masashi, *et al.*[82] have established the size-controlled synthesis of nanocrystalline $LiCoO_2$ through a hydrothermal reaction and clarified the structural and electrochemical properties of this intercalation cathode material. Electrochemical

Table 4.1   Li ion diffusion lengths based on different discharging rates and diffusion coefficients.

Rate/Length/ Diffusion coefficient	0.1 C	0.2 C	0.5 C	1 C	2 C	5 C	10 C
10^{-12} cm^2 s^{-1}	3.8 μm	2.7 μm	1.7 μm	1.2 μm	0.8 μm	0.5 μm	0.4 μm
10^{-11} cm^2 s^{-1}	12 μm	8.5 μm	5.4 μm	3.8 μm	2.7 μm	1.7 μm	1.2 μm
10^{-10} cm^2 s^{-1}	38 μm	27 μm	17 μm	12 μm	8.5 μm	5.4 μm	3.8 μm
10^{-9} cm^2 s^{-1}	120 μm	85 μm	54 μm	38 μm	27 μm	17 μm	12 μm

measurements and theoretical analyses on nanocrystalline $LiCoO_2$ revealed that extreme size reduction below 15 nm was not favorable for most applications. An excellent high-rate capability (65% of the 1 C rate capability at 100 C) was observed in nanocrystalline $LiCoO_2$ with an appropriate particle size of 17 nm. Lu *et al.*[83] also synthesized nanosized and well-dispersed $LiCoO_2$ powders and the electrochemical analysis reveals that the discharge capacity of $LiCoO_2$ significantly depends on the particle size and agglomeration state of the synthesized powders.

$LiMn_2O_4$: Spinel $LiMn_2O_4$ is a low-cost, environmentally friendly, and highly abundant material for Li ion battery cathodes and its theoretical specific capacity is 148 mAh/g. Similar to $LiCoO_2$, normally microsize powder is used for commercial battery applications. Several groups[86,87] have synthesized nanosize $LiMn_2O_4$ and obtained higher rate performance than common microsize powder. Cui and coworkers[84] synthesized nanosize single-crystalline $LiMn_2O_4$ nanorods, which have an average diameter of 130 nm and length of 1.2 μm. Galvanostatic battery testing showed that $LiMn_2O_4$ nanorods have a high charge storage capacity at high power rates compared with commercially available powders and more than 85% of the initial charge storage capacity was maintained for over 100 cycles (Figure 4.20).

$LiFePO_4$: Olivine $LiFePO_4$ has received widely interests because of its high capacity (170 mAh/g), low cost and low toxicity since it was first proposed by Padhi and coworkers in 1997.[88–106] $LiFePO_4$ does not include noble elements such as cobalt, the price of raw material is lower and both phosphorus and iron are abundant on Earth, which lowers raw material availability issues. Originally, $LiFePO_4$ is only for low rate application because of the low electron and ionic diffusion. Chiang *et al.*[87] have successfully synthesized doped nanosize $LiFePO_4$ to increase the electron conductivity and reduce the Li ion diffusion length to increase high rate performance.[87] Recently, Ceder *et al.*[89] synthesized nanosize $LiFePO_4$ with unique surface phase, which is shown in Figure 4.21.[89] From the figure, we can see that nanosize $LiFePO_4$ can be discharged at 400 °C. Later they also explain why nanosize $LiFePO_4$ can have such a high rate, which is shown in Figure 4.22.[104] Li ion is 1D diffusion in $LiFePO_4$, which can be impeded by the presence of immobile and low-mobility defects residing in the diffusion path. The presence of point defects will have a drastic effect on the

Figure 4.20 Nanosize $LiMn_2O_4$ and the electrochemical performance, (a) $LiMn_2O_4$ nanorod, (b) and (c) electrochemical performance.[91]

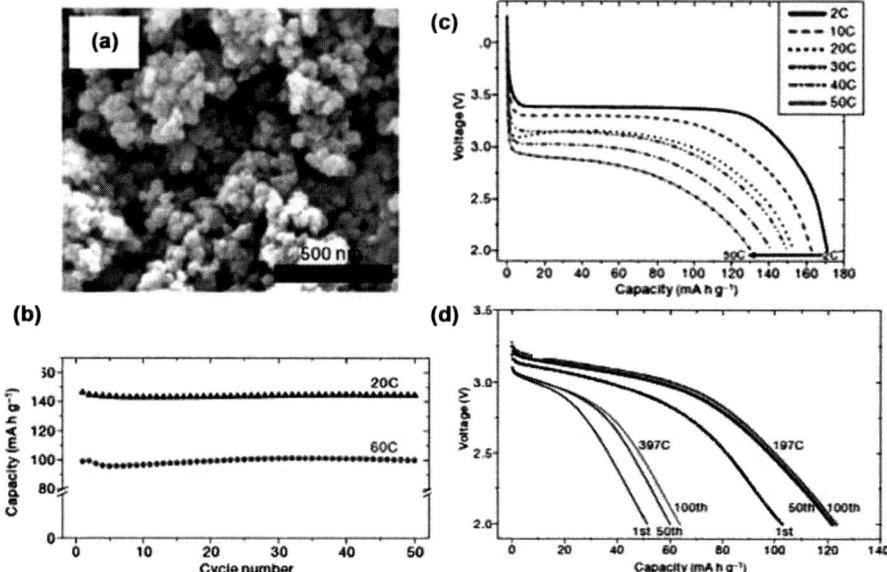

Figure 4.21   Nanosize LiFePO$_4$ and electrochemical performance, (a) synthesized
nanosize LiFePO$_4$, (b) discharge curve at different rates, (c) cycle life at
different rates, (d) discharge curve at very high rates.[96]

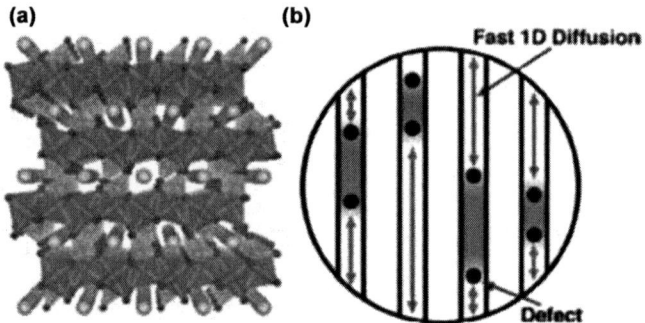

Figure 4.22   (a) crystal structure of LiFePO$_4$ with 1D Li ion diffusion channels,
(b) schematic illustration of Li ion diffusion.[111]

rate at which ions can move through the crystal. Therefore, adopting nanosize
powder can significantly reduce the chance that Li ions are impeded, which is
shown in Figure 4.23.

4.4.3   Anode

The conventional anode material is graphite and its theoretical capacity
is 372 mAh/g. Microsize graphite powders are used for Li ion batteries.

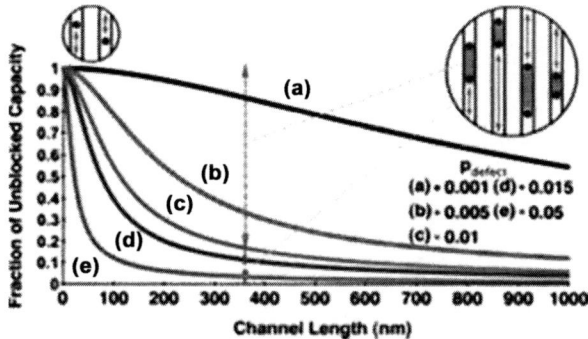

Figure 4.23 Expected unblocked capacity *vs.* channel length in LiFePO$_4$.[111]

The major rate-limiting factor in current Li ion batteries is that the charging rate of graphite is low compared to the cathode materials due to the low voltage (0.15 V *vs.* Li/Li$^+$). Li will be deposited on graphite if the overpotential is too high during charging, which causes ineffective charging or shorting of the cell. All kinds of nanostructured carbon or graphite, including nanopower, nanofiber, nanosheet, grapheme, *etc.* have been fabricated to reduce the Li ion diffusion length in the solid phase to increase the rate performance.

Si: In current Li ion battery technology, the anode is mainly graphite and its specific capacity is 372 mAh/g. The development of portable electronics, hybrid and electric cars has produced demand for even higher capacities and energy densities. Si is a very attractive anode material with a low discharge potential (~0.5 V *vs.* Li/Li$^+$) and high theoretical capacity (4200 mAh/g). However, the huge volume expansion (300%) for a Si electrode limits its practical application in Li ion battery. The stress induced by volume changes causes cracking and pulverization of Si anode and Si particles can lose electronic contact with each other and the current collector, which results in fast fading.[107]

In order to overcome the volume-expansion problem, several approaches have been proposed. One approach is to use 3D Si nanowire or nanotube, which is commonly deposited onto a substrate by a chemical vapor deposition (CVD) method, with a result from the literature shown in Figure 4.24.[108] Si nanowires of small diameter better accommodate large volume changes and the 1D geometry provides an efficient electronic charge transport pathway. Another group developed a hierarchical bottom-up approach to grow the nanosize Si particles on the carbon black dendrite particles as the electrode (Figure 4.25),[109] which shows very stable cycle life with high capacity. Although results from nanostructured electrodes are promising, high synthesis costs with the CVD method can prevent their adoption by industry and long cycle life needs to be proven in full cells.

The main alternative approach is to use a C-Si composite to overcome the volume expansion problem.[110] Carbon acts as the buffer for the large volume expansion of Si during charging. However, the large volume changes of Si

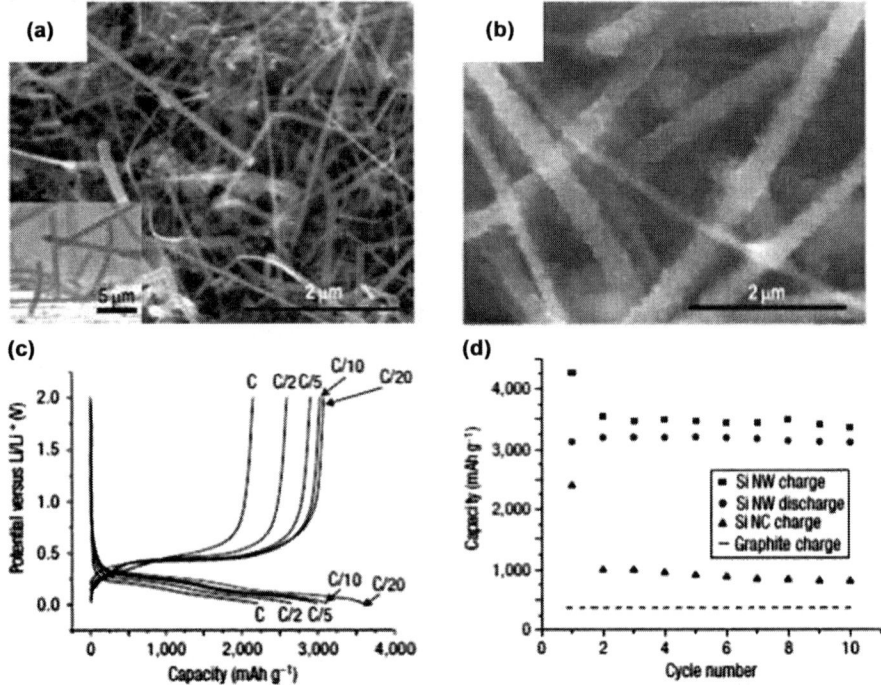

Figure 4.24   Silicon nanowire and its electrochemical performance, (a) before cycling,
(b) after cycling, (c) rate performance, (d) cycle test.[113]

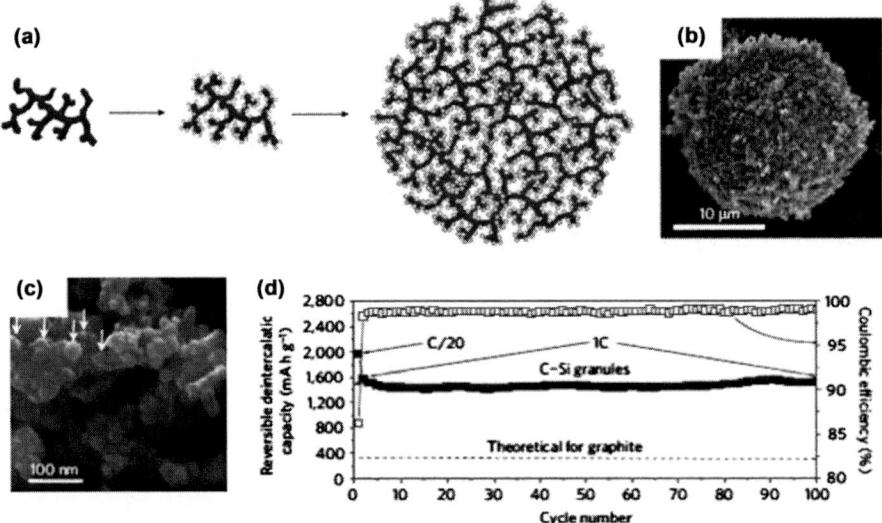

Figure 4.25   Si electrode based on bottom-up approach and its electrochemical
performance, (a) hierarchical bottom-up approach, (b) Morphology in
low magnification, (c) Morphology in high-magnification performance,
(d) cycle test at different C rates.[114]

Table 4.2 Examples of metal and their theoretical capacity and volume expansion.

Metal	Limiting composition	Volume expansion	Theoretical capacity (mAh/g)
Sn	$Li_{4.4}Sn$	259%	994
Ge	$Li_{4.4}Ge$	272%	1624
Sb	Li_3Sb	135%	536
Bi	Li_3Bi	115%	385

during Li insertion can be accommodated by carbon only to a limited degree, thus offering limited stability and cycle life. There is also some research work on Si thin film with or without coatings.[111] Although Si thin film showed improved cycle life, it still cannot meet practical requirement for industrial applications due to the cost and low capacity associated with the thin-film structure. Igor *et al.*,[112] has utilized alginate as the binder with nano-Si to fabricate the electrodes for Li ion batteries and very stable and high capacity has been obtained in their recent report.[112]

$Li_4Ti_5O_{12}$: The theoretical capacity of $Li_4Ti_5O_{12}$ is 175 mAh/g and its voltage is 1.5 V *vs.* Li/Li+. The energy density is low due to its high voltage *versus* Li metal. For example, the working voltage of $LiCoO_2$ and $Li_4Ti_5O_{12}$ is only about 2 V. However, there is no SEI layer formed on $Li_4Ti_5O_{12}$ and irreversible capacity is very low. The volume change of fully discharged $Li_4Ti_5O_{12}$ to fully charged $Li_7Ti_5O_{12}$ is less than 0.2%, which assures that $Li_4Ti_5O_{12}$ has a much longer cycle life than graphite. Many researchers have studied the synthesis of $Li_4Ti_5O_{12}$ with various nanostructures, including nanowires,[113–115] nanotubes,[116] flower-like nanosheets,[117] hollow spheres[118] and 0D–1D heterogeneous nanostructures.[119]

Other materials: Many metals or metal oxides can be used as the anode materials for Li ion batteries and they offer very high capacity, which is shown in Table 4.2. Similar to Si, huge volume expansion causes the fast capacity fading. All kinds of nanostructures including nanoparticles have been fabricated in order to suppress the volume expansion and increase the cycle life.

4.5 Summary

One of the most daunting challenges we face is sustainable development in the 21st century. With the population increasing at 1.4% annual rate and energy demand increasing at a 1.7% annual rate, the need for renewable energy resources is pivotal. QD solar cells fall into the third generation of solar cells, which can allow for both greatly enhanced efficiencies and low cost. The decreased manufacturing costs arise from the facile, solution processing of QDs. Enhanced efficiency can be attributed to better solar cell absorption spectral matching with the sun's radiation spectrum, including the capacity to harvest IR photons. Additional effects, such as hot-carrier extraction and multiple exciton generation could potentially allow for QD solar cell efficiency values greater than the Shockley–Quessier limit of ~31%. Unfortunately, neither of these effects has been measurable in the external and internal quantum efficiency values reported for any solar cells to date. Four different QD solar-cell configurations have been presented. Two such configurations, vacuum-deposited solar cells and liquid-junction solar cells, are interesting from a research standpoint; however, challenges in their complicated and costly fabrication must be overcome before they can compete with traditional silicon photovoltaics. Hybrid organic-QD solar cells have undergone extensive research, but have achieved only modest power-conversion efficiencies. Fully inorganic, solution-processed QD solar cells, such as the Schottky–QD and depleted heterojunction structures, have shown significant promise with the highest QD solar cell efficiency of 5.1%. In all configurations, the large number of deep interfacial traps limits device efficiencies. In order to become competitive with current photovoltaic technology, it is now necessary to investigate new device architectures to take advantage of hot carriers and multiple exciton generation effects.

In the drive towards improved electrical energy storage for applications ranging from electric vehicles to grid stabilization to renewable energy storage, the important role of electrochemical storage, and batteries specifically, will continue due to their high energy density, simplicity, reliability, and potential for favorable performance/cost ratio. Li rechargeable chemistry is likely to be widely adopted in, first, hybrid electric vehicles (HEVs), and subsequently, plug-in hybrids (PHEVs) and limited-range battery electric vehicles (BEVs, here referring to fully electric vehicles). Recent evaluations of high-power Li batteries in grid stabilization (frequency regulation) applications where an entire system has 20–40 MW power and 5–10 MWh stored energy suggests that Li chemistry may have further technological reach than previously anticipated.

With the fast development of personnel electronics, hybrid or electric vehicles, high power and energy density Li ion batteries are needed. Nanotechnology and chemistry advances in electrode design are the key research topics that academy and industry need to push for developing Li ion batteries. However, nanomaterials can also cause some challenges that we do not faced before. The main challenges mainly include the high cost of manufacturing, purification of materials, and environmental concerns.

Acknowledgement

The authors thank the University of Kansas and Worcester Polytechnic Institute for its startup financial supporting.

References

1. A. P. Alivisatos, *Science*, 1996, **271**, 933.
2. W. A. Deheer, *Rev. Mod. Phys.*, 1993, **65**, 611.
3. M. A. Kastner, *Rev. Mod. Phys.*, 1992, **64**, 849.
4. W. P. McCray, *Nature Nano.*, 2007, **2**, 259.

5. C. Toumey, *Nature Nano.*, 2010, **5**, 239.
6. L. E. Brus, *Nano Lett.*, 2010, **10**, 363.
7. R. Ashoori, *Nature*, 1996, **379**, 413.
8. T. Oosterkamp, *Nature*, 1998, **395**, 873.
9. C. B. Murray, D. J. Norris and M. G. Bawendi., *J. Am. Chem. Soc.*, 1993, **115**, 8706.
10. X. G. Peng, J. Wickham and A. P. Alivisatos., *J. Am. Chem. Soc.*, 1998, **120**, 5343.
11. A. J. Nozik, *et al.*, *Chem. Rev.*, 2010, **110**, 6873.
12. R. L. Whetten, J. T. Khoury and M. M. Alvarez, *et al.*, *Adv. Mater.*, 1996, **8**, 428.
13. C. Chen, A. Herhold and C. Johnson, *et al.*, *Science*, 1997, **276**, 398.
14. K. Eberl, *Phys. World*, 1997, **10**, 47.
15. V. Klimov, A. Mikhailovsky, S. Xu, A. Malko, J. Hollingsworth, C. Leatherdale, H. Eisler and M. G. Bawendi, *Science*, 2000, **290**, 314.
16. S. Coe, W. K. Woo, M. G. Bawendi and V. Bulovic, *Nature*, 2002, **420**, 800.
17. J. Tang and E. H. Sargent, *Adv. Mater.*, 2011, **23**, 12.
18. A. Denton and N. Ashcroft, *Phys. Rev. A*, 1991, **43**, 3161.
19. L. Brus, *J. Chem. Phys.*, 1983, **79**, 5566.
20. L. Brus, *J. Chem. Phys.*, 1984, **80**, 4403.
21. X. Peng, M. Schlamp, A. Kadavanich and A. P. Alivisatos, *J. Am. Chem. Soc.*, 1997, **119**, 7019.
22. M. Steigerwald, A. Alivisatos, J. Gibson, T. Harris, R. Kortan, A. Muller, A. Thayer, T. Duncan, D. Douglass and L. Brus, *J. Am. Chem. Soc.*, 1988, **110**, 3046.
23. A. P. Alivisatos, T. Harris, L. Brus and A. Jayaraman, *J. Chem. Phys.*, 1988, **89**, 5979.
24. M. Bawendi, M. L. Steigerwald and L. E. Brus, *Ann. Rev. Phys. Chem.*, 1990, **41**, 477.
25. Z. Peng and X. Peng, *J. Am. Chem. Soc.*, 2001, **123**, 183.
26. R. Debnath, *et al.*, *J. Am. Chem. Soc.*, 2010, **132**, 5952.
27. H. J. Lee, *et al.*, *Nano Lett.*, 2009, **9**, 4221.
28. H. Lee, J. Yum, H. Leventis, S. Zakeeruddin, S. Haque, P. Chen, S. Seok, M. Grätzel and M. Nazeeruddin, *J. Phys. Chem. C*, 2008, **112**, 11600.
29. J. Kim, O. Benson, H. Kan and Y. Yamamoto, *Nature*, 1999, **397**, 500.
30. A. Mohan, *et al.*, *Small*, 2010, **6**, 1268.
31. A. Goetzberger, C. Hebling and H. Schock, *Mater. Sci. Eng. R*, 2003, **40**, 1.
32. H. Gerischer and M. Liibke, *J. Electroanal. Chem. Interfac. Electrochem.*, 1986, **204**, 225.
33. R. Plass, S. Pelet, J. Krueger, M. Grätzel and U. Bach, *J. Phys. Chem., B*, 2002, **106**, 7578.
34. A. J. Nozik, *Physica E*, 2002, **14**, 115.
35. A. J. Nozik, *Ann. Rev. Phys. Chem.*, 2001, **52**, 193.
36. J. Garcıa, G. Medeiros-Ribeiro, K. Schmidt, T. Ngo, J. L. Feng, A. Lorke, J. Kotthaus and P. Petroff, *Appl. Phys. Lett.*, 1997, **71**.

37. M. Green, *J. Mater. Sci., Mater. Electron.*, 2007, **18**, S15.
38. K. Cheng, *IEEE Proc.*, 1997, **85**, 1694.
39. P. Kamat, *J. Phys. Chem. C*, 2008, **112**, 18737.
40. I. Robel, B. A. Bunker and P. V. Kamat, *Adv. Mater.*, 2005, **17**, 2458.
41. N. Greenham, X. Peng and A. P. Alivisatos, *Phys. Rev. B*, 1996, **54**, 17628.
42. C. Choi, K. Koski, S. Sivasankar and A. P. Alivisatos, *Nano Lett.*, 2009, **9**, 3544.
43. D. Norris, M. Nirmal, C. Murray, A. Sacra and M. Bawendi, *Z. Phys. D: At., Mol. Clusters*, 1993, **26**, 355.
44. G. Konstantatos and E. H. Sargent, *Nature Nano.*, 2010, **5**, 391.
45. D. Qi, E. Fishbein, M. Drndic and S. Selmic, *Appl. Phys. Lett.*, 2005, **86**, 093103.
46. S. Zhang, P. W. Cyr, S. A. Mcdonald, G. Konstantatos and E. H. Sargent, *Appl. Phys. Lett.*, 2005, **87**, 233101.
47. J. Tang, L. Brzozowski, D. Barkhouse, X. Wang, R. Debnath, R. Wolowiec, E. Palmiano, L. Levina, A. G. Abraham, D. Jamakosmanovic and E. H. Sargent, *ACS Nano*, 2010, **4**, 869.
48. S. Günes, H. Neugebauer and N. S. Sariciftci, *Chem Rev.*, 2007, **107**, 1324.
49. H. Lee, H. Leventis, S. Moon, P. Chen, S. Ito, S. Haque, T. Torres, F. Nuesch, T. Geiger, S. Zakeeruddin, M. Grätzel and M. Nazeeruddin, *Adv. Func. Mater.*, 2009, **19**, 2735.
50. G. Larramona, C. Chone, A. Jacob, D. Sakakura, B. Delatouche, D. Pere, X. Cieren, M. Nagino and R. Bayon, *Chem. Mater.*, 2006, **18**, 1688.
51. W. Huynh, X. G. Peng and A. P. Alivisatos, *Adv. Mater.*, 1999, **11**, 923.
52. A. Alivisatos, *Science*, 1996, **271**, 933.
53. S. Ren, L. Chang, S. Lim, J. Zhao, M. Smith, N. Zhao, V. Bulovic, M. Bawendi and S. Gradecak, *Nano Lett.*, 2011, **11**, 3998.
54. J. Clifford, K. Johnston, L. Levina and E. H. Sargent, *Appl. Phys. Lett.*, 2007, **91**, 253117.
55. J. Bang and P. V. Kamat, *ACS Nano*, 2009, **3**, 1467.
56. K. Johnston, A. Abraham, J. Clifford, S. Myrskog, D. MacNeil, L. Levina and E. H. Sargent, *Appl. Phys. Lett.*, 2008, **92**.
57. E. Klem, D. MacNeil, L. Levina and E. H. Sargent, *Adv. Mater.*, 2008, **20**, 3433.
58. M. Hanna and A. J. Nozik., *J. Appl. Phys.*, 2006, **100**, 074510.
59. W. Ma, J. M. Luther, H. Zheng, Y. Wu and A. P. Alivisatos, *Nano Lett.*, 2009, **9**, 1699.
60. K. Leschkies, T. Beatty, M. Kang, D. Norris and E. S. Aydil, *ACS Nano*, 2009, **3**, 3638.
61. A. Pattantyus-Abraham, I. Kramer, A. Barkhouse, X. Wang, G. Konstantatos, R. Debnath, L. Levina, I. Raabe, M. Nazeeruddin, M. Graetzel and E. H. Sargent, *ACS Nano*, 2010, **4**, 3374.
62. K. S. Leschkies, M. S. Kang, E. Aydil and D. Norris, *J. Phys. Chem. C*, 2010, **114**, 9988.

63. A. Abraham, I. Kramer, A. Barkhouse, X. Wang, G. Konstantatos, R. Debnath, L. Levina, I. Raabe, M. Nazeeruddin, M. Gratzel and E. H. Sargent, *ACS Nano*, 2010, **18**, 144.
64. J. Choi, Y. Lim, M. Santiago-Berrios, M. Oh, B. Hyun, L. Sun, A. C. Bartnik, A. Goedhart, G. Malliaras, H. D. Abrun, F.W. Wise and T. Hanrath, *Nano Lett.*, 2009, **9**, 3749.
65. S. Ghosh, S. Hoogland, V. Sukhovatkin, L. Levina and E. H. Sargent, *Appl. Phys. Lett.*, 2011, **99**, 101102.
66. A. J. Nozik, *Physica E*, 2002, **14**, 115.
67. J. Sambur, *et al.*, *Science*, 2010, **330**, 63.
68. A. E. Shabaev and A. J. Nozik, *Nano Lett.*, 2006, **6**, 2856.
69. J. Luther, M.C. Beard, Q. Song, M. Law, R. Ellingson and A. J. Nozik, *Nano Lett.*, 2007, **7**, 1779.
70. A. J. Nozik, *Chem. Phys. Lett., Front. Chem.*, 2008, **457**, 3.
71. J. Tarascon and M. Armand, *Nature*, 2001, **414**, 359.
72. M. Armand and J. M. Tarascon, *Nature*, 2008, **451**, 652.
73. J. Ma, C. Wang and S. Wroblewski, *J. Power Sources*, 2007, **164**, 849.
74. H. Yan, X. Huang, H. Li and L. Chen, *Solid State Ion.*, 1998, **11**, 113.
75. M. Levi, G. Salitra, B. Markovsky, H. Teller, D. Aurbach, U. Heider and L. Heider, *J. Electrochem. Soc.*, 1999, **146**, 1279.
76. J. McGraw, C. Bahn, P. Parilla, J. Perkins, D. Readey and D. Ginley, *Electrochim. Acta*, 1999, **45**, 187.
77. S. B. Tanga, M. O. Laia and L. Lu, *J. Alloys Compd.*, 2008, **449**, 300.
78. A. Honders, J. M. der Kinderen, A. H. wan Heeren, J. H. W. de Wit and G. H. J. Broers, *Solid State Ion.*, 1985, **15**, 265.
79. H. Xia, L. Lu and G. Ceder, *J. Power Sources*, 2006, **159**, 1422.
80. C. N. Polo da Fonsecaa, J. Davalosa, M. Kleinkea, M. C. A. Fantinib and A. Gorenstein, *J. Power Sources*, 1999, **81–82**, 575.
81. Y.-M. Choi, S.-I. Pyun, J.-S. Bae and S. Moon, *J. Power Sources*, 1995, **56**, 25.
82. D. Aurbach, M. D. levi, E. Levi, H. Teller, B. Markovsky and G. Salitra, *J. Electrochem. Soc.*, 1998, **145**, 3024.
83. Y. I. Jang, B. J. Neudecker and N. J. Dudney, *Electrochem. Solid State Lett.*, 2001, **4**, A74.
84. M. Okubo, E. Hosono, J. Kim, M. Enomoto, N. Kojima, T. Kudo, H. Zhou and I. Honma, *J. Am. Chem. Soc.*, 2007, **129**, 7444.
85. C.-H. Lu, H.-H. Chang and Y.-K. Lin, *Ceram. Int*, 2004, **30**, 1641.
86. D.-K. Kim, P. Muralidharan, H.-W. Lee, R. Ruffo, Y. Yang, C. K. Chan, H. Peng, R. A. Huggins and Y. Cui, *Nano Lett.*, 2008, **8**, 3948.
87. M. Okubo, Y. Mizuno, H. Yamada, J. Kim, E. Hosono, H. Zhou, T. Kudo and I. Honma, *ACS Nano*, 2010, **4**, 741.
88. A. K. Padhi, K. S. Nanjundaswamy and J. B. Goodenough, *J. Electrochem. Soc.*, 1997, **144**, 4 1188.
89. S. Y. Chung, J. T. Bloking and Y. M. Chiang, *Nature Mater.*, 2002, **1**, 123.
90. B. Ellis, W. H. Kan, W. R. M. Makahnouk and L. F. Nazar, *J. Mater. Chem.*, 2007, **17**, 3248.

91. B. Kang and G. Ceder, *Nature*, 2009, **458**, 190.
92. S. Yang, P. Y. Zavalij and M. S. Whittingham, *Electrochem. Commun.*, 2001, **3**, 505.
93. S. Yang, Y. Song, P. Y. Zavalij and M. S. Whittingham, *Electrochem. Commun.*, 2002, **4**, 239.
94. S. Franger, F. Le Cras, C. Bourbon and H. Rouault, *J. Power Sources*, 2003, **119–121**, 252.
95. N. C. Y. Ravet, M. J. Fagnan, S. Besner, M. Gauthier and M. Armand, *J. Power Sources*, 2001, **97–8**, 503.
96. P. S. Herle, B. Ellis, N. Coombs and L. F. Nazar, *Nature Mater.*, 2004, **3**, 147.
97. H. Huang, S.-C. Yin and L. F. Nazar, *Electrochem, Solid-State Lett.*, 2001, **4**, A170.
98. R. Dominiko, M. Bele, M. Gaberscek, M. Remskar, S. Pejovnik and J. Jamnik, *J. Electrochem. Soc.*, 2005, **152**, A607.
99. A. Yamada, S. C. Chung and K. Hinokuma, *J. Electrochem. Soc.*, 2001, **148**, A224.
100. M. S. Islam, D. J. Driscoll, C. A. J. Fisher and P. R. Slater, *Chem. Mater.*, 2005, **17**, 5085.
101. J. Chen and M. S. Whittingham, *Electrochem. Commun.*, 2006, **8**, 855.
102. D. Morgan, A. Van der Ven and G. Ceder, *Electrochem. Solid State Lett.*, 2004, **7**, A30.
103. C. Delacourt, P. Poizot, S. Levasseur and C. Masquelier, *Electrochem. Solid State Lett.*, 2006, **9**, A352.
104. D. H. Kim and J. Kim, *Electrochem. Solid State Lett.*, 2006, **9**, A439.
105. G. Y. Chen, X. Y. Song and T. J. Richardson, *Electrochem. Solid State Lett.*, 2006, **9**, A295.
106. R. Malik, D. Burch, M. Bazant and G. Ceder, *Nano Lett.*, 2010, **10**, 4123.
107. R. Teki, M. K. Datta, R. Krishnan, T. C. Parker, T. Lu, P. N. Kumta and N. Koratkar, *Small*, 2009, **5**, 2236.
108. C. Chan, H. Peng, G. Liu, K. McIlwrath, X. Zhang, R. Huggins and Y. Cui, *Nature Nano.*, 2008, **3**, 3.
109. A. Magasinki, P. Dixon, B. Hertzberg, A. Kvit, J. Ayala and G. Yushin, *Nature Mater.*, 2010, **9**, 353.
110. M. Li, M. Qu, X. He and Z. Yu, *J. Electrochem. Soc.*, 2009, **156**, A294.
111. H. Wolf, Z. Pajkica, T. Gerdesa and M. Willert-Poradaa, *J. Power Sources*, 2009, **190**, 157.
112. I. Kovalenko, B. Zdyrko, A. Magasinski, B. Hertzberg, Z. Milicev, R. Burtovyy, I. Luzinov and G. Yushin, *Science*, 2011, **334**, 75.
113. J. R. Li, Z. L. Tang and Z .T. Ahang, *J. Electrochem. Soc.*, 1995, **142**, 1431.
114. J. Li, Z. Tang and Z. Zhang, *Electrochem. Commun.*, 2005, **7**, 894.

115. J. Kim and J. Cho, *Electrochem Solid-State Lett.*, 2007, **10**, A81.
116. Y. Li, K. Xi and X.P. Gao, *Mater. Lett.*, 2009, **63**, 304.
117. Y. F. Tang, L. Yang, Z. Qiu and J.S. Huang, *Electrochem. Commun.*, 2008, **10**, 1513.
118. J. Huang and Z. Jiang, *Elecrochem, Solid-State Lett.*, 2008, **11**, A116.
119. D.-K. Lee, H.-W. Shim, J.-Sul. An, C.-M. Cho, I.-S. Cho, K.-S. Hong and D.-W. Kim, *Nanoscale Res Lett.*, 2010, **5**, 1585.

CHAPTER 5

The Green Synthesis and Environmental Applications of Nanomaterials[†]

CHANGSEOK HAN,[a] MIGUEL PELAEZ,[a]
MALLIKARJUNA N. NADAGOUDA,[b]
SHERINE O. OBARE,[c] POLYCARPOS FALARAS,[d]
PATRICK S.M. DUNLOP,[e] J. ANTHONY BYRNE,[e]
HYEOK CHOI[f] AND DIONYSIOS D. DIONYSIOU*[a]

[a] Environmental Engineering and Science Program, University of Cincinnati, Cincinnati, Ohio 45221, USA; [b] The US Environmental Protection Agency, ORD, NRMRL, WSWRD, 26 West MLK Drive, Cincinnati, Ohio 45268, USA; [c] Department of Chemistry, Western Michigan University, Kalamazoo, Michigan 49008, USA; [d] Division of Physical Chemistry, Institute of Advanced Materials, Physicochemical Processes, Nanotechnology and Microsystems (IAMPPNM), National Center for Scientific Research "Demokritos", 15310 Aghia Parakevi Attikis, Athens, Greece; [e] Nanotechnology and Integrated BioEngineering Centre, School of Engineering, University of Ulster, Northern Ireland, BT37 0QB, UK; [f] Department of Civil Engineering, The University of Texas at Arlington, Arlington, Texas 76019, USA
*Email: dionysios.d.dionysiou@uc.edu

[†] All authors have contributed equally in this chapter.

RSC Green Chemistry No. 19
Sustainable Preparation of Metal Nanoparticles: Methods and Applications
Edited by Rafael Luque and Rajender S Varma
© The Royal Society of Chemistry 2013
Published by the Royal Society of Chemistry, www.rsc.org

5.1 Green Chemistry of the Synthesis of Nanomaterials

The ample use of nanomaterials has led to increasing demand in the chemical industry. Hence, many chemicals such as solvents, raw materials, reagents, and template materials, have been utilized in order to produce nanomaterials. In addition, the production of toxic or hazardous intermediates and products, as well as wastes has increased through reactions aiming at preparing desirable chemicals. The concept of green chemistry was introduced to chemical science and industry[1] to reduce or eliminate the generation of undesirable products and to decrease the use of hazardous materials in chemical processes. This chapter will discuss green routes that have been applied in chemical processes to synthesize various nanomaterials.

5.1.1 TiO$_2$ Nanomaterials

Titanium dioxide (titania, TiO$_2$) has been widely used in pigments, semi-conductors, ceramic materials, cosmetics, and photocatalysts.[2–7] Industrial processes have used many chemicals (solvents, template materials, and precursors) to synthesize TiO$_2$ nanomaterials. Therefore, it is of great priority to apply the concept of green chemistry toward the synthesis of TiO$_2$ photocatalyst nanomaterials in order to reduce or eliminate the use of toxic or hazardous chemicals as well as the production of toxic or hazardous intermediates, products, and wastes. There are different green chemical approaches in the synthesis of TiO$_2$ nanomaterials. These approaches include the use of green solvents as primary solvents and natural materials as templates and precursors, the employment of microorganisms, and the simplification of the synthesis procedure.

Several investigations on nontoxic green solvents and natural template materials have been reported.[8–12] Venkataramanan et al.[8] used ionic liquid as a nontoxic green solvent and natural cellulose fibers as green templates for the synthesis of titania nanowires. Using cellulosic substrate and a green solvent like ionic liquid successfully led to TiO$_2$/cellulose nanowires with specific surface area of 14.65 m^2 g^{-1}. Zeng et al.[9] also synthesized fiber-like TiO$_2$ nanomaterials using cellulose fibers. They reported the synthesis of pure, fiber-like TiO$_2$ nanomaterials containing different ratios of anatase/rutile structural modifications by varying the calcination temperature. Figure 5.1 shows the synthesized TiO$_2$ composite fiber and pure TiO$_2$ fibers calcined at 400, 600, and 800 °C, respectively. TiO$_2$ fibers calcined at 400 °C contained 100% anatase phase, had high BET surface area (39.73 m^2 g^{-1}) and showed the best photocatalytic activity on the degradation of methyl orange, compared to fibers calcined at higher temperatures.

In another study,[10] sucrose was used as a green template to develop mesoporous TiO$_2$ nanomaterials with high surface area and pore volume without any surfactant templates. Wormhole-like mesostructured TiO$_2$/sucrose nanocomposites were formed following drying at room temperature. After calcination at 300–800 °C, wormhole-like mesoporous titania with BET surface

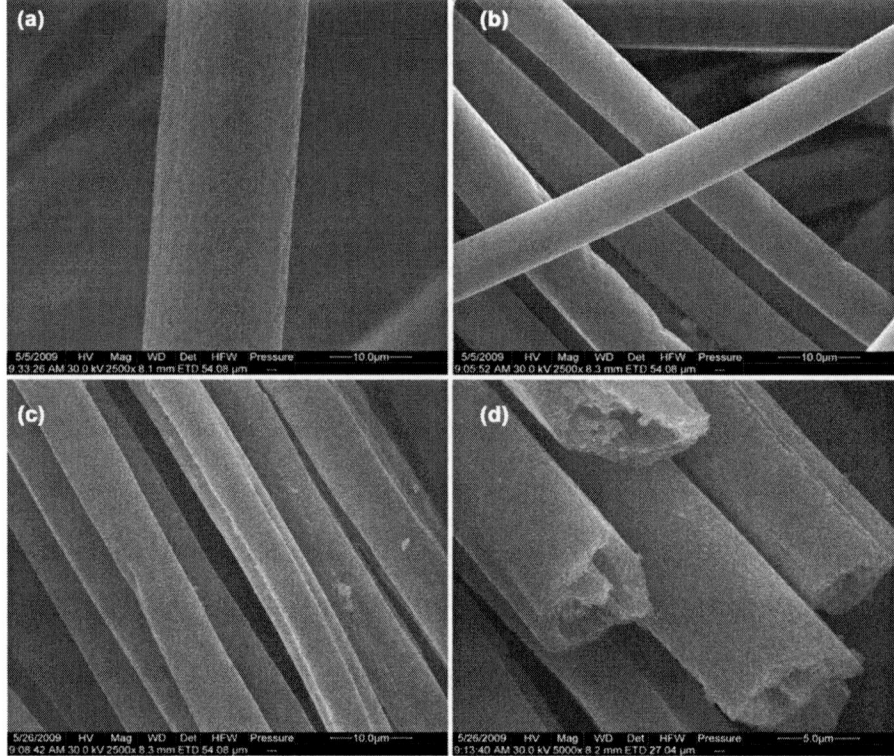

Figure 5.1 SEM images of the TiO$_2$ composite fibers (a), fibers calcined 400 °C (b), fibers calcined 600 °C (c), and fibers calcined 800 °C (d).
Reproduced with permission from ref. 9, copyright 2011 American Chemical Society.

area of 101–193 m^2/g and pore volume of 0.12–0.42 cm^3/g was obtained by the elimination of the sucrose. Sundrarajan and Gowri[11] mixed nyctanthes arbortristis leaves extract and titanium tetraisoproxide and following calcination at 500 °C, stable TiO$_2$ nanoparticles of 100–150 nm size were rapidly synthesized through this ecofriendly green chemistry approach. Moreover, Li *et al.*[12] reported the use of green-leaf biotemplates to enhance the light-harvesting and photocatalytic properties in *morph*-TiO$_2$. Based on the unique physical structure of green-leaf providing an excellent light-harvesting efficiency to plants, green-leaf structured *morph*-TiO$_2$ was successfully synthesized. The nitrogen contained in green leaves was also self-doped into *morph*-TiO$_2$ during the synthesis process.

Scheme 5.1 shows the whole procedure for synthesis of green-leaf structured *morph*-TiO$_2$ nanomaterials. The microstructure of *morph*-TiO$_2$ completely replicated the structure of green-leaf from macro- to nanoscale (see Figure 5.2). The light-harvesting efficiency of synthesized *morph*-TiO$_2$ has been increased extremely from 103–258% within the visible light range and a red shift of

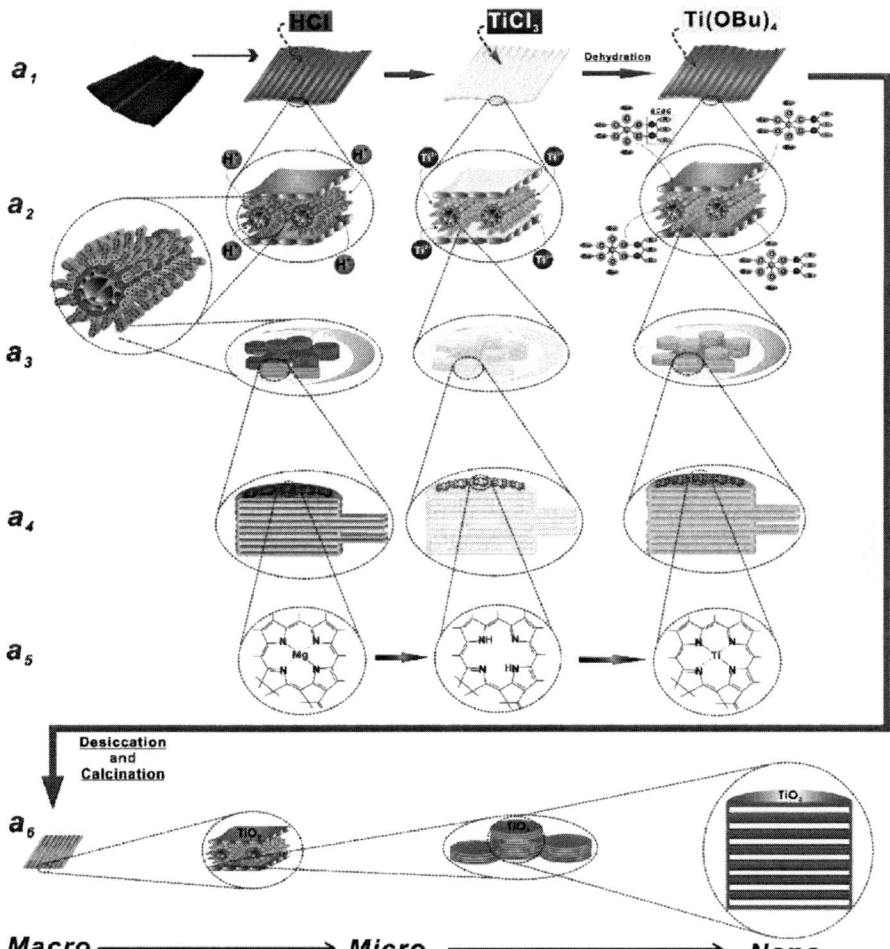

Scheme 5.1 The synthesis procedure of *morph*-TiO$_2$. The rows a$_1$ to a$_5$ are illustrations of the leaf structure on different scales: a$_1$) *Saccharum officinaru
Linn.* leaf observed by the naked eye; a$_2$) microstructure of a cross section
of the leaf (Kranz-type anatomy); a$_3$) magnification of a chloroplast
contained in a mesophyll cell; a$_4$) Granum – the layered nanostructure of
the thylakoid membranes; a$_5$) porphyrin ring of the chlorophyll after
different treatments; a$_6$ shows illustrations of the resulting *morph*-TiO$_2$
on different length scales.
Reproduced with permission from ref. 12, Copyright 2009 WILEY-VCH
Verlag GmbH & Co. KGaA, Weinhein.

absorption also observed at the edge of UV and visible-light range. In addition,
the photocatalytic activity of *morph*-TiO$_2$ on degradation of rhodamine dye
was enhanced under both UV and visible light illumination.

There are also very interesting studies regarding the synthesis of TiO$_2$
nanoparticles by microorganisms. The pH, partial pressure of gaseous

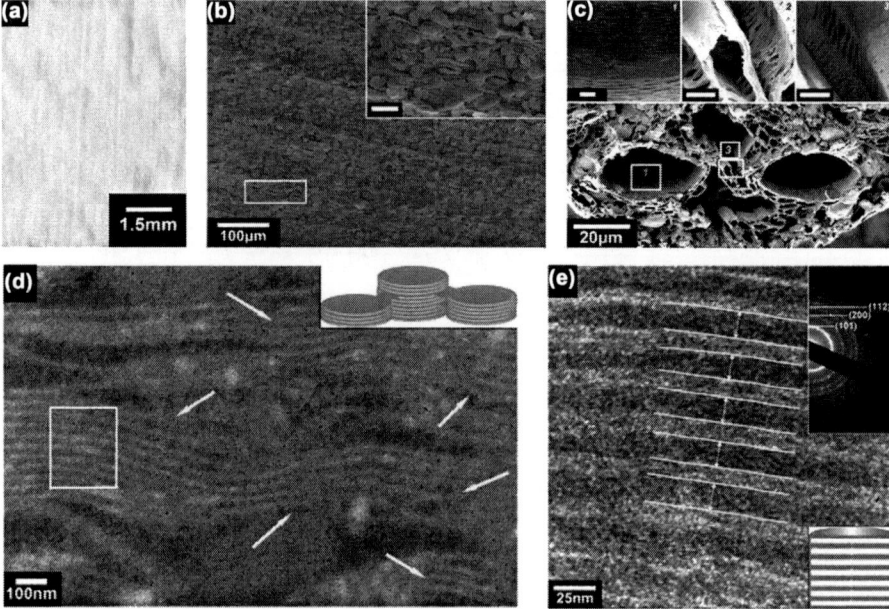

Figure 5.2   Structures from the macro- to the nanoscale of *Saccharum officinarum
              Linn.* – leaf morph-TiO$_2$: a) digital image of the sample; b) FESEM image
              of the surface structure; the inset shows a magnification of the area con-
              tained the white pane: the scale bar is 25 µm; c) cross-sectional FESEM
              image; the inset shows a magnification of the corresponding areas con-
              tained in the white panes: the scale bars are 2 µm; d) TEM image of the
              layered nanostructure; e) magnification of the area contained in the white
              pane in (d), the inset shows the SAED pattern.
              Reproduced with permission from ref. 12, copyright 2009 WILEY-VCH
              Verlag GmbH & Co. KGaA, Weinheim.

hydrogen (rH$_2$) or redox potential of the culture solution could be important
parameters in the TiO$_2$ synthesis by microorganisms. Jha *et al.*[13] reported that
many organisms can produce inorganic materials either on the intra- or
extracellular level, and that fungus and bacteria can be employed to synthesize
TiO$_2$ nanoparticles. The Baker's yeast (*Saccharomyces cerevisiae*) and *Lactobacillus* species were selected for transformation of TiO$_2$ nanoparticles. Thus,
well crystallized TiO$_2$ nanoparticles of about 13 nm and 25 nm size were
synthesized from *Lactobacillus* cells and yeast, respectively, containing both
anatase and rutile phase.

Several studies were carried out to simplify the chemical process of the
synthesis of TiO$_2$ nanomaterials.[3,14,15] Dong *et al.*[14,15] suggested a facile, energy
efficient, and environmentally friendly approach to synthesize in water visible
light-activated carbon-doped TiO$_2$ nanomaterials using low-cost Ti(SO$_4$)$_2$ and
glucose/sucrose as titania and carbon sources, respectively. After hydrothermal
treatment with or without or postheat treatment, carbon-doped TiO$_2$

nanoparticles (10–12 nm) with high BET surface area (122.52 $m^2 g^{-1}$ for glucose and 79.11 $m^2 g^{-1}$ for sucrose) were produced. The C-TiO$_2$ nanoparticles prepared using glucose or sucrose decomposed gaseous toluene efficiently under visible light illumination. Han *et al.*[3] developed visible light-activated sulfur-doped TiO$_2$ films for the degradation of an emerging water contaminant, cyanotoxin microcystin-LR (MC-LR). The use of sulfuric acid as a sulfur source, as well as an in situ water forming agent in titania solution, simplified the chemical process and stabilized the titania solution for the synthesis of nanostructured sulfur-doped TiO$_2$ films. The sulfur content, BET surface area, and photocatalytic activity of of S-TiO$_2$ for the degradation of MC-LR were significantly influenced by the calcination temperature. The primary particle size was increased from 3 nm to 12 nm when calcination temperature increased from 350 °C to 500 °C. Otherwise, the sulfur content, BET surface area, and photocatalytic activity of S-TiO$_2$ materials decreased by increasing the calcination temperature. The best photocatalytic activity of the degradation of MC-LR corresponds to S-TiO$_2$ materials with sulfur content and BET surface area of 4.1% and 179 $m^2 g^{-1}$, respectively.

5.1.2 Other Semiconductors

Other metal oxides nanoparticles, such as ZnO and CdS, have been employed as photocatalysts for environmental remediation, in particular for the removal of recalcitrant pollutants in water and air. The introduction of environmentally friendly methodologies to establish effective synthesis of nanoparticles without substantial harmful byproducts has also been explored for these semiconductor nanomaterials.

Biological processes have been applied as green chemistry methods to synthesize nanostructured materials. Biomolecules have unique features including high water solubility, high surface area (from swelling) that allows them to act as efficient templates and presence of several functional groups that permit their use as stabilizers. ZnO-based hollow microspheres can be prepared through a low temperature approach with biomolecules. In particular, nanocomposite ZnO microspheres decorated with Ag nanoparticles were prepared using sodium alginate through a facile, one-pot hydrothermal method.[16] The proposed synthesis mechanism indicates that a negatively charged carboxylic group from sodium alginate binds to a Zn^{2+} precursor, which is then hydrolyzed by ammonia, and forms ZnO after hydrothermal treatment. Enhanced photocatalytic and disinfection efficiency were obtained at the optimum Ag loading. A Ag/ZnO composite was also obtained by hydrothermal synthesis using the bovine serum albumin. The as-formed nanoparticles showed hierarchical micro/nano structure and enhanced photocatalytic activity towards the decolorization of Orange G dye under UV light.[17] Another template-assisted green approach for the synthesis of ZnO hollow spheres was proposed by Patrinoiu *et al.*,[18] who used a starch-derived composite precursor to form hollow spheres followed by impregnation and heat treatment. The synthesis was performed in aqueous solution without any hazardous reagents and the main products during calcination were water and CO_2. A similar strategy with other biomolecules was applied in terms of bonding of Zn with functional groups present in the starch-based spheres. The degradation of phenol under UV and visible light was carried out and showed the enhancement on the photocatalytic efficiency compared to a commercially available ZnO photo-catalyst[18] reference. The use of various biological entities in the production of nanoparticles has also been investigated.[19] CdS is a metal sulfide that can be used as a photocatalyst in aqueous solutions. Different variety of yeast and fungus were employed in the biosynthesis of CdS nanocrystals from cadmium salts.[19] A microwave-assisted method was developed to decrease the reaction time and energy consumption on the production of CdS nanocrystals,[20] where thioglycolic acid (TA) was employed as capping agent and $CdSO_4$ and $Na_2S_2O_3$ were used as precursors. The mechanism of this method is based on the dissociation of $Na_2S_2O_3$ after microwave irradiation for a control release of S^{2-} ions along with the capping effect of TA to produce CdS. Modifications on the irradiation time can lead to crystal sizes between 2.7 and 3.7 nm that correspond to different bandgap values (3.4–2.9 eV).[20]

5.1.3   Metal and Metal Oxides Nanoparticles

Green chemistry or sustainable chemistry is based on processes that minimize the use and generation of hazardous substances.[21] Recently, there has been renewed interest in using green chemistry principles to synthesize metal nano-particles. For their production green chemistry gives emphasis (i) in the choice of solvent, (ii) in the reducing agent employed, and (iii) in the capping agent (or dispersing agent) used, *etc.* Within this framework, there has also been growing interest in identifying environmentally friendly reagents that are multi-functional, *i.e.* those which can serve the role of a reducing as well as a capping agent. Greener production of nanomaterials can find applications in catalysis and organic transformations, with a special emphasis on environmental remediation.

For example, the coffee/tea extracts (polyphenols) function as both reducing and capping agents for Ag and Pd metals.[22] In addition to its high water solubility, biodegradability and low toxicity compared to other reducing agents sodium borohydride ($NaBH_4$), sodium citrate, hydroxylamine hydrochloride, *etc.*, caffeine is the most widely used, behaviorally active drug in the world. By using natural available resources like tea leaf extract, it is possible to prepare nanospheres (see Figure 5.3). It is also possible to control the shape of the particles by changing the tea extract concentration/source.[22] Similarly, Vitamin B1[23] was used as the reducing and capping agent for the reduction of noble salts.[23] The method is one-pot, greener in nature, and bulk quantities of nanoballs of aligned nanobelts and nanoplates of the noble metal palladium in water can be synthesized without the need of any external capping or surfactant agents and/or large amounts of insoluble templates commonly deployed.

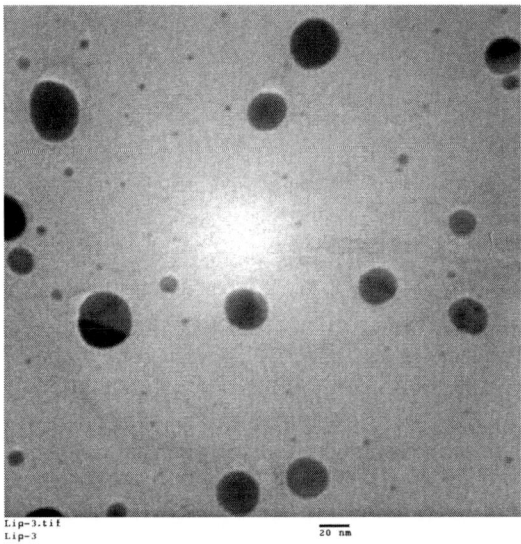

Lip-3.tif
Lip-3   20 nm

Figure 5.3   TEM image of silver nanoparticles produced using lipton tea extract.

Further, Vitamin C[24] was used to fabricate the novel core–shell metal nanocrystals.

Metal salts of Cu and Fe were reduced using ascorbic acid in solution and then by simultaneous addition of noble salts yielding core–shell type nanostructures. This method is benign and uses a naturally available antioxidant.

A green, single-step synthesis of iron nanoparticles using tea (*Camellia sinensis*) polyphenols is described that uses no additional surfactants/polymers as capping or reducing agents. The expedient reaction between polyphenols and ferric nitrate occurs within a few minutes at room temperature and is indicated by color changes from pale yellow to dark greenish/black in the formation of iron nanoparticles.[25] The obtained nanoparticles were utilized to catalyze hydrogen peroxide for treatment of organic contamination and results were compared with Fe-EDTA and Fe-EDDS. Bromothymol blue, a commonly deployed pH indicator, was used as a model contaminant for free radical reactions (catalyzed decomposition of hydrogen peroxide to form the hydroxyl radical), due to its stability in the presence of H_2O_2 and its absorbance in the visible range at pH 6. The concentration of bromothymol blue was conveniently monitored using ultraviolet-visible (UV-Vis) spectroscopy during treatment with non sources used in the study and H_2O_2. Various concentrations of iron were tested to allow for the determination of initial rate constants for the different iron sources.

Following a trans-metallic reaction between copper and silver, dendritic Ag structures were produced using a copper grid without the use of any other chemical reducing agents.[26] These results demonstrate a facile, aqueous, room-temperature synthesis of a range of noble metal nano- and mesostructures

Figure 5.4   TEM image of silver dendrites produced using copper TEM grid.

(see Figure 5.4) that have widespread technological potential in the design and development of next-generation fuel cells, catalysts, and antimicrobial coatings.

Bulk production of silver (Ag) nanorods was achieved using polyethylene glycol (PEG) under microwave irradiation conditions.[27] The formation of nanorods depended upon the concentration of PEG used in the reaction with Ag salt.

Gold (Au) nanostructures were synthesized using naturally occurring bio-degradable plant surfactants such as VeruSOL-3 (mixture of D-limonene and plant-based surfactants), VeruSOL-10, VeruSOL-11, and VeruSOL-12 (individual plant-based surfactants derived from coconut and castor oils) without any special reducing agent/capping agents. This green method uses water as a benign solvent and surfactant/plant extract as a reducing agent.[28] The morphology of the Au crystallites is a function of the concentration of the corresponding gold salt and thus, different shapes (spheres, prisms, and hexagonal) can be formed with sizes varying from the nanometer to micrometer scale level, depending on the plant extract used for preparation (see Figure 5.5).

Several green approaches were developed to prepare silver coated goethite and the synthesized goethites were tested for arsenic removal and antibacterial activity.[29] Bayoxide® E33 (E-33) is a widely used commercial material for arsenic adsorption composed of a mixture of iron oxyhydroxide and oxides. Primarily used to remove arsenic from water or wastewater stream, this non-magnetic, iron oxyhydroxide/oxide is incorporated in fixed bed pressure filters that generally lack multifunctionality. Similar to other adsorptive media, it is subject to surface fouling by precipitates, including iron and manganese, and biofilms that can create diffusion limitations. Surface modification of E-33 with silver nanoparticles to enhance the material properties could add multi-functionality and be beneficial, provided that the unique arsenic adsorption

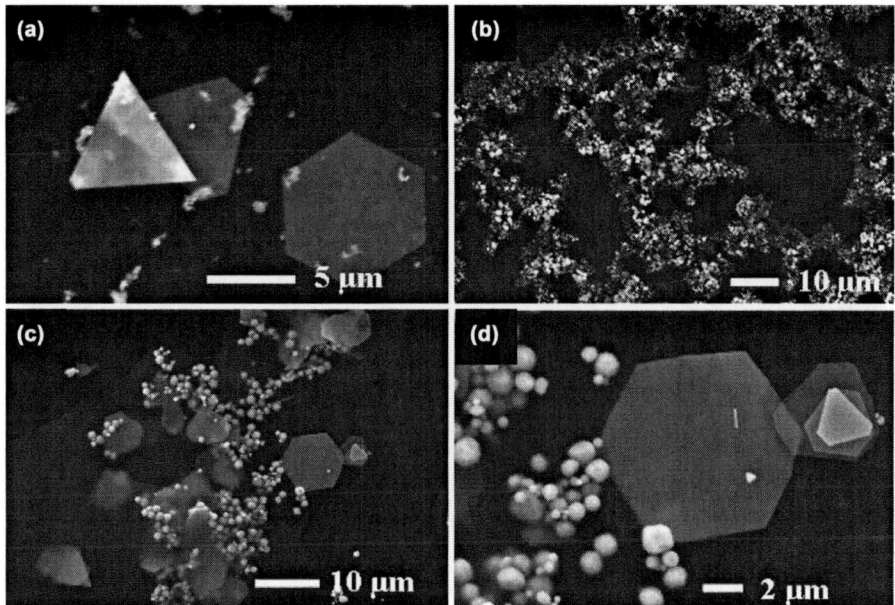

Figure 5.5 SEM images of gold particles synthesized using different surfactants.

properties are not compromised. Commercially available arsenic adsorption media (E-33) was combined to create core–shell composites, *i.e.* silver coated goethite (in magnetic and nonmagnetic forms), that are capable of removing arsenic and present additional antibacterial, and magnetic properties.[29]

Iron oxide (hematite, α-Fe$_2$O$_3$) is an abundant material found in nature with photocatalytic properties due to the capacity to absorb in the visible region (2.1 eV).[30] Fe$_2$O$_3$ nanoparticles have been synthesized using starch as the capping agent in a one pot green procedure.[31] Ferric chloride, in combination with an aqueous starch solution, can form ferric hydroxide after precipitation with an amine. After centrifugation and calcination, the resulting nanoparticles were monodispersed and exhibited well-defined morphology and narrow size distribution due to the structuring effect of the starch. Nevertheless, changes in pH can lead to different size and shape ranging from spherical to tubular.[31] A different green synthesis approach to prepare hematite is the use of hydrogen peroxide (H$_2$O$_2$), in a photochemical reaction, under solar light.[30] Iron acetate was used as precursor and irradiated under simulated solar light in the presence of H$_2$O$_2$. After aging, the solution was dried and sintered at high temperature (500 °C) to obtain pure phase α-Fe$_2$O$_3$ submicrometer range ($\sim$200 nm) particles.[30]

Monodisperse MFe$_2$O$_4$ (M = Ni, Co, Mn) and γ-Fe$_2$O$_3$ nanoparticles synthesis at a water – toluene interface under microwave hydrothermal conditions as well as conventional conditions using readily available nitrate or chloride salts and oleic acid as the dispersing agent has also been explored.[32]

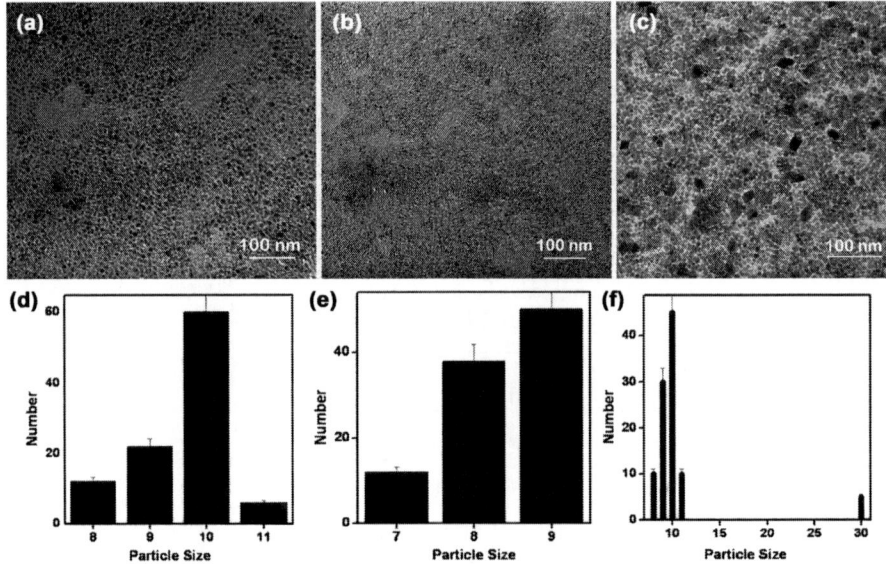

Figure 5.6   TEM micrographs of (a) $CoFe_2O_4$ and (b) $MnFe_2O_4$ particles and
(c) γ-Fe_2O_3. (d–f) Corresponding particle size distributions calculated over
100 particles.

The resultant particles present a narrow size distribution ranging from 5 to
10 nm (see Figure 5.6). The procedure is atom-efficient and can be scaled up
to a level of more than a few grams without affecting the size and shape.
The modulation of water solubility profile of the particles was accomplished
via functionalization using organic moieties. The particles were found to
be superparamagnetic with negligible coercivities equal to 11–12 Oe at
room temperature. The saturation magnetization values were in the range of
20–25 emu/g.

Nanoferrite dendritic structures (see Figure 5.7) with micro-pine morphology
have been synthesized under microwave irradiation conditions without using
any reducing or capping reagent.[33] The nanoferrites were then functionalized
with amine groups and attached with Pd metal, which catalyzes various C–C
coupling and hydrogenation reactions with high yields. The ease of recovery
and high efficiency, combined with the intrinsic stability of this material, make
this method robust and economic.

Self propagation combustion synthesis of iron oxide/iron coated carbons
such as activated carbon, anthracite, cellulose fiber, and silica was explored.[34]
The reactions were carried out in alumina crucibles using a Panasonic kitchen
microwave with inverter technology, and the reaction process was completed
within a few minutes. The method used no additional fuel and nitrate, which is
present in the precursor itself, to drive the reaction. The size of the iron oxide/
iron nanoparticle-coated activated carbon, anthracite, cellulose fiber, and silica
samples were found to be in the 50–400 nm range. The XRD pattern indicated

Figure 5.7 TEM image of dendritic α-Fe$_2$O$_3$.

the presence of iron oxide/iron nanoparticles. The iron oxide/iron nanoparticles were crystallized into cubic symmetry and confirmed the existence of four major phases: γ-Fe$_2$O$_3$, α-Fe$_2$O$_3$, Fe$_3$O$_4$, and Fe. These iron-coated activated carbon, anthracite, cellulose fiber, and silica samples were tested for arsenic removal (adsorption) through batch experiments, revealing that some samples had significant arsenic adsorption capacity.

5.1.4 Metallic and Bimetallic Nanoparticles

The formation of nanoparticles involves two important processes: (1) nucleation and (2) growth.[35] The initial nucleation process occurs through homogenous, heterogeneous or secondary nucleation in a supersaturated solution. The nucleation process is followed by growth of the nuclei *via* molecular addition. Molecular addition causes the concentration of the target element to drop below the critical point, resulting in an end to the nucleation process. However, the growth process continues until the concentration of the precipitates reaches equilibrium. Nanoparticles of uniform size begin to form due to the difference in free energy (ΔG) of the driving force within nanoparticles of different size. As smaller particles have larger free energy relative to larger particles, this difference in free energies results in faster growth for the smaller particles. It should be noted that under these conditions, the reaction should be stopped or more reactant should be added to the system in order to retrieve the depletion of its concentration due to particle growth. Alternatively, Ostwald ripening occurs, whereby the larger particles continue growing, as a result of generation of smaller particles. Formation of smaller particles is kinetically favored while the formation of larger particles is thermodynamically favored and the process continues until the smaller particles completely dissolve in the solvent.

In 1992 Bönneman[36] expressed that the synthesis of metal nanoparticles requires at least three parameters: a metal precursor, a stabilizer, and a reducing agent. Elemental metals, their inorganic salts, and inorganic complexes are examples of metal precursors that are used for metallic nanoparticle synthesis. The role of the stabilizer is to cap the particle surface, control the particle size and shape, and prevent or reduce particle aggregation. Metal nanoparticles of desired size, shape, and morphology can be obtained by carefully choosing and manipulating these three substrates.[37] The synthesis method of metal nanoparticles is categorized by the media in which the nanoparticles are formed. According to this classification, there are two major methods for synthesis of metallic nanoparticles: solution synthesis or a *water-in-oil emulsion* synthesis.[38-40]

Solution synthesis involves the synthesis of metal nanoparticles by chemical reduction of the metal salt in solution.[41] Generally, stirring the solution after addition of the reducing agent is needed to ensure even distribution of the reducing agent throughout the solution. Furthermore, to prevent possible oxidation of the synthesized nanoparticles, excess amount of reducing agent is added into the reaction solution or the reaction may be carried out under oxygen free conditions. Careful control of the reaction conditions, concentrations of the reactants, and solvent volume are required to control particle size and size distribution.

The *water-in-oil emulsion* method provides nanoparticles with improved size-control.[42] In this method, an aqueous solution of metal ions is added into an oil phase, and a water-soluble cosolvent is added into the formed emulsion. The reducing agent is usually added into the system as the last step in the reaction. Metal nanoparticles synthesized by this method are well-capped, show a narrow size distribution, and have fewer tendencies to aggregate.

Bimetallic alloy nanoparticles (*i.e.* metallic entities comprising atoms of two different metallic elements) have been found to be quite effective toward dehalogenation processes, relative to their single-metal counterparts.[43-50] The most common methods for the preparation of bimetallic nanoparticle alloys include (1) coreduction, (2) successive reduction, and (3) thermal decomposition.

Co-reduction involves the simultaneous reduction of the two corresponding metal ions in the presence of a stabilizing ligand. Often the particles are produced by chemical reduction, however, electrochemical reduction can also be employed. Depending on the composition of the metallic or bimetallic nanoparticles, they may be susceptible to oxidation, and therefore Schlenk line techniques should be employed during the synthesis procedure and once prepared, if needed, the nanoparticles should be stored under nitrogen to avoid oxidation.

Successive reduction is a suitable procedure for synthesizing bimetallic alloy nanoparticles where the metal ions have a large difference in reduction potential. The metal ion with the more negative reduction potential is reduced first, followed by the second metal.

Thermal decomposition involves reacting the corresponding metal precursors, of varied ratios, at increased temperature in the presence of the stabilizing ligands. The ligands coordinated to the metal precursors dissociate, leading to the nucleation and growth of the metal nanoparticles.

The size, shape, and morphology of nanoscale materials have a major influence on their chemical and physical properties and their reactivity. Thus, characterization of nanoparticles is important toward understanding their reactivity with various substrates including environmental pollutants. Typically, the characterization of metallic and bimetallic nanoparticles involve microscopic, X-ray, gas adsorption and spectroscopic techniques including, transmission electron microscopy (TEM), scanning electron microscopy (SEM), atomic force microscopy (AFM),[51] Brunauer-Emmette-Teller method (BET),[52,53] X-ray photoelectron spectroscopy (XPS),[54] X-ray diffraction (XRD),[55] Raman spectroscopy,[56] energy–dispersive X-ray spectroscopy (EDX) and extended X-ray absorption fine structure (EXAFS).[57]

5.2 Environmental Applications of Nanomaterials

5.2.1 Photocatalytic Degradation of Organic Pollutants in Air, Water, and Soil

The high effectiveness of TiO_2 to generate extremely reactive chemical species (*e.g.*, hydroxyl radicals) upon UV light illumination along with its environmentally benign properties (*i.e.* nontoxicity, absence of dissolution in water, photostability) and relatively low cost, rendered TiO_2 a key material for the complete destruction of recalcitrant organic pollutants in water and air that make it ideally suited for remediation of heavily polluted industrial environments or remote areas lacking adequate power supplies.[3,56,58–76] Lab-scale photocatalytic experiments are performed under UV-A, solar, and visible light in the presence of aqueous solutions of the target pollutants. As the efficiency of the photocatalytic process is heavily dependent on the chemical nature of the pollutants being treated, which relates to the stability of radicals that were formed during the degradation process,[77,78] model herbicides, *i.e.* atrazine, propanil, and diuron used worldwide are the target compounds in optimization of photocatalytic efficiency of the best performing TiO_2 nanocatalysts to reduce its presence down to 0.1 ppb. Analytical methods including LC-MS/MS, LC-HRMS and NMR are used for the process monitoring and the identification of degradation products, while the mineralization extent is assessed by total organic carbon (TOC) and inorganic carbon (IC) measurements. Quantitative analysis is based on LC-MS/MS in multireaction monitoring mode and internal standard calibration that provides confirmatory analysis. For toxicity, different bioassays are used and ion chromatography is employed to determine small organic acids, inorganic end products, and degradation pathways.

Recent literature reports on the evaluation of photocatalytic performance and better understanding of the photocatalytic process. In fact, comparative evaluation of the TiO_2 nanomaterials is performed by the measurement of the quantum yield of reaction products for model organic compounds [Methylene

Blue (MB), Acid Orange 7 (AO7), Rhodamine B (RhB), 4-Chlorphenol, *etc.*] and thus a standard protocol for the evaluation of materials performance under visible light is established. The outperforming photocatalytic materials are assessed by a multiactivity test according to the protocol proposed for water treatment.[79] This assessment is based on the examination of model pollutants having completely different molecular structure such as: phenol, dichloroacetic acid, tetramethyl ammonium, and trichloroethylene, which represent the aromatic, anionic, cationic, and chlorohydrocarbon compounds, respectively.

The photocatalytic activity and its correlation with the TiO_2 nanomaterials selectivity toward the formation of distinct intermediates is also explored by using suitable organic probe compounds that are prone to oxidative, reductive, or both oxidative/reductive degradation pathways (*e.g.* phenols, 4-chlorophenol).[80] The ensuing activity-selectivity relationship is evaluated with respect to the structural and physicochemical properties and most importantly the surface structure of the TiO_2 nanomaterials toward the deep understanding of the materials reactivity and its judicious manipulation for photocatalytic applications.

Important research efforts aim at investigating the mechanism of visible light activated photocatalysis. The adjustment of the materials bulk and surface properties gives evidence in the photocatalytic mechanism,[81] by evaluating the oxidative chemistry of the modified TiO_2, and distinguishing the hydroxyl-type chemistry from the single-electron transfers from the organic molecule to the photogenerated holes.[82] Such mechanistic models are further supported by physicochemical (light dependent EPR spectroscopy and spin trapping) as well as photoelectrochemical studies.[83]

Apart from testing the materials under sequel photocatalytic cycles, chemical stability (after exposure of the films in acidic or basic environments) as well as mechanical stability (possible formation of cracks) and the influence of film support,[84] which can be detrimental for the long term use of the nanomaterials and the films, are also studied. This is accompanied by tests under extreme pollution conditions that may cause permanent fouling of the films and accelerated weathering tests, simulating the exterior conditions (sun light, humidity, rain).

To enhance the photocatalytic efficiency against organic pollutants (in air, water and soil), the behaviour of new visible light-activated TiO_2-based photocatalysts are being investigated. The destruction of selected recalcitrant pollutants (pharmaceuticals and cyanotoxins) in aquatic systems[85] mainly involves target water pollutants ranging from classical water contaminants such us phenols, herbicides, pesticides, and azo-dyes to the extremely hazardous cyanobacterial toxins and related contaminants (*e.g.* microcystin-LR, geosmin, 2-methylisoborneol) as well as emerging endocrine disrupting compounds (*e.g.* 17β-estradiol, and bisphenol[4,86–90]).

Harmful air agents, *e.g.* volatile organic compounds (VOCs), aldehydes, *etc.* are released into the environment during industrial activities. Photocatalytic degradation leads to the decomposition of pollutants towards nonharmful compounds in one step at ambient temperature. There are two main domains of photocatalytic applications depending on the degree of air pollution: high-polluted gas from industrial processes (agents in concentrations in order of hundreds ppm) and low-contaminated air present in urban areas (order of units ppm). Acetaldehyde, *n*-hexane, and toluene are used as model air pollutants and special attention is given to the relation between surface area, adsorption, relative humidity, degradation kinetics, intermediates, and mineralization.[91–94]

Polycyclic aromatic hydrocarbons (PAHs) from anthropogenic activities are a group of major soil contaminants that are ubiquitous in the environment. PAHs, especially those with four or more rings and their metabolites, are considered as hazardous pollutants due to their toxicity, mutagenicity, and carcinogenicity and are classified as compounds with significant health risk. The photodegradation of selected PAHs on soil surfaces in the presence of new sensitized photocatalysts and their comparison with commercial TiO_2 nano-catalysts (*e.g.*, Degussa P25, Kronos) are investigated, taking into account the effects of various parameters such as photocatalyst concentration, irradiation wavelength, and soil thickness.[95,96]

Innovative photocatalysts are incorporated into highly performing photo-reactors that are adequately designed for water and air remediation from organic pollutants. The active elements of the purification devices are highly porous nanomaterials (*e.g.* carbon nanotubes) bearing on their bulk matrix effectively dispersed the photocatalytic TiO_2 nanomaterials. The reactors operate in the flow through membrane mode, forcing all of the fluid to pass through the pores of the fiber and ensuring turbulent flow that will enhance mixing and transfer rates of the organic pollutants to the surface of the embedded photocatalyst nanoparticles and nanocomposites. Typical air pollutants and aldehydes or organic solvents will apply for the device characterization and testing at different operational conditions such as pollutant concentration and gas flow rate.[97–100] Recently, a photocatalytic membrane purification device was designed and optimized to photo degrade organic contaminants existing in polluted water sources. The purpose is to integrate membrane separation and photocatalytic degradation technology in a continuous single pass water purification process. In a scheduled embodiment the reactor includes simultaneously, in vertical arrangement, one tubular ceramic nanofiltration (NF) membrane with TiO_2, undoped or doped active nanoporous layers on its internal and external surfaces. The layers are deposited by physical (*e.g.* vapor deposition) or chemical method (*e.g.* sol-gel dip coating). Moreover, photocatalytically active mixed matrices from light transparent polymer (*e.g.* polymethyl methacrylate, polysulfonic acid, alginate) in the form of asymmetric porous fibers with embedded active photocatalysts or photocatalytic nanocomposites are accommodated in the retentate side of the NF membrane to act as a pre-treatment photodegradation stage before the pollutants reach the external photocatalytically active surface of the membrane, thus considerably increasing the photodegradation efficiency.

Besides TiO_2, the use of other nanostructured metal oxides semiconductors (*e.g.*, ZnO, CdS, ZnS and α-Fe_2O_3) for environmental remediation has been widely investigated. These metal oxides can be synthesized under mild, environmentally friendly conditions through sol-gel procedures. They posses relevant photocatalytic activity and have been employed for the decontamination of water and air treatment. Decolorization of wastewater containing dye, degradation of pesticides, and volatile organic compounds has been effective when employing ZnO under UV light irradiation.[101] ZnO has also been explored for sensing considering the size-dependent emission properties in the visible region.[102] This emission is sensitive to aromatic compounds, such as chlorinated phenols, present in water. Nevertheless, an evident limitation of ZnO is its high solubility in water that could increase sensitivity in aquatic organisms. Also, long-term stability of ZnO has been questioned since it has a tendency to photo-corrode under exposure to UV light.[103] Similar to ZnO, CdS and ZnS have been employed as photocatalysts but the same effect is observed for photo-corrosion.[103]

Iron oxide (in particular hematite, α-Fe_2O_3) is stable, low cost, nontoxic, and with high resistance to photo-corrosion with the ability to absorb visible light. It has been synthesized with a wide variety of methods and applied for the removal of organic dyes, 4-nitrotoluene, phenols, and other organic contaminants, heavy metals and the oxidation of CO.[104,105] Coupled semiconductors have also been proven to have an enhanced performance in the removal of recalcitrant pollutants. For instance, mixed CdS/TiO_2, α-Fe_2O_3, or CdS deposited on ZnO improves the charge separation and enhances the photocatalytic activity due to a higher quantum yield in the catalytic reactions in the system. The degradation of trinitrotoluene and other phenolic compounds has been demonstrated.[106,107]

5.2.2 Dehalogenation using Metallic and Bimetallic Nanoparticles

5.2.2.1 *Metallic Nanoparticles*

Zero-valent iron nanoparticles have proven to be effective toward the dehalogenation of a number of organohalides. The most notable examples include treatment of lindane and atrazine, two well known pesticides,[108] trichloroethylene,[109] and hexachlorobenzene (HCB).[110]

Liu *et al.* reported hydrodechlorination of monochlorobenzene in the presence of colloidal platinum nanoparticles stabilized on polyvinylpyrrolidone (PVP).[111] Unlike the hydrodechlorination of chlorobenzene using other metallic nanoparticles which result in formation of additional environmental pollutants like benzene, aniline and naphthalene, Pt nanoparticles resulted in the complete degradation of chlorobenzene to produce aliphatic products.

Zero-valent palladium nanoparticles have proven to be effective toward the dechlorination of chlorobenzene.[112] In most cases, the colloidal nanoparticles are used in the solution phase for degradation, however this leads to separation challenges. One step forward to overcome this challenge has been accomplished by attaching the metallic nanoparticles to the surface of solids supports (*e.g.* nanoscale magnetite). Magnetite being magnetic, allows catalyst separation using a magnetic field.

5.2.2.2 Bimetallic Nanoparticles

While the use of zero-valent iron nanoparticles for remediation of organohalides has been common, a major limitation lies in the decreased reactivity due to the formation of an oxide layer on the surface of zero-valent iron nanoparticles. This layer acts as barrier and blocks the surface active sites of nanoparticles.[113] Introducing a second metal to zero-valent iron, for example, nickel, platinum, palladium, or zinc to forming bimetallic nanoparticles enhances the dehalogenation rates relative to using iron nanoparticles alone.[113–117] Most reports that utilize bimetallic nanoparticles have made use of supportive membranes to reduce the tendency of the nanoparticles toward aggregation. The supportive membranes also aid in the recyclability of the bimetallic nanoparticles from water. In order to make the nanoparticles compatible with the supportive membranes, specific stabilizers should be used; for example, carboxymethyl cellulose (CMC),[118,119] polyvinylpyrrolidone (PVP), PEG2000, and octa(di-aceticaminophenyl)silsesquioxa-ne (OAAPS).[120]

5.2.2.2.1 Iron/Palladium Bimetallic Nanoparticles. Fe/Pd particles have been effective toward the degradation of organohalides in the presence of H_2. The combination of Fe and Pd metals is advantageous because palladium serves as a metal on which hydrogen gas is adsorbed. Hydrogen gas results from the corrosion reaction of iron. The H_2 is then transformed to form H^+ or H^- and aids in the dehalogenation reaction to combine with the halogen atom or to replace the halogen atom, respectively.[121]

Successive reduction was used to synthesize iron/palladium (Fe/Pd) nanoparticles because of their position in electrochemical series. In this case, iron nanoparticles were first synthesized by solution synthesis using sodium borohydride as the reducing agent. This was followed by addition of the palladium salt to the solution containing the iron nanoparticles. Equation (1) shows formation of Fe/Pd nanoparticles by the successive reduction method.[118,122–124]

$$Fe^0 + Pd^{2+} \rightarrow Fe^{2+} + Pd^0 \tag{1}$$

In this system, the Fe component of the nanoparticle acts as the electron source, while the Pd component of the nanoparticles plays the role of the catalyst. In this case, a three step reaction occurs. As shown in equations (2 to 7), the reaction begins with the generation of hydrogen gas, followed by Fe corrosion. In the second step, the generated hydrogen gas combines with Pd and is embedded into the Pd crystal lattice.[125] The last step involves the dehalogenation reaction (equation 7).

$$Fe + 2H^+ \rightarrow Fe^{2+} + H_2 \quad (\textit{in acidic solutions}) \tag{2}$$

$$Fe + 2H_2O \rightarrow Fe^{2+} + H_2 + 2OH^- \quad (\textit{in basic solutions}) \tag{3}$$

$$Fe + RCl + H^+ \rightarrow RH + Fe^{2+} + Cl^- \qquad (4)$$

$$Pd + H_2 \rightarrow Pd.H_2 \qquad (5)$$

$$Pd + RCl \rightarrow Pd \ldots Cl \ldots R \qquad (6)$$

$$Pd.H_2 + Pd \ldots Cl \ldots R \rightarrow RH + H^+ + Cl^- + 2Pd \qquad (7)$$

Chlorinated methanes, γ-hexachlorocylohexane (Lindane),[113] poly-chlorocyclohexanes, PCBs,[121] p-NCB, chlorinated aromatics,[126,127] and chloro-acetic acids,[128] are key examples for dehalogenation using Fe/Pd bimetallic nanoparticles.

5.2.2.2.2 Iron/Nickel Bimetallic Nanoparticles. Fe/Ni bimetallic nano-particles are attractive catalysts for organohalide degradation due to the cor-rosion stability of Ni, its low cost, and availability.[115]

As shown in equation (8) addition of sodium borohydride to a solution of ferrous sulfate (FeSO$_4$) and nickel (II) chloride (NiCl$_2$) with a 4:1 Fe:Ni ratio resulted in the coreduction of the metal ions leading to the formation of Fe/Ni bimetallic nanoparticles.[129]

$$2Fe^{2+}(Ni^{2+}) + BH_4^- + 2H_2O \rightarrow 2Fe^0(Ni^0) + 2H_2 + BO_2^- + 4H^+ \qquad (8)$$

Ni is also effective when combined with Fe because it can reactivate the surface. For example, Cheng[130] and coworkers showed that physical addition of microscale Ni particles resulted in the reactivation of zero-valent Fe particles that had lost their surface activity.

Cwiertny *et al.* showed that for Fe/Ni particles to display enhanced activity, the Ni and Fe atoms need to be in electronic contact with each other within the bimetallic particle.[131] Another important factor is that having a layer of Ni nanoparticles on the iron surfaces (in a core–shell arrangement) is advanta-geous. The dehalogenation mechanism carried out by Fe/Ni nanoparticles follows a similar pathway to that of Fe/Pd nanoparticles. Fe/Ni nanoparticles have been effective toward the dehalogenation of p-NCB,[132] p-chlorophenol (p-CP),[133] Fe/Ni bimetallic nanoparticles have also been effective toward the remediation of other organohalides including pentachlorophenol,[134] mono-chlorobenzene,[135] and brominated diphenyl ethers,[136] trichloroethylene,[42] carbon tetrachloride,[137] chloroform,[138] p-NCB,[132] and brominated methanes.[129] In all reports, the Ni loading, pH, and temperature play an important role and govern the reaction efficiency. Increased reaction tem-perature and low pH have typically led to favorable results.

5.2.2.2.3 Iron/Copper Bimetallic Nanoparticles. Fe/Cu bimetallic nanoparticles have also been effective toward dehalogenation and their catalytic activity, low cost and biocompatibility (relative to Ni and Pd) make Fe/Cu bimetallic nanoparticles interesting candidates for organohalide remediation.[138,139] Cao *et al.* used Fe/Cu nanoparticles for the degradation of 1,2,4-trichlorobenzene (1,2,4-TCB).[138] The Cu atoms in the bimetallic Fe/Cu nanoparticles play the role of the catalyst. 1,2,4-TCB was dechlorinated by Fe/Cu bimetallic nanoparticles to form 1,2-dichlorobenzene. Further reaction of 1,2-dichlorobenzene with Fe/Cu nanoparticles led to its hydrodechlorination to form chlorobenzene and finally benzene. 90% of 1,2,4-TCB was dechlorinated after a 24 hr treatment with Fe/Cu bimetallic nanoparticles. Zhu *et al.* also investigated the efficiency of Fe/Cu bimetallic nanoparticles for the dechlorination of hexacholorobenzene.[139] Their results showed that 98% of hexachlorobenzene was reduced after 48 hrs of treatment leading to the formation of pentachlorobenzene, tetrachlorobenzene, trichlorobenzene and dichlorobenzene as byproducts with no selectivity toward a stepwise dechlorination process.

5.2.2.2.4 Other Bimetallic Nanoparticles. While iron-based bimetallic nanoparticles have been most commonly used for improving the dehalogenation process of organohalides relative to zero-valent iron technology, there have been recent developments in using noniron based bimetallic nanoparticles as well. Lo and coworkers synthesized palladium/tin (Pd/Sn) bimetallic nanoparticles stabilized on a resin, and demonstrated their effectiveness toward the degradation of trichloroethylene (TCE) in aqueous media.[140] Boronina *et al.* showed that tin metal is an effective electron donor for mediating organohalide dehalogenation.[140] An added advantage to using tin is its stability in ground water relative to zero-valent iron.

Silver/gold (Ag/Au) bimetallic nanoparticles have also been reported as being effective for dehalogenation processes. Using an underpotential deposition-redox placement technique, Zhou used Ag/Au bimetallic nanoparticles as the cathode in the electrochemical reduction of benzyl chloride.[123]

Platinum/palladium (Pt/Pd) bimetallic nanoparticles,[141] iron/silver (Fe-Ag)[142] and gold/platinum (Au-Pt)[143] are increasingly being investigated for dehalogenation and it is expected that with optimization of reaction conditions, such nanoparticles will prove to be more effective and more stable than iron-based bimetallic systems.

5.2.3 Photocatalysis for the Disinfection of Drinking Water

In 2008 it was estimated that 884 million people were without access to "improved sources of drinking water".[144] Many more are forced to rely on supplies that are microbiologically unsafe, resulting in a higher risk of water-borne diseases, including typhoid, hepatitis A and E, polio and cholera.[145–147] On July 28, 2010, the UN General Assembly adopted Resolution 64/292 recognizing that safe and clean drinking water and sanitation is a human right essential to the full enjoyment of life and all other human rights. It called on United Nations Member States and international organizations to offer funding, technology and other resources to help poorer countries scale up their efforts to provide clean, accessible and affordable drinking water and sanitation for everyone.[148] Household water treatment and safe storage (HWTS) is one option for improving the quality of water for consumption within the home, especially where water handling and storage is necessary and recontamination is a real risk between the point of collection and point of use. "Appropriate" household water treatment (HWT) methods include boiling, filtration, adding chlorine or bleach, and *solar disinfection.*

Sunlight is widely and freely available on Earth and the combined effects of IR, visible, and UV energy from the sun can inactivate pathogenic organisms present in water. Solar disinfection of water (SODIS) is practiced by around 5 million people world-wide. The basic protocol is to fill a UV transparent container (*e.g.* 2 L PET bottle) with water and place it in direct sunlight for 6 hours. This significantly reduces the number of viable pathogenic microorganisms in the water. There are, however, a number of parameters which affect the efficacy of the SODIS process including the solar irradiance, ambient temperature, the quality of the water to be treated (turbidity, suspended solids, *etc.*), and the nature of the contamination (as some pathogens are more resistant to SODIS than others). SODIS enhancement technologies may improve the effectiveness without increasing the cost substantially. One approach to SODIS enhancement is the use of heterogeneous photocatalysis, which may provide a relatively low cost solution to the purification of water in developing regions. Solar photocatalysis is truly a *Clean Technology* and could provide a sustainable and low cost approach to the remediation of contaminated water.

For titanium dioxide photocatalysis, the absorption of UV light results in the formation of electron hole pairs which participate in redox reactions at the surface leading to the formation of reactive oxygen species (ROS). ROS are very active, indiscriminate oxidants, especially the hydroxyl radical, and can degrade organic chemical contaminants in water and cause fatal damage to microorganisms. Matsunaga *et al.*[149] reported the first application of TiO_2 photocatalysis for the inactivation of bacteria. Since then, there have been a large number of research publications dealing with the inactivation of microorganisms including bacteria, viruses, protozoa, fungi and algae. Blake *et al.*[150] carried out an extensive review of the microorganisms reported to be inactivated by photocatalysis. McCullagh *et al.*[151] reviewed the published work on photocatalysis for the disinfection of water contaminated with pathogenic microorganisms. In 2009, Malato *et al.* published an extensive review on the decontamination and disinfection of water by solar photocatalysis[152] and, in 2010, Dalrymple *et al.* reviewed the modeling and mechanisms of photocatalytic disinfection.[153]

In most studies concerning photocatalytic disinfection, the hydroxyl radical is suggested to be the primary species responsible for microorganism inactivation. Some papers reported other reactive oxygen species, such as H_2O_2, $O_2^{\bullet-}$ to be responsible for inactivation process.[154–157] These reactive species can cause fatal damage to microorganisms by disruption of the cell membrane or by attacking DNA and RNA.[150] Other modes of action of TiO_2 photocatalysis have been proposed including damage to the respiratory system within the cells[158] or loss of fluidity and increased ion permeability in the cell membrane.[155] Many researchers have attributed cell death to lipid peroxidation of bacterial cell membrane.[154–156] The peroxidation of the unsaturated phospholipids that are contained in the bacterial cell membrane causes loss of respiratory activity[158] and/or leads to a loss of fluidity and increased ion permeability.[155] This is suggested to be the main reason for cell death. Other researchers suggested that the cell membrane damage can open the way for further oxidative attack of internal cellular components, ultimately resulting in cell death.[159]

Alrousan *et al.*[160] reported on the photocatalytic inactivation of *E. coli* in surface water using immobilized nanoparticle TiO_2 films. It was found that the rate of photocatalytic inactivation of *E. coli* was more efficient with UVA-TiO_2 than direct photolytic inactivation with UVA alone, both for distilled water and real surface water. Bacterial cells have been described as a rather easy target for disinfectants, with bacterial spores and protozoa suggested as more robust target organisms. *Clostridium perfringens* spores have been reported to be chlorine resistant at levels used in potable water supplies. Dunlop *et al.* reported on the photocatalytic inactivation of *Clostridium perfringens* spores on TiO_2 electrodes.[161] The rate of photocatalytic inactivation of *E. coli* was found to be one order of magnitude greater than that of *C. perfringens* spores, demonstrating the greater resistance of the spores to environmental stress. In this work, it was shown that the application of an external electrical bias (electrochemically assisted photocatalysis) significantly increased the rate of photocatalytic disinfection of *C. perfringens* spores.

Cryptosporidium species are waterborne, protozoan parasites that infect a wide range of vertebrates. The life cycle involves the production of an encysted stage (oocyst) which is discharged in the faeces of their host. The disease, Cryptosporidiosis, in humans usually results in self-limited, watery diarrhoea but has far more devastating effects on immune-compromised patients (*e.g.* AIDS patients) and can be life threatening as a result of dehydration caused by chronic diarrhoea. Owing to their tough outer walls, the oocysts are highly resistant to disinfection and can survive for several months in standing water. *Cryptosporidium* oocysts therefore present as an excellent challenge for disinfection technologies. Sunnotel *et al.* reported on the photocatalytic inactivation of *Cryptosporidium parvum* oocysts on nanostructured titanium dioxide films.[162] The photocatalytic inactivation of the oocysts was shown to occur in Ringers buffer solution (78.4% after 180 min) and surface water (73.7% after 180 min). Scanning electron microscopy (SEM) confirmed cleavage at the suture line of oocyst cell walls, revealing large numbers of empty (ghost) cells after exposure to photocatalytic treatment (see Figure 5.8). Importantly, no significant inactivation was observed in the oocysts exposed to UVA radiation alone.

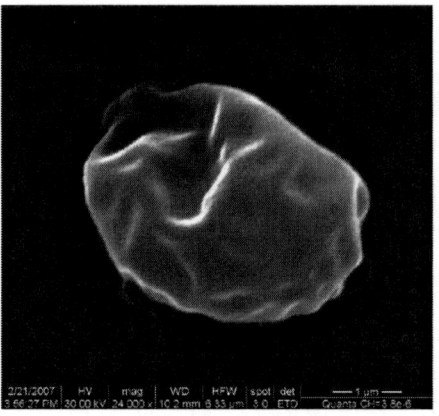 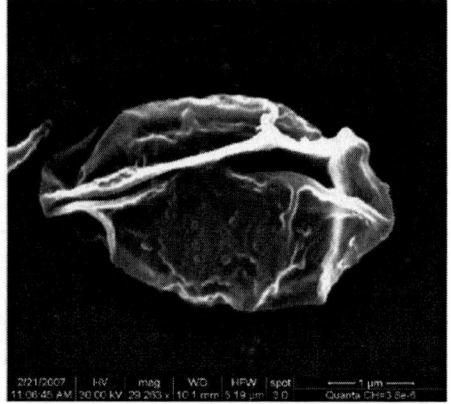

Figure 5.8   Showing SEM images showing intact oocyst (left) and empty oocyst shell
following photocatalytic treatment (right).
Reproduced with permission from ref. 162, copyright IWA Publishing 2010.

Gelover *et al.* studied the small scale batch disinfection of spring water
naturally polluted with coliform bacteria in plastic bottles with and without
TiO_2.[163] TiO_2 was coated onto small Pyrex-glass cylinders, using a sol-gel
method, and these were placed inside each bottle. It was found that photocatalytically enhanced SODIS was by far more effective than SODIS alone for
the inactivation of both the total coliforms and the faecal coliforms. They
tested bacterial re-growth following treatment and found that re-growth was
observed with SODIS alone, but not with photocatalytically enhanced SODIS.
This is an important finding, as bacteria have repair mechanisms which allow
recovery following stress, and demonstrates the differences in the kill
mechanisms involved.

With respect to larger scale systems, Fernandez *et al.* reported on the pilot
scale photocatalytic disinfection of water under real sun conditions using a
photoreactor with compound parabolic collectors (CPC).[164] The experiments
were carried out under sunlight at the Plataforma Solar de Almeria in southern
Spain and the photoreactor volume was 5.4 L and the total plant volume 11 L.
Experiments were performed with suspended TiO_2 and with supported TiO_2. In
both cases, *Degussa P25* was employed as the photocatalyst. For the immobilized study, the catalyst was immobilized on glass fiber using SiO_2 as an
inorganic binder. *E. coli* was used as the model microorganism suspended in
distilled water. It was found that the photocatalytic suspension reactor was
most efficient, followed by the immobilized photocatalytic reactor, with the
solar irradiation alone (no TiO_2) being the least efficient. Further work
undertaken by the group at Plataforma Solar de Almeria (PSA) investigated the
effect of UV intensity and dose on SODIS and photocatalytic SODIS of bacteria and fungi.[165] The aim of the work was to study the dependence on solar
irradiation conditions under natural sunlight. This dependency was evaluated
for solar photocatalysis with TiO_2 and solar-only disinfection of three

microorganisms, a pure *E. coli* K-12 culture and two wild strains of the *Fusarium* genus, *F. solani* and *F. anthophilum*. Photocatalytic disinfection experiments were carried out with TiO_2 supported on a paper matrix around concentric tubes with CPCs or with TiO_2 as slurry in bottle reactors, under natural solar irradiation at PSA. The experiments were performed with different illuminated reactor surfaces, in different seasons of the year, and under changing weather conditions (*i.e.*, cloudy and sunny days). All results showed that once the minimum solar dose had been received, the photocatalytic disinfection efficacy was not particularly enhanced by any further increase. The solar-only disinfection turned out to be more susceptible to changes in solar irradiation, and therefore, only took place at higher irradiation intensities.

For application in remote locations and developing regions, the treatment system must be robust, noncomplex, and require only low-level maintenance. Therefore, photocatalyst regeneration stages are undesirable. There is need for more research to study the longevity of the photocatalyst under real working conditions. To reduce the complexity of the treatment system, immobilized photocatalyst systems are preferred; however; catalyst stripping may be a problem if the immobilization protocol does not produce a robust hard wearing coating. Also, catalyst fouling by inorganic species present in the water can lead to a reduction in the photocatalytic efficiency over time.

The overall efficiency of TiO_2 under natural sunlight is limited to the UV-driven activity (for anatase $\lambda \leq 387$ nm), accounting only to ca. 5% of the incoming solar energy on the Earth's surface. Therefore, there has been a substantial amount of research effort towards shifting of the absorption spectrum of TiO_2 towards the visible region of the electromagnetic spectrum. According to the literature, one of the more promising approaches to achieve visible light activity is doping with nonmetal elements including N and S. Since Asahi *et al.*[166] reported the visible-light photo-activity of TiO_2 with nitrogen doping, many groups have demonstrated that anion doping of TiO_2 extends the optical absorbance of TiO_2 into the visible-light region. However, the number of publications concerning the photocatalytic activity of these materials for the inactivation of microorganisms is limited. Li *et al.*[167] based at the University of Illinois, reported on the inactivation of the MS2 phage under visible light irradiation using a palladium-modified nitrogen-doped titanium oxide (TiON/PdO) photocatalytic fiber, synthesized on a mesoporous activated carbon fiber template by a sol–gel process. Dark adsorption led to virus removal and subsequent visible light illumination resulted in additional virus removal of 94.5–98.2% within 1 h of additional contact time. By combining adsorption and visible-light photocatalysis, TiON/PdO fibers reached final virus removal rates of 99.75–99.94%. EPR measurements confirmed the production of $^{\bullet}OH$ radicals by TiON/PdO under visible light illumination. Wu *et al.*[168] from the University of Illinois also reported on the visible-light-induced photocatalytic inactivation of bacteria by composite photocatalysts of palladium oxide and nitrogen-doped TiO_2. The PdO/TiON catalysts were tested for visible-light-activated photocatalysis using gram-negative organisms *i.e.*, *E. coli* and *Pseudomonas aeruginosa*, and a gram-positive organism *Staphylococcus aureus*. The

PdO/TiON photocatalysts had a much better visible photocatalytic activity than either palladium-doped (PdO/TiO$_2$) or nitrogen-doped titanium oxide (TiON). The photocatalytic reactor was rather basic, utilizing a petri dish, which was stirred periodically. The light source was a metal halogen desk lamp with a low UV output. While these photocatalysts show promise for truly "visible" light activity, there was no comparison with undoped TiO$_2$ for photocatalytic activity under solar simulated light (which has around 5% UV).

In many cases, the UV activity of undoped TiO$_2$ is much greater than the visible light activity of the doped material. Therefore, for solar applications, the photocatalysts should be tested under simulated solar irradiation or under real sun conditions. Along these lines, Rengifo-Herrera and Pulgarin recently reported on the photocatalytic activity of N, S codoped and N-doped commercial anatase (Tayca TKP 102) TiO$_2$ powders towards phenol oxidation and *E. coli* inactivation under simulated solar light irradiation.[169] However, these novel materials did not present an enhancement for the photocatalytic degradation of phenol or the photocatalytic inactivation of *E. coli* under simulated solar light, as compared to *Degussa P-25* (undoped TiO$_2$). They suggest that while the N or N, S codoped TiO$_2$ may show a visible light response, the localized states responsible for the visible light absorption do not play an important role in the photocatalytic activity (see Figure 5.9 for proposed mechanism). More research is required to determine if visible light active materials can deliver an increase in the efficiency of photocatalysis under solar irradiation.

Semiconductor photocatalysis has been shown to be effective for the inactivation of a wide range of microorganisms at lab scale and under real sun conditions for both small scale and large scale applications. Nevertheless, there are a number of issues to be addressed before photocatalytically enhanced solar disinfection can be effectively deployed in developing regions. These include improvements in photoreactor design and the assessment of photocatalyst

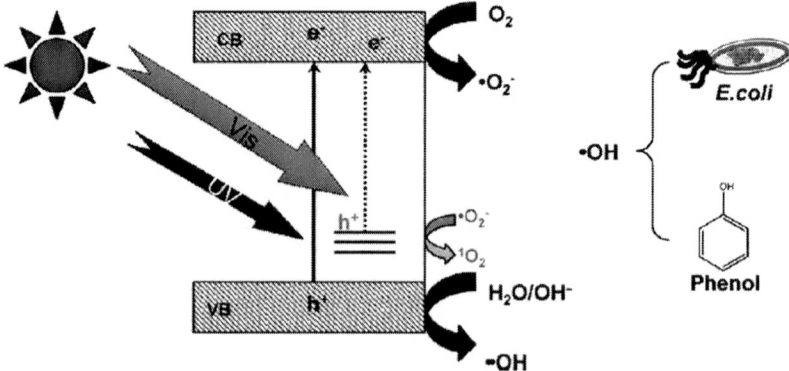

Figure 5.9   UV and visible light photocatalytic oxidation of phenol and inactivation of *E. coli* using nonmetal-doped TiO$_2$.
Reproduced with permission from ref. 169, copyright Elsevier 2010.

longevity under real operating conditions. Future developments in relation to visible light active photocatalytic materials may lead to more efficient solar photocatalysis for the disinfection of water.

5.3 Immobilization of Nanoparticles for Sustainable Environmental Applications

5.3.1 Need for Particle Immobilization

Metal and metal oxide nanoparticles, including titanium oxide, zinc oxide, silver, iron, gold, cadmium sulfide, and bimetallic nanoparticles (Fe/Pd, Fe/Cu and Fe/Ni), show their own reactivity for decomposition of organic contaminants in the environment. When practical reaction conditions relevant in natural systems, where technologies employing the nanoparticles are ultimately implemented, are considered, the remediation technologies pose a challenging question on sustainable environmental applications, *i.e.*, can we apply the metal nanoparticles directly to contaminated sites for water and wastewater treatment and groundwater remediation?[170] TiO_2 nanoparticles show a high photocatalytic oxidation capability to decompose organic chemicals in the presence of ultraviolet (UV) radiation. Their suspension is often utilized to demonstrate the fast degradation kinetics of organic contaminants in water since the nanoparticles are characterized by large catalytic surface area and low mass transfer limitations. However, the suspension system requires filtering of the effluent to remove TiO_2 nanoparticles before the discharge of the treated water.[171] The postfiltration process is tedious and costly. Meanwhile, zerovalent iron (ZVI) nanoparticles exhibit a unique reduction capability to electrochemically dechlorinate chlorinated chemicals, particularly trichloroethylene and perchloroethylene widely found in groundwater. In most cases, ZVI nanoparticles tend to aggregate to form a large cluster, significantly reducing their surface area and reactivity.[172] In addition, when they are deployed in actual contaminated groundwater, the delivery and dispersion of ZVI colloidal particles is a critical hurdle. As a result, it is more demanded to immobilize nanoparticles onto a stable substrate for their sustainable and practical environmental applications.

5.3.2 Goals and Strategies of Particle Immobilization

When nanoparticles are immobilized onto a variety of substrates, the following aspects should be addressed: i) mechanical stability (partial loss of nanoparticles by attrition diminishes the purpose of immobilization), ii) chemical stability (leaching of metal ions to the aqueous phase creates another environmental problem), iii) uniform dispersion (nanoparticles should be uniformly distributed onto the surface of a substrate to maximize their reactivity), and iv) metal content control (metal content should be controlled to comply with the application goal). It is true that an immobilized system generally exhibits a low reactivity due to the limited loading of metal particles onto a supporting material, restricted activation of metal particles present mostly at the near surface, and possible mass transfer limitations.[173] However, this drawback can be significantly overcome by introducing novel nanotechnological materials processing and synthesis routes.[173,174]

Substrates for immobilizing metal nanoparticles vary, including secondary metal particles, glass/metal plates, and organic and inorganic porous materials.[175–179] Metal nanoparticles can be immobilized onto secondary metal bulk particles so that their removal is much easier later by filtration. They can also be incorporated into a ferromagnetic material, which can be removed by magnetic-based separations.[175] Nanoparticles can be immobilized onto glass and stainless steel substrates to form a thin/thick film, which can be utilized in immobilized film reactors.[176,177] Most importantly, nanoparticles can be integrated into the surface of porous membranes to combine the physical separation function of the membranes with the unique chemical reactivity of the nanoparticles[176,179] and may constitute the active elements of very efficient water purification devices.[99,100] They can also be impregnated into the porous structure of zeolite and activated carbon to utilize the organic adsorption capability of the substrates.[178] An incipient wetness impregnation method that is followed by heat treatment is one of the simplest and popular methods for the immobilization of a variety of nanoparticles. During the thermal treatment, metal-support interactions occur and thus their mechanical and chemical stability is tunable. Chemical binders can be added to enhance the mechanical strength of the bindings between metal particles and substrates. In more advanced techniques, electron-beam-induced and UV-assisted immobilization of nanoparticles onto a substrate has been also proposed to enhance the selective immobilization of target metal nanoparticles.[180,181]

5.3.3 Application Examples of Immobilized Systems using Nanoparticles

There have been significant advances in the development of immobilized/incorporated metal nanoparticles onto/into substrates. Cobalt oxide nanoparticles have been proposed to effectively activate peroxymonosulfate for the heterogeneous generation of sulfate radicals, which can oxidatively decompose organic contaminants in water. A novel approach of using Co oxide attached to Fe oxide was proposed.[175] As shown in Figure 5.10, the ferromagnetic properties of the mixed oxide nanoparticles in form of $CoFe_2O_4$ facilitate their separation from the solution using a magnetic bar after its environmental applications, as compared to other metal nanoparticles such as Co_3O_4 and Fe_2O_3.

Meanwhile, TiO_2 nanoparticles can be successfully immobilized onto a glass substrate, as shown in Figure 5.11.[176] The TiO_2 film was synthesized by following a sol-gel method/dip-coating/calcination procedure, which is common for the fabrication of inorganic films. In this case, in order to overcome the

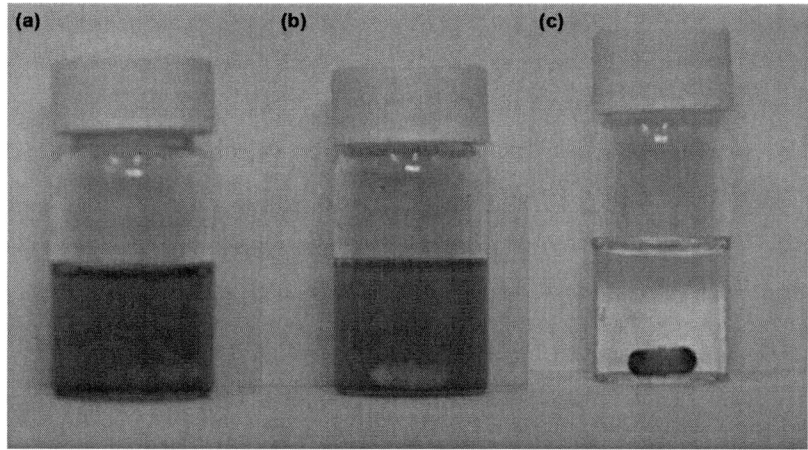

Figure 5.10 Separation of Co-Fe mixed oxide nanoparticles based on their ferro-magnetic properties after their use in advanced oxidation technologies. (a) Co_3O_4 and (b) Fe_2O_3 are not separated while (c) the mixed oxide nanoparticles in form of $CoFe_2O_4$ are easily removed by a magnetic bar. Reproduced with permission from ref. 175, copyright 2009 Elsevier.

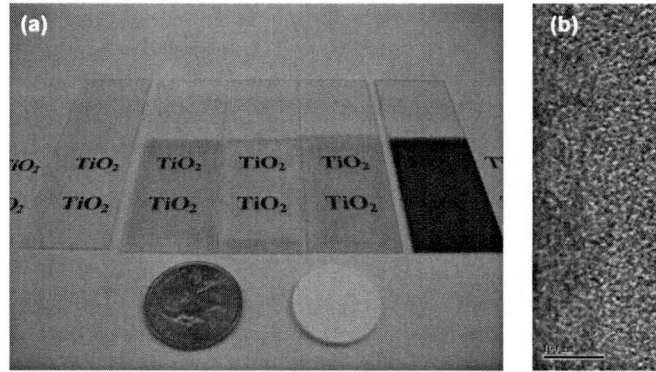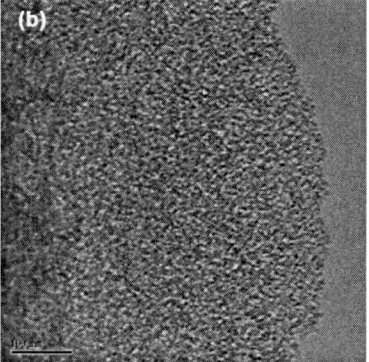

Figure 5.11 Images of a TiO_2 thin film: (a) a transparent thin film (100–300 nm thickness) composed of TiO_2 particles is uniformly immobilized onto a glass substrate and (b) a porous structure in the TiO_2 film is clearly seen. Reproduced with permission from ref. 176, copyright 2006 Elsevier.

drawbacks of the immobilized system (*e.g.*, mass transfer limitation), the film is designed to possess a porous structure by modifying the sol-gel method with a pore-templating technique employing surfactant self-assembly as a pore template. The TiO_2 film shows a high reactivity to decompose organic contaminants while maintaining high mechanical and chemical stability.

More recently, the surface of granular activated carbon (GAC) with a well developed mesoporous structure was impregnated with palladized ZVI

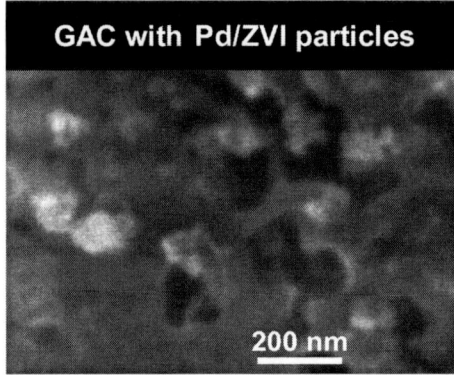

Figure 5.12   Electron microscopic image of activated carbon impregnated with palladized ZVI nanoparticles. The mesoporous structure of activated carbon is occupied by the metal nanoparticles.
Reproduced with permission from ref. 178, copyright 2008 American
Chemical Society.

nanoparticles, as shown in Figure 5.12.[178] Palladized ZVI particles show a
much higher reactivity to dechlorinate chlorinated chemicals such as
polychlorinated biphenyls (PCBs), compared to ZVI particles, while GAC
effectively adsorbs such a hydrophobic organic contaminant.[182] As a result, the
so-called reactive activated carbon (RAC) combines the chemical decomposition of PCBs with their physical adsorption. GAC plays multiple roles as: i) a
substrate for ZVI particle immobilization (a permeable reactive barrier composed of RAC granules is proposed), ii) a controller for the reactivity of ZVI
particles (iron oxides formed in the boundary of a RAC grain protect ZVI
particles in the grain core from immediate oxidation), and iii) an adsorbent for
organic contaminants (the adsorption capability of GAC to place PCBs into its
pores in close proximity to the palladized ZVI particles makes the treatment
strategy attractive). To increase the mechanical stability of RAC and facilitate
nanoscaling of ZVI particles, the key synthesis events include in situ formation
of Fe particles in the GAC pores from their molecular precursor and crystal
growth limitation of Fe oxides within the GAC pores.[178]

5.4   Conclusions

In conclusion, this book chapter summarizes green routes for the sustainable
preparation of nanoparticles. Different nanomaterials can be successfully synthesized, and the use of hazardous chemicals can be decreased by using nontoxic
solvents and natural template materials, such as ionic liquid, leaf and tea extract,
natural cellulose, sucrose, starch, and leaves. In addition, microorganisms
could be employed to synthesize nanomaterials as a green synthesis route.

In the application of metallic and bimetallic nanomaterials for environmental
remediation there are several areas of interest: photocatalytic degradation,
dehalogenation of organohalides, and photocatalytic disinfection. In the field

of photocatalysis, various nanomaterials have shown effective photocatalytic degradation of organic contaminants. The improvement of the light-harvesting ability of the photocatalyst, the enhancement of the charge separation efficiency, and the elucidation/optimization of the corresponding interfacial charge transfer reaction mechanisms are critical challenges in this field. For the dehalogenation, metallic and bimetallic nanoparticles have emerged as viable materials for degrading aliphatic and aromatic halogenated compounds. Bimetallic systems show a clear advantage over their monometallic counterparts. The degradation pathway has primarily involved electron transfer and has resulted in the formation of dehalogenated byproducts. For photocatalytic disinfection, semiconductor photocatalysis has been shown to be effective for the inactivation of a wide range of microorganisms at lab scale and under natural sunlight conditions for both small scale and large scale applications. Finally, the immobilization of nanoparticles onto a variety of substrates could be a more environmentally sustainable and versatile approach for environmental remediation.[183]

"The U.S. Environmental Protection Agency, through its Office of Research and Development, funded and managed, or partially funded and collaborated in, the research described herein. It has been subjected to the Agency's peer and administrative review and has been approved for external publication. Any opinions expressed are those of the author(s) and do not necessarily reflect the views of the Agency, and therefore, no official endorsement should be inferred. Any mention of trade names or commercial products does not constitute endorsement or recommendation for use."

Acknowledgements

D. D. Dionysiou acknowledges financial support by the National Science Foundation (US–Ireland collaborative research CBET (033317)) and the Cyprus Research Promotion Foundation through Desmi 2009–2010 which was co-funded by the Republic of Cyprus and the European Regional Development Fund of the EU under contract number NEA IPODOMI/STRATH/0308/09.

P. Falaras acknowledges funding through the Clean Water Project which is a Collaborative Project (Grant Agreement number 227017) co-funded by the Research DG of the European Commission within the joint RTD activities of the Environment and NMP Thematic Priorities.

S. O. Obare is grateful to the National Science Foundation for funding under Grant CHE 1005456.

J. A. Byrne acknowledges the Department for Employment and Learning Northern Ireland for funding under the US-Ireland R&D Partnership.

References

1. M. Doble and A. K. Kruthiventi, *Green Chemistry & Engineering*, Academic Press, Elsevier, MA, 2007.

2. G. P. Dransfield, *Radiat. Prot. Dosim.*, 2000, **91**, 271–273.
3. C. Han, M. Pelaez, V. Likodimos, A. G. Kontos, P. Falaras, K. O'Shea and D. D. Dionysiou, *Appl. Catal., B*, 2011, **107**, 77–87.
4. M. Pelaez, P. Falaras, V. Likodimos, A. G. Kontos, A. A. de la Cruz, K. O'Shea and D. D. Dionysiou, *Appl. Catal., B*, 2010, **99**, 378–387.
5. H. Haugen, J. Will, A. Köhler, U. Hopfner, J. Aigner and E. Wintermantel, *J. Eur. Ceram. Soc.*, 2004, **24**, 661–668.
6. S. L. Isley, S. David, R. Jordan and L. Penn, *Mater. Res. Bull.*, 2008, **44**, 1–7.
7. F. M. Wang, Z. S. Shi, F. Gong and J. T. Jiu, *Chinese J. Chem. Eng.*, 2007, **15**, 754–759.
8. N. S. Venkataramanan, K. Matsui, H. Kawanami and Y. Ikushima, *Green Chem.*, 2007, **9**, 18–19.
9. J. Zeng, R. Li, S. Liu and L. Zhang, *ACS Appl. Mater. Interfaces*, 2011, **3**, 2074–2079.
10. D.-W. Lee and K.-H. Lee, *Micropor. Mesopor. Mat.*, 2011, **142**, 98–113.
11. M. Sundrarajan and S. Gowri, *Chalcogenide Lett.*, 2011, **8**, 447–451.
12. X. Li, T. Fan, H. Zhou, S.-K. Chow, W. Zhang, D. Zhang, Q. Guo and H. Ogawa, *Adv. Funct. Mater.*, 2009, **19**, 45–56.
13. A. K. Jha, K. Prasad and A. R. Kulkarni, *Colloid. Surface. B*, 2009, **71**, 226–229.
14. F. Dong, H. Wang and Z. Wu, *J. Phys. Chem. C*, 2009, **113**, 16717–16723.
15. F. Dong, S. Guo, H. Wang, X. Li and Z. Wu, *J. Phys. Chem. C*, 2011, **115**, 13285–13292.
16. S. Gao, Z. Li and H. Zhang, *Current Nanoscience*, 2010, **6**, 452.
17. S. Gao, X. Jia, S. Yang, Z. Li and K. Jian, *J. of Solid State Chem.*, 2011, **184**, 764.
18. G. Patrinoiu, M. Tudose, J. M. Calderon-Moreno, R. Birjega, P. Budrugeac, R. Ene and O. Carp, *J. of Solid State Chem.*, 2012, **186**, 17.
19. P. Mohanpuria, N. K. Rana and S. K. Yadav, *J. Nanopart. Res.*, 2008, **10**, 507.
20. M. Molaei, E. S. Iranizad, M. Marandi, N. Taghavinia and R. Amrollahi, *Appl. Surf. Sci.*, 2011, **257**, 9796.
21. M. N. Nadagouda, T. F. Speth and R. S. Varma, *Acc. Chem. Res.*, 2011, **44**, 469.
22. M. N. Nadagouda and R. S. Varma, *Green Chem.*, 2008, **10**, 859.
23. M. N. Nadagouda, V. Polshettiwar and R. S. Varma, *J. Mater. Chem.*, 2009, **19**, 2026.
24. M. N. Nadagouda and R. S. Varma, *Cryst. Growth Des.*, 2007, **7**, 2582.
25. G. E. Hoag, J. B. Collins, J. L. Holcomb, J. R. Hoag, M. N. Nadagouda and R. S. Varma, *J. Mater. Chem.*, 2009, **19**, 8671.
26. M. N. Nadagouda and R. S. Varma, *Aust. J. Chem.*, 2009, **62**, 260.
27. M. N. Nadagouda and R. S. Varma, *Cryst. Growth Des.*, 2008, **8**, 291.
28. M. N. Nadagouda, G. Hoag, J. Collins and R. S. Varma, *Crys. Growth Des.*, 2009, **9**, 4979.

29. M. N. Nadagouda, D. Lytle, C. Bennet-Stamper and C. White, *RSC Advances*, 2012, DOI: 10.1039/C2RA01306A.
30. R. K. Gupta, K. Ghosh, L. Dong and P. K. Kahol, *Mater. Lett.*, 2010, **64**, 2132.
31. P. S. Chowdhury, P. R. Arya and K. Raha, *Synth. React. Inorg. Met.*, 2007, **37**, 447.
32. B. Baruwati, M. N. Nadagouda and R. S. Varma, *J. Phys. Chem. C*, 2008, **112**, 18399.
33. V. Polshettiwar, M. N. Nadagouda and R. S. Varma, *Chem. Commun.*, 2008, 6318.
34. M. N. Nadagouda and D. R. Lytle, *Journal of Nanotechnology*, 2011, **doi: 10.1155/2011/972486**.
35. C. B. Murray, D. J. Norris and M. G. Bawendi, *J. Am. Chem. Soc.*, 1993, **115**, 8706.
36. H. Bönneman, W. Brijoux, R. Brinkmann, E. Dinjus, T. Fretzen, B. Joussen and J. Korall, *Angew. Chemie.*, 1992, **31**, 323.
37. R. G. Freemantle, M. Liu, W. Guo and S. O. Obare in *Metallic Nanomaterials for Life Sciences*, ed. C. S. S. R. Kumar, Wiley-VCH, Weinheim, 2009, p. 305.
38. C. B. Wang and W. X. Zhang, *Environ. Sci. Technol.*, 1997, **31**, 2154.
39. A. Martino, M. Stoker, M. Hicks, C. H. Bartholomew, A. G. Sault and J. S. Kawola, *Appl. Catal. A Gen.*, 1997, **161**, 235.
40. F. Li, C. Vipulanandan and K. K. Mohanty, *Colloids Surf. A Physicochem. Eng. Asp.*, 2003, **223**, 103.
41. L. Wu, M. Shamsuzzoha and S. M. C. Ritchie, *J. Nanoparticle. Res.*, 2005, **7**, 469.
42. L. Wu and S. M. C. Ritchie, *Chemosphere*, 2006, **63**, 258.
43. Z. H. Meng, H. L. Liu, Y. Liu, J. Zhang, S. Yu, F. Y. Cui, N. Q. Ren and J. Ma, *J. Membr. Sci.*, 2011, **372**, 165.
44. J. Xu and D. Bhattacharyya, *Ind. Eng. Chem. Res.*, 2007, **46**, 2348.
45. J. H. Sinfelt, *J. Catal.*, 1973, **29**, 308.
46. J. H. Sinfelt, *Acc. Chem. Res.*, 1987, **20**, 134.
47. G. Meitzner, G. H. Via, F. W. Lytle and J. H. Sinfelt, *J. Chem. Phys.*, 1983, **78**, 2533.
48. G. Meitzner, G. H. Via, F. W. Lytle and J. H. Sinfelt, *J. Chem. Phys.*, 1983, **78**, 882.
49. G. Meitzner, G. H. Via, F. W. Lytle and J. H. Sinfelt, *J. Chem. Phys.*, 1985, **83**, 4793.
50. N. Toshima and T. Yonezawa, *New J. Chem.*, 1998, **22**, 1179.
51. F. Giessibl, *Rev. Modern Phys.*, 2003, **75**, 949.
52. S. Brunauer, P. H. Emmett and E. Teller, *J. Am. Chem. Soc.*, 1938, **60**, 309.
53. I. Langmuir, *J. Am. Chem. Soc.*, 1918, **40**, 1361.
54. Z. Fang, X. Qiu, J. Chen and X. Qiu, *J. Haz. Mater.*, 2011, **185**, 958.
55. A. W. Hull, *J. Am. Chem. Soc.*, 1919, **41**, 1168.

56. I. M. Arabatzis, S. Antonaraki, T. Stergiopoulos, A. Hiskia, E. Papakonstantinou and P. Falaras, *J. Photochem. Photob. A: Chem.*, 2002, **149**, 237.

57. B. K. Teo, *EXAFS: Basic Principles and Data Analysis – Inorganic Chemistry Concepts*, Springer, Verlag, 1986.

58. A. Fujishima and K. Honda, *Nature*, 1972, **238**, 37.

59. M. R. Hoffmann, S. T. Martin, W. Y. Choi and D. W. Bahnemann, *Chem. Rev.*, 1995, **95**, 69.

60. A. Fujishima, X. Zhang and D. A. Tryk, *Surf. Sci. Rep.*, 2008, **63**, 515.

61. H. Zhang, G. Chen and D. W. Bahnemann, *J. Mater. Chem.*, 2009, **19**, 5089.

62. A. P. Xagas, E. Androulaki, A. Hiskia and P. Falaras, *Thin Solid Films*, 1999, **357**, 173.

63. P. Falaras and I. M. Arabatzis, *Environ. Sci. Pollut. R.*, 2002, **3**, 57.

64. I. M. Arabatzis, T. Stergiopoulos, G. Katsaros, M. C. Bernard, D. Labou, S. G. Neofytides and P. Falaras, *Appl. Catal. B: Environ.*, 2003, **42**, 187.

65. P. Falaras, I. M. Arabatzis, T. Stergiopoulos and M. C. Bernard, *Int. J. Photoenergy*, 2003, **5**, 123.

66. I. M. Arabatzis, T. Stergiopoulos, D. Andreeva, S. Kitova, S. G. Neophytides and P. Falaras, *J. Catal.*, 2003, **220**, 127.

67. P. Falaras and I. M. Arabatzis, *Trends in Physical Chemistry*, 2004, **10**, 79.

68. A. I. Kontos, I. M. Arabatzis, D. S. Tsoukleris, A. G. Kontos, M. C. Bernard, D. E. Petrakis and P. Falaras, *Catal. Today*, 2005, **101**, 275.

69. D. S. Tsoukleris, A. I. Kontos, P. Aloupogiannis and P. Falaras, *Catal. Today*, 2007, **124**, 110.

70. A. O. Ibhadon, I. M. Arabatzis and P. Falaras, *Chem. Eng. J.*, 2007, **133**, 317.

71. A. O. Ibhadon, G. M. Greenway, Y. Yue, P. Falaras and D. Tsoukleris, *J. Photochem. Photob. A: Chem.*, 2008, **197**, 321.

72. A. Kyrkou, A. I. Kontos, G. Papavasileiou and P. Falaras., *J. Adv. Oxid. Technol.*, 2008, **11**, 402.

73. A. G. Kontos, A. I. Kontos, D. S. Tsoukleris, V. Likodimos, J. Kunze, P. Schmuki and P. Falaras, *Nanotechnology*, 2009, **20**, 045603.

74. D. D. Dionysiou and P. Falaras, *Reviews in Environmental Science and Bio/Technology*, 2010, **9**, 87.

75. A. G. Kontos, M. Pelaez, V. Likodimos, N. Vaenas, D. D. Dionysiou and P. Falaras, *Photochem. Photobiol. Sci.*, 2011, **10**(3), 350.

76. G. Liu, C. Han, M. Pelaez, D. Zhu, S. Liao, V. Likodimos, N. Ioannidis, A. G. Kontos, P. Falaras, P. S. M. Dunlop, J. A. Byrne and D. D. Dionysiou, *Nanotechnology*, 2012, accepted.

77. T. Aungpradit, P. Sutthivaiyakit, D. Martens, S. Sutthivaiyakit and A. A. F. Kettrup, *J. Hazard. Mater.*, 2007, **146**, 204.

78. W. Bahnemann, M. Muneer and M. M. Haque, *Catal. Today*, 2007, **124**, 133.

79. J. Ryu and W. Choi, *Environ. Sci. Technol.*, 2008, **42**, 294.

80. A. V. Emeline, X. Zhang, M. Jin, T. Murokami and A. Fujishima, *J. Photochem. Photobiol.*, 2009, **207**, 13.
81. A. I. Kontos, A. G. Kontos, Y. S. Raptis and P. Falaras, *Phys. Stat. Sol. (RRL)*, 2008, **2**, 83.
82. E. M. Rockafellow, L. K. Stewart and W. S. Jenks, *Appl. Catal. B: Environ.*, 2009, **91**, 554.
83. S. Kment, P. Kluson, V. Stranak, P. Virostko, J. Krysa, M. Cada, P. Adamek and Z. Hubicka, *Electrochim. Acta*, 2009, **54**, 3352.
84. J. Krysa, P. Novotna, S. Kment and A. Mills, *J Photochem. Photobiol. A: Chem.*, 2011, 81.
85. K. Kruwal, F. Sacher, A. Werner, J. Müller and T. R. Knepper, *Sci. Total Environ.*, 2005, **340**, 57.
86. V. A. Sakkas, I. M. Arabatzis, I. K. Konstantinou, A. D. Dimou, T. A. Albanis and P. Falaras, *Appl. Catal. B: Environ.*, 2004, **49**, 195.
87. A. O. Ibhadon, G. M. Greenway, Y. Yue, P. Falaras and D. Tsoukleris, *Appl. Catal. B: Environ.*, 2008, **84**, 351.
88. M. Pelaez, A. A. de la Cruz, E. Stathatos, P. Falaras and D. D. Dionysiou, *Catal. Today*, 2009, **144**, 19.
89. M. Pelaez, A. A. de la Cruz, K. O'Shea, P. Falaras and D. D. Dionysiou, *Water Res.*, 2011, **45**, 3787.
90. T. M. Triantis, T. Fotiou, T. Kaloudis, A. G. Kontos, P. Falaras, D. D. Dionysiou, M. Pelaez and A. Hiskia, *J. Hazard. Mater.*, 2012, **211–212**, 196.
91. D. S. Tsoukleris, T. Maggos, C. Vassilakos and P. Falaras, *Catal. Today*, 2007, **129**, 96.
92. A. G. Kontos, A. Katsanaki, V. Likodimos, T. Maggos, D. Kim, C. Vasilakos, D. D. Dionysiou, P. Schmuki and P. Falaras, *Chem. Eng. J.*, 2012, **179**, 151.
93. I. Karatasios, M. S. Katsiotis, V. Likodimos, A. I. Kontos, G. Papavassiliou, P. Falaras and V. Kilikoglou, *Appl. Catal. B: Environ.*, 2010, **95**, 78.
94. A. G. Kontos, A. Katsanaki, T. Maggos, V. Likodimos, A. Ghicov, D. Kim, J. Kunze, C. Vasilakos, P. Schmuki and P. Falaras, *Chem. Phys. Lett.*, 2010, **490**, 58.
95. H. F. Lin and K. T. Valsaraj, *J. Hazard. Mater.*, 2003, **99**, 203.
96. O. T. Woo, W. K. Chung, K. H. Wong, A. T. Chow and P. K. Wong, *J. Hazard. Mater.*, 2009, **168**, 1192.
97. I. M. Arabatzis, N. Spyrellis, Z. Loizoz and P. Falaras, *J. Mater. Process. Tech.*, 2005, **161**, 224.
98. G. Romanos, C. P. Athanasekou, F. K. Katsaros, N. K. Kanellopoulos, D. D. Dionysiou, V. Likodimos and P. Falaras, *J. Hazard. Mater.*, 2012, **211–212**, 304.
99. C. P. Athanasekou, G. E. Romanos, F. K. Katsaros, K. Kordatos, V. Likodimos and P. Falaras, *J. Membrane Sci.*, 2012, **392–393**, 192.
100. S. K. Papageorgiou, F. K. Katsaros, E. P. Favvas, G. E. Romanos, C. P. Athanasekou, K. G. Beltsios, O. I. Tzialla and P. Falaras, *Water Res.*, 2012, **46**, 1858.

101. E. Yassıtepe, H. C. Yatmaz, C. Öztürk, K. Öztürk and C. Duran, *J. Photoch. Photobio. A.*, 2008, **198**, 1.
102. P. V. Kamat and D. Meisel, *C. R. Chimie*, 2003, **6**, 999.
103. J. C. Colmenares, R. Luque, J. M. Campelo, F. Colmenares, Z. Karpinski and A. A. Romero, *Materials*, 2009, **2**, 2228.
104. M. Mohapatra and S. Anand, *Int. J. Environ. Sci. Te.*, 2010, **2**, 127.
105. C. Pulgarin and J. Kiwi, *Langmuir*, 1995, **11**, 519.
106. S. Sakthivel, S.-U. Geissen, D.W. Bahnemann, V. Murugesan and A. Vogelpohl, *J. Photoch. Photobio. A.*, 2002, **148**, 283.
107. S. O. Obare and G. J. Meyer, *J. Environ. Sci. Heal. A.*, 2004, **A39**, 2549.
108. S. H. Joo and D. D. Zhao, *Chemosphere*, 2008, **70**, 418.
109. Y. H. Shih, C. Y. Hsu and Y. F. Su, *Sep. Purif. Technol.*, 2011, **76**, 268.
110. W. X. Zhang, *J. Nanopart. Res.*, 2003, **5**, 323.
111. M. Liu, M. Han and W. W. Yu, *Environ. Sci. Technol.*, 2009, **43**, 2519.
112. H. Hildebrand, K. MacKenzie and F. D. Kopinke, *Global Nest. J.*, 2008, **10**, 47.
113. V. Nagpal, A. D. Bokare, R. C. Chikate, C. V. Rode and K. M. Paknikar, *J. Hazard. Mater.*, 2010, **175**, 680.
114. C. B. Wang and W. X. Zhang, *Environ. Sci. Technol.*, 1997, **31**, 2154.
115. B. Schrick, B. W. Hydutsky, J. L. Blough and T. E. Mallouk, *Chem. Mater.*, 2004, **16**, 2187.
116. L. Liang, N. E. Korte, G. D. Goodlaxson, J. Clausen, Q. Fernando and R. Muftikian, *Ground Water Monit. Rem.*, 1997, **17**, 122.
117. W. X. Zhang, C. B. Wang and H. L. Lien, *Catal. Today.*, 1998, **40**, 387.
118. T. Dong, H. Luo, Y. Wang, B. Hu and H. Chen, *Desalination*, 2011, **271**, 11.
119. J. Cao, R. Xu, H. Tang, S. Tang and M. Cao, *Sci. Tot. Environ.*, 2011, **409**, 2336.
120. X. Li, B. Li, M. Cheng, Y. Du, X. Wang and P. Yang, *J. Mol. Catal. A.*, 2008, **284**, 1.
121. K. Venkatachalam, X. Arzuaga, N. Chopra, V. G. Gavalas, J. Xu, D. Bhattacharyya, B. Henning and L. G. Bachas, *J. Hazard. Mater.*, 2008, **159**, 483.
122. Z. Zhang, Q. Shen, N. Cissoko, J. Wo and X. Xu, *J. Hazard. Mater.*, 2010, **182**, 252.
123. G. Zhang, Y. Kuang, J. Liu, Y. Cui, J. Chen and H. Zhou, *Electrochem. Commun.*, 2010, **12**, 1233.
124. C. H. Lin, Y. H. Liou and S. L. Lo, *Chemosphere*, 2009, **74**, 314.
125. X. Wang, C. Chen, Y. Chang and H. Liu, *J. Hazard. Mater.*, 2009, **161**, 815.
126. Y. H. Shih, Y. C. Chen, M. Y. Chen, Y. T. Tai and C. P. Tso, *Coll. Surf. A.*, 2009, **332**, 84.
127. T. Zhou, Y. Li and T. T. Lim, *Sep. Purif. Technol.*, 2010, **76**, 206.
128. X. Wang, P. Ning, H. Liu and J. Ma, *Appl. Catal. B.: Envion.*, 2010, **94**, 55.

129. T. T. Lim, J. Feng and B. W. Zhu, *Water Res.*, 2007, **41**, 875.
130. S. F. Chen and S. C. Wu, *Chemosphere*, 2000, **41**, 1263.
131. D. M. Cwiertny, S. J. Branssfield and A. L. Robert, *Environ. Sci. Technol.*, 2007, **41**, 3734.
132. X. Xu, J. Wo, J. Zhang, Y. Wu and Y. Liu, *Desalination*, 2009, **242**, 346.
133. Z. Wei-hua, Q. Xie and Z. Zhuo-yong, *J. Environ. Sci.*, 2007, **19**, 362.
134. R. Cheng, W. Zhou, J. Wang, D. Qi, L. Guo, W. Zhang and Y. Qian, *J. Hazard. Mater.*, 2010, **180**, 70.
135. Y. Han, W. Li, M. Zhanh and K. Toa, *Chemosphere*, 2008, **72**, 53.
136. Z. Fang, X. Qiu, J. Chen and X. Qiu, *J. Hazard. Mater.*, 2011, **185**, 958.
137. J. Feng and T. T. Lim, *Chemosphere*, 2005, 1267.
138. J. Cao, R. Xu, H. Tang, S. Tang and M. Cao, *Sci. Tot. Environ.*, 2011, **409**, 2336.
139. N. Zhu, H. Luan, S. Yuan, J. Chen, X. Wu and L. Wang, *J. Hazard. Mater.*, 2010, **176**, 1101.
140. T. Boronina, K. J. Klabunde and G. Sergeev, *Environ. Sci. Technol.*, 1995, **29**, 1511.
141. Y. Shiraishi, Y. Takeda, Y. Sugano, S. Ichikawa, S. Tanaka and T. Hirai, *Chem. Commun.*, 2011, **47**, 7863.
142. S. Luo, S. Yang, X. Wang and C. Sun, *Chemosphere*, 2010, **79**, 672.
143. M. O. Lutt, J. B. Hughes and M. S. Wong, *Environ. Sci. Technol.*, 2005, **39**, 1346.
144. WHO-UNICEF (2010) Progress on Sanitation and Drinking-water: 2010 Update, WHO.
145. T. Clasen and P. Edmondson, *Int. J. Hyg. Envir. Heal.*, 2006, **209**, 173.
146. WHO (2007) Economic and health effects of increasing coverage of low cost household drinking-water supply and sanitation interventions to countries off-track to meet MDG target 10.
147. J. D. Burch and K. E. Thomas, *Sol. Energy*, 1998, **64**, 87.
148. UNICEF and World Health Organization, Drinking Water Equity, safety and sustainability, JMP Thematic Report on Drinking Water 2011.
149. T. Matsunaga, R. Tomoda, T. Nakajima and H. Wake, *FEMS Microbiol. Lett.*, 1985, **29**, 211.
150. D. M. Blake, P. C. Maness, Z. Huang, E. J. Wolfrum, J. Huang and W. A. Jacoby, *Separ. Purif. Methods*, 1999, **28**, 1.
151. C. McCullagh, J. M. Robertson, D. W. Bahnemann and P. J. K. Robertson, *Res. Chem. Intermediat.*, 2007, **33**, 359.
152. S. Malato, P. Fernandez-Ibanez, M. I. Maldonado, J. Blanco and W. Gernjak, *Catal. Today*, 2009, **147**, 1.
153. O. K. Dalrymple, E. Stefanakos, M. A. Trotz and D. Y. Goswami, *Appl. Catal. B.: Environ.*, 2010, **98**, 27.
154. Z. Huang, P. Maness, D. M. Blake, E. J. Wolfrum, S. L. Smolinski and W. A. Jacoby, *J. Photoch. Photobio. A*, 2000, **130**, 163.
155. M. Wainwright, *Int. J. Antimicrob. Ag.*, 2000, **16**, 381.
156. K. Sunada, T. Watanabe and K. Hashimoto, *J. Photoch. Photobio. A*, 2003, **156**, 227.

157. M. Cho, H. Chung, W. Choi and J. Yoon, *Water Res.*, 2004, **38**, 1069.
158. A. G. Rincon, C. Pulgarin, N. Adler and P. Peringer, *J. Photoch. Photobio. A*, 2001, **139**, 233.
159. A. Rincon and C. Pulgarin, *Sol. Energy*, 2004, **77**, 635.
160. D. M. A. Alrousan, P. S. M. Dunlop, T. A. McMurray and J. A. Byrne, *Water Res.*, 2009, **43**, 47.
161. P. S. M. Dunlop, T. A. McMurray, J. W. J. Hamilton and J. A. Byrne., *J. Photoch. Photobio. A*, 2008, **196**, 113.
162. O. Sunnotel, R. Verdoold, P. S. M. Dunlop, W. J. Snelling, C. J. Lowery, J. S. G. Dooley, J. E. Moore and J. A. Byrne, *J. Water Health*, 2010, **8**, 83.
163. S. Gelover, L. A. Gomez, K. Reyes and M. T. Leal, *Water Res.*, 2006, **40**, 3274.
164. P. Fernandez, J. Blanco, C. Sichel and S. Malato, *Catal. Today*, 2005, **101**, 345.
165. C. Sichel, J. Tello, M. de Cara and P. Fernández-Ibáñez, *Catal. Today*, 2007, **129**, 152.
166. R. Asahi, T. Morikawa, T. Ohwaki, T. Aoki and K. Taga, *Science*, 2001, **293**, 269.
167. Q. Li and M. A. Page, B. J. Marinas and J. K. Shang, *Environ. Sci. Technol.*, 2008, **42**, 6148.
168. P. Wu, R. Xie, J. A. Imlay and J. K. Shang, *Appl. Catal. B: Environ.*, 2009, **88**, 576.
169. J. A. Rengifo-Herrera and C. Pulgarin, *Sol. Energy*, 2010, **84**, 37.
170. M. R. Wiesner, G. V. Lowry, P. Alvarez, D. D. Dionysiou and P. Biswas, *Environ. Sci. Technol.*, 2006, **40**, 4336.
171. S.-A. Lee, K.-H. Choo, C.-H. Lee, H.-I. Lee, T. Hyeon, W. Choi and H.-H. Kwon, *Ind. Eng. Chem. Res.*, 2001, **40**, 1712.
172. N. Saleh, K. Sirk, Y. Liu, T. Phenrat, B. Dufour, K. Matyjaszewski, R. T. Tilton and G. V. Lowry, *Environ. Eng. Sci.*, 2007, **24**, 45.
173. H. Choi, S. R. Al-Abed, D. D. Dionysiou, E. Stathatos and P. Lianos in *Sustainability Science and Engineering Volume 2: Sustainable Water for the Future*, ed. I. I. Escobar, A. I. Schafer, Elsevier, Amsterdam, 2010, p. 229.
174. A. K. Goyal, E. S. Johal and G. Rath, *Current Nanoscience*, 2011, **7**, 640.
175. Q. Yang, H. Choi, S. R. Al-Abed and D. D Dionysiou, *Appl. Catal. B: Environ.*, 2009, **88**, 462.
176. H. Choi, E. Stathatos and D. D. Dionysiou, *Appl. Catal. B: Environ.*, 2006, **63**, 60.
177. Y.-J. Chen and D. D. Dionysiou, *Appl. Catal. B: Environ.*, 2006, **62**, 255.
178. H. Choi, S. R. Al-Abed, S. Agarwal and D. D. Dionysiou, *Chem. Mater.*, 2008, **20**, 3649.
179. S. Yang, J. S. Gu, H. Y. Yu, J. Zhou, S. F. Li, X. M. Wu and L. Wang, *Sep. Purif. Technol.*, 2011, **83**, 157.
180. D. J. Burbridge, S. Crampin, G. Viau and S. N. Gordeev, *Nanotechnology*, 2008, **19**, 445302.

181. Z. Li, Y. Li, X.-F. Qian, J. Yin and Z.-K. Zhu, *Appl. Surf. Sci.*, 2005, **31**, 109.
182. H. Choi, S. Agarwal and S. R. Al-Abed, *Environ. Sci. Technol.*, 2009, **43**, 488.
183. M. I. Litter, W. Choi, D. D. Dionysiou, P. Falaras, A. Hiskia, G. Li Puma, T. Pradeep and J. Zhao, *J. Hazard. Mater.*, 2012, **211–212**, 1–2 (Editorial).

CHAPTER 6

Green Nanotechnology – a Sustainable Approach in the Nanorevolution

AJIT ZAMBRE,[§a] ANANDHI UPENDRAN,[§b,f]
RAVI SHUKLA,[a] NRIPEN CHANDA,[a] KAVITA K. KATTI,[a]
CATHY CUTLER,[c,d,e] RAGHURAMAN KANNAN[a,f] AND
KATTESH V. KATTI*[a,b,f]

Departments of [a]Radiology and [b]Physics; [c]Nuclear Science and
Engineering Institute; [d]Chemistry; [e]Missouri University Research Reactor,
University of Missouri, Columbia, MO 65212, USA; [f]Nanoparticle Biochem
Inc., Columbia, MO 65211, USA
*Email: KattiK@health.missouri.edu

6.1 Introduction

Nanotechnology, often referred to as "disruptive science" is inherently inter-disciplinary and has the power to break traditional boundaries that exist between various areas of science, agriculture, medicine and engineering.[1–7] Nanotechnology has opened up new opportunities for the creation and means of miniaturization of existing products, provides new materials with exceptional performance properties, and enriches our knowledge toward new insights and deep understanding of nature and life. Nanotechnology is poised

---

[§]Authors contributed equally for the chapter.

RSC Green Chemistry No. 19
Sustainable Preparation of Metal Nanoparticles: Methods and Applications
Edited by Rafael Luque and Rajender S Varma
© The Royal Society of Chemistry 2013
Published by the Royal Society of Chemistry, www.rsc.org

Figure 6.1 Green nanotechnology: Market potential and applications in various fields.

for rapid growth with myriad of applications in pharmaceutical research, smart electronic materials, alternative energy generation, environmental restoration, and allied fields (Figure 6.1).[8–11]

This interdisciplinary field is now being discussed in terms of its potential to deliver a life-saving second industrial revolution, radically disrupting existing manufacturing processes and thereby bringing about a paradigm shift in the way chemicals, materials and drug intermediates are developed for medical, health, hygiene, agricultural and automotive industry applications.[12,13] It has been estimated that over 85% of all future drugs and pharmaceutical processes will utilize one or the other form of nanotechnology. As the nanotechnology revolution continues to penetrate technological, medical, hygienic, environmental, and energy sectors, the combined market potential has been touted to exceed 3 Trillion dollars.[14,15] The growth and prospects of nanotechnology depend on new advances that allow efficient production of various types of nanoparticles. Development of large-scale production technologies for both metallic and nonmetallic nanoparticles with little or no collateral environmental pollution is becoming a basic necessity, with significant scientific and societal relevance. Therefore, it is imperative to develop sustainable processes that embrace minimization of toxic reactants, life-threatening chemical precursors and environmentally toxic variables within the overall nanoparticulate production platforms. The production and utilization of phytochemical-based nanoparticles have become an attractive area of interest in the field of nanotechnology. Various herbs and plant sources occlude powerful antioxidants that are present as phytochemical constituents in seeds, stems, fruits and in leaves.[16,17] These naturally occurring antioxidants have existed in human food chain for thousands of years, and are known to be nontoxic to living organisms and to the environment.[18,19] Therefore, the utility of plant-based phytochemicals in the overall synthesis and architecture of nanoparticles creates an important symbiosis between natural/plant sciences and nanotechnology. This connection provides an inherently green approach to nanotechnology, referred to as "*green nanotechnology*".

Gold nanoparticles have gained significance recently as they have a wide variety of applications in the field of nanomedicine.[20–27] We have carried out extensive research exploring the application of green nanotechnology routes for the synthesis of biocompatible gold nanoparticles for use in biomedical applications.[21,25–27] Specifically, we have recently reported the application of phytochemicals available within tea, cinnamon and cumin as dual reducing and stabilizing agents for the synthesis and stabilization of gold nanoparticles. This chapter discusses the application of redox-active phytochemicals, from a plethora of naturally available herbs including tea, cinnamon and cumin, for the production of gold nanoparticles in a 100% green and ecofriendly approach.

6.2   Synthesis of Gold Nanoparticles Using Phytochemicals

The synthesis of gold nanoparticles using the rich variety of phytochemicals in various plants and herbs is built on a scientifically sound premise that naturally available chemicals in leaves, bark, fruits and roots possess ideal redox properties that allow efficient reduction of metal precursors for conversion into the corresponding nanoparticles. The ability of a reducing agent to reduce gold salts to the corresponding gold nanoparticles is partly related to its standard one-electron reduction potential ($E^{\circ\prime}$), which is also a measure of the reactivity of an antioxidant as hydrogen or electron donor under standardized conditions. A lower $E^{\circ\prime}$ indicates that less energy is required for hydrogen or electron donation and is an important factor in determining reduction abilities of chemical compounds. The variety of phytochemicals in tea, cinnamon, cumin and soy singularly or collectively contribute toward the efficient reduction and stabilization of gold nanoparticles as outlined in Table 6.1. We initiated our green nanotechnology based gold nanoparticle synthesis using phytochemicals in soy beans (Figure 6.2) and with this initial success, we have explored the application of the rich variety of redox-active phytochemicals from a wide range of naturally occurring herbs such as tea, cinnamon and cumin. The synthetic details of the individual phytochemical based nanoparticles are described below.

6.2.1   Gold Nanoparticles from Cinnamon Phytochemicals (Cin-AuNPs)

The phytochemicals within cinnamon reveals that the herb comprises of essential oil (trans-cinnamaldehyde, eugenol, linalool, *trans*-cinnamic acid,

Table 6.1 Phytoconstituents in cinnamon, cumin and tea.

	Phytoconstituents	Major reducing phytochemical constituent	Redox Potential
Cinnamon	Cinnamaldehyde Linalool Eugenol Limonene Trans-Cinnamyl acetate Cinnamyl alcohol Linalyl acetate Cumin aldehyde Dihydrocinnamaldehyde Terpinene Camphene Eugenyl acetate Benzyl benzoate Carbohydrates	Cinnamaldehyde	−1.5 V
Cumin	Volatile essential oils, free amino acids, flavonoid glycosides, including derivatives of apigenin and luteolin	Cocktail of phytoconstitutents	–
Tea	Catechins (Catechin, Epicatechin, Epicatechin gallate, Epigallocatechin, Epigallocatechin gallate) Theaflavins (Theaflavin, Theaflavin 3-gallate, Theaflavin 3'-gallate, Theaflavin 3,3'-gallate, Thearubigins (Polymeric polyphenols)	Epigallocatechin gallate (EGCg)	0.42 V

terpenes and others; 1–4% by weight), polyphenols (catechin, epicatechin, anthocyanidin, catechin/epicatechin oligomers, kaempferitrin and others; 5–10% by weight) and carbohydrates (starch, polysaccharides, ash; 80–90% by weight).[28–33] The production of cinnamon phytochemical based nanoparticles (Cin-AuNPs) was explored in detail (Figure 6.3). The roles of individual phytochemicals present in cinnamon for the generation and stabilization of Cin-AuNPs by interactions with $NaAuCl_4$ in aqueous media have been systematically investigated. It was observed that the redox processes that allow conversion of $NaAuCl_4$ to the corresponding gold nanoparticles using individual components are less efficient as compared to the utility of cocktail of phytochemicals present in herbs or plants species. After detailed investigations, it was concluded that *trans*-cinnamaldehyde and linalool present in cinnamon are the major constituents that reduce $NaAuCl_4$ in aqueous media to

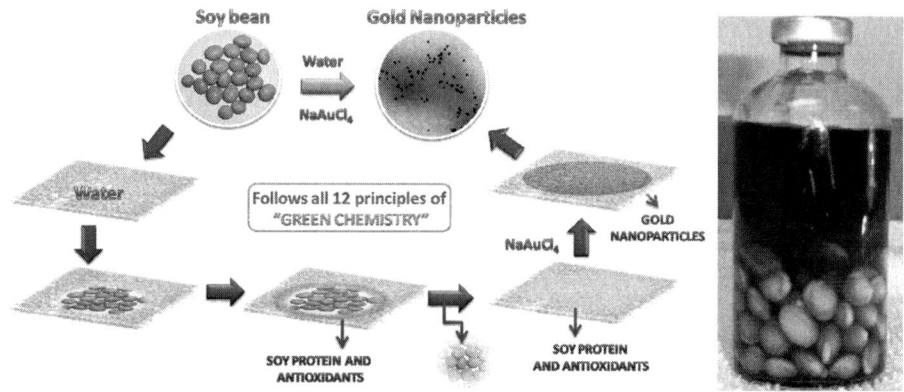

Figure 6.2   Synthesis of Soy-AuNPs.
Reprinted with permission from ref. 23 © 2008 Wiley-VCH Verlag GmbH
& Co. KGaA, Weinheim.

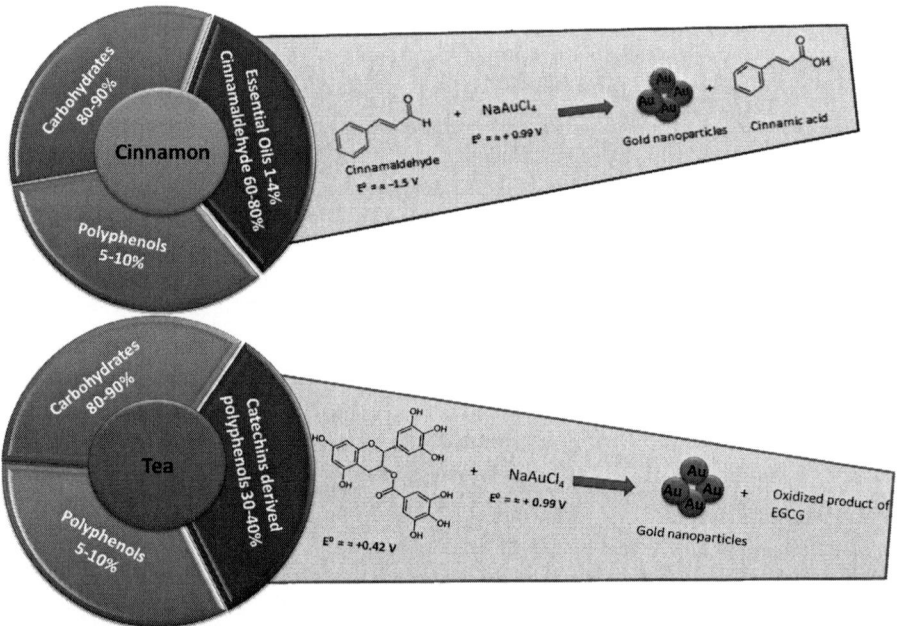

Figure 6.3   Generation of Cin-AuNPs and T-AuNPs.

produce a purple-red solution of Cin-AuNPs. The oxidized product of
trans-cinnamaldehyde, cinnamic acid precipitated as a crystalline solid on the
walls of the reaction vessel as evidenced by ^{1}H NMR.[34] The redox potential of
different organic aldehydes was examined and it was concluded that the
reduction potential of *trans*-cinnamaldehyde is in the vicinity of –1.5 V, a value
that is far below the reduction potential of [AuCl4]$^-$/Au (+0.99 V).[35,36]

Therefore, a redox couple of NaAuCl$_4$-cinnamaldehyde is thermodynamically feasible, resulting in the formation of stable gold nanoparticles of well-defined size and shape. Linalool, another major constituent was also found to be effective to reduce NaAuCl$_4$ to generate AuNPs but they failed to stabilize nanoparticles in aqueous solution. From the series of experiments performed using various phytochemicals present in cinnamon, it was inferred that only *trans*-cinnamaldehyde provides both reducing and stabilizing properties simultaneously during the nanoparticle formation. The chemical constitutions of benzaldehyde, linalool, are comprised of alcoholic, aldehyde functional groups that may be responsible for the reduction of NaAuCl$_4$.

6.2.2 Gold Nanoparticles from Cumin Phytochemicals (Cum-AuNPs)

Cumin seeds occlude redox active phytochemicals present including volatile essential oils (5%), numerous free amino acids and a variety of flavonoid glycosides, derivatives of apigenin and luteolin.[37] In order to understand the critical roles of the various phytochemicals present in cumin seeds on the overall reduction of NaAuCl$_4$ to the corresponding gold nanoparticles, a series of independent experiments using commercially available chemicals that are present in cumin seeds have been performed.[37] Results from such experiments have unambiguously confirmed that none of the individual constituents, when used independently, are effective in reducing gold salts to corresponding gold nanoparticles. The optimum mixture naturally composed within cumin seeds provides the ideal combined reduction potential to reduce gold salt for conversion to gold nanoparticles in aqueous medium. Therefore, the "cocktail" effect as seen in cinnamon-based phytochemicals is also in operation in cumin to provide the optimum reduction power for the conversion of gold salts to nanoparticles without the intervention/production of any toxic chemicals.

6.2.3 Gold Nanoparticles from Tea Phytochemicals (T-AuNPs)

Polyphenols in tea have $E^{\circ\prime}$ values comparable to that of α-tocopherol (vitamin E), but higher than ascorbate (vitamin C). The one-electron reduction potential of epigallocatechin gallate (EGCg), the second major constituent of polyphenol in tea under standard conditions is 550 mV, a value lower than that of glutathione (920 mV) and comparable to that of α-tocopherol (480 mV).[38] This superior hydrogen donor (antioxidant) of tea polyphenols can be advantageously utilized for the production of biocompatible gold nanoparticles. Recent results demonstrate that simple mixing of tea leaves or EGCg, with NaAuCl$_4$ resulted in the production of gold nanoparticles of uniform size distribution (Figure 6.3). The redox reaction occurs even at 0–5 °C, suggesting excellent kinetic propensity of polyphenols-based reducing agents in tea for the transformation of gold salt into gold nanoparticles. Tea-stabilized gold nanoparticles (T-AuNPs) generated through this process do not agglomerate, suggesting that the combination of thearubugins, theaflavins, catechins and various phytochemicals present in tea also serve as excellent stabilizers on nanoparticles and thus, provide robust shielding from agglomeration. EGCg-stabilized nanoparticles (EGCG-AuNPs) also have been synthesized and studied in detail.[39] EGCg-AuNPs show high stability, uniformity in size distribution and have been explored for their therapeutic application in the treatment of cancer (see section on biomedical applications).

6.3 Dual Roles of Reduction and Stabilization

Phytochemicals from cinnamon, cumin and tea are highly effective in reducing gold salt to the corresponding gold nanoparticles. However, not all the individual phytochemical constituents will provide stability to the resulting nanoparticles against agglomeration. For example, Linalool one of the phyto-chemical constituents in cinnamon, when interacted individually with $NaAuCl_4$ resulted in the efficient reduction to generate AuNPs, however, this phyto-chemical failed to stabilize nanoparticles against agglomeration in aqueous media. Further studies on the possible dual roles of reduction and stabilization of phytochemicals in cinnamon have revealed that *trans*-cinnamaldehyde (phytochemical constituent in cinnamon) provides both reducing and stabilizing properties simultaneously during the nanoparticle formation. The presence of carbohydrates for gold nanoparticle stabilization during the reduction process of $NaAuCl_4$ by linalool in aqueous media has been further investigated. These studies have clearly shown that coating of carbohydrates onto AuNPs provide good stability, whereas alcohol or aldehyde components of cinnamon are directly involved in $NaAuCl_4$ reduction. These detailed studies provide credible evidence on the dual reduction and stabilization influence of cocktail of phyto-chemicals in cinnamon for the production of well-defined and homogeneously dispersed nanoparticles. Extensive investigations involving phytochemicals from cumin and tea have demonstrated the importance of naturally available cocktail mixture of phytochemicals in dictating optimum redox chemistry required for the reduction of gold precursors and concurrent stabilization of the generated nanoparticles. The detailed physicochemical properties of gold nanoparticles synthesized using phytochemicals of cinnamon, cumin and tea are listed in Table 6.2. The UV-visible spectra and transmission electron microscopic (TEM) images of the nanoparticles synthesized by green process are shown in Figure 6.4. These studies reveal that the particles are uniform in size and shape.

Table 6.2 Size and zeta potential parameters of nanoparticles synthesized by green process.

	Size			
Nanoconstruct	Core Size by TEM [nm]	Hydrodynamic Radius (DLS) [nm]	DCS (CPS) [nm]	Zeta Potential [mV]
Cin-AuNPs	13 ± 5	155 ± 2	32 ± 2	-31 ± 2
Cum-AuNPs	13 ± 4	77 ± 1	12 ± 2	-15 ± 2
Tea-AuNPs	30 ± 3	165 ± 1	25 ± 1	-25 ± 1

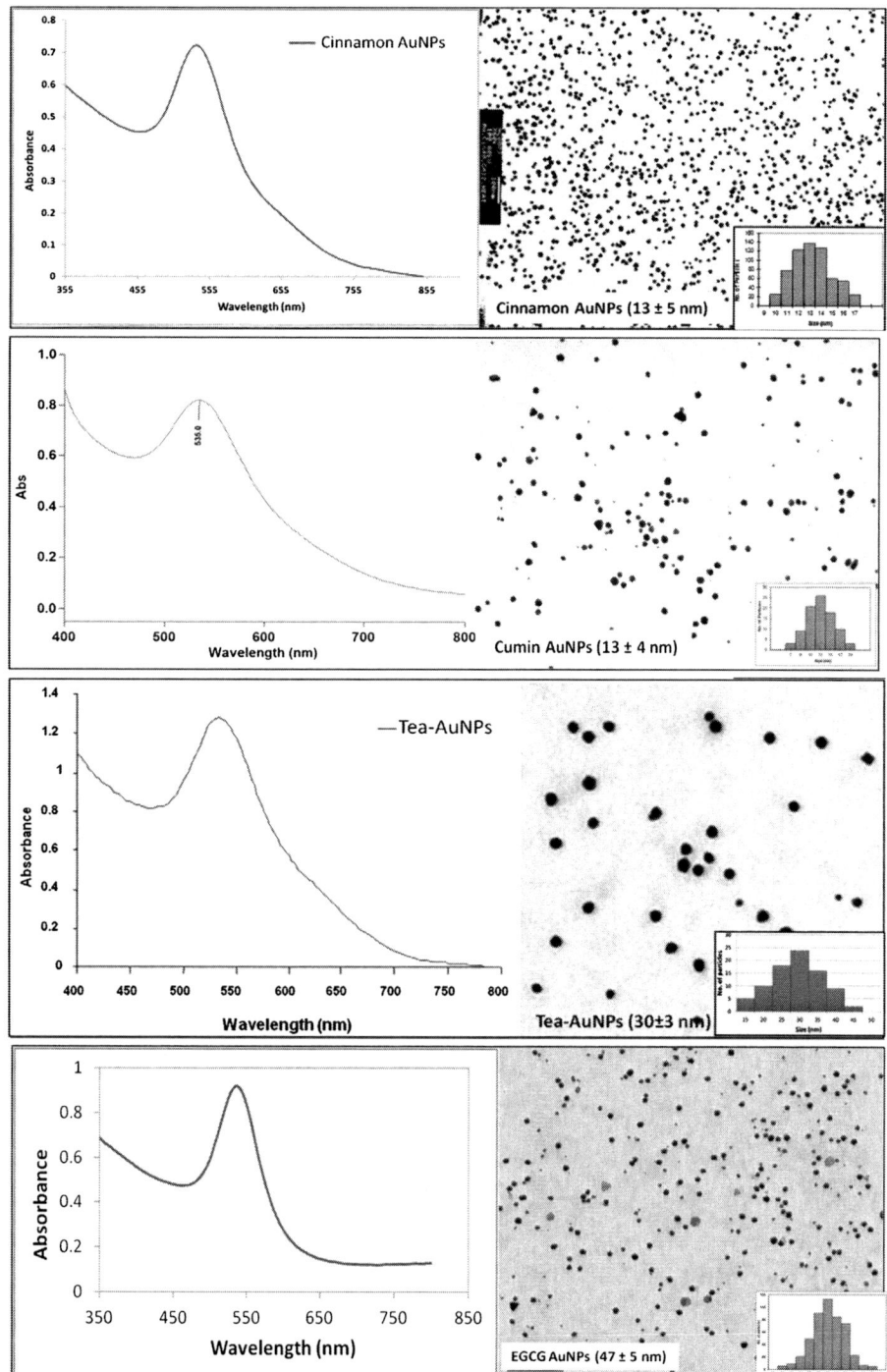

Figure 6.4   UV-Visible spectra and TEM images of Cin-AuNPs, Cum-AuNPs, Tea-AuNPs and EGCG-AuNPs.

6.4   Biomedical Applications

The biocompatibility of Cin-AuNPs, Cumin-AuNPs and Tea-AuNPs allow their utility in a number of biomedical applications.[34,37,38] MTT assays of these nanoparticles have demonstrated that they are nontoxic to normal fibroblastic cells.[34,37,38] In order to explore the biomedical applications of Cin-AuNPs, Cum-AuNPs and T-AuNPs, we have performed detailed *in vitro* studies on cancer cells using prostate tumour cells of human origin (PC-3).[34] The TEM images revealed uptake of gold nanoparticles within cellular membranes, suggesting that the internalization of nanoparticles within cells could have occurred *via* phagocytosis, fluid-phase endocytosis, and receptor-mediated endocytosis. Cin-AuNPs have shown extraordinary propensity for internalization within PC-3 cells.[34] The effective internalization of AuNPs within tumor cells offers excellent opportunities for biomedical imaging and therapy. The diagnostic potential of Cin-AuNPs has been validated from their ability in generating photoacoustic signals of individual nanoparticle-embedded cancer cells. Photoacoustic (PA) detection of PC-3 cells using Cin-AuNPs can provide a novel approach of utilizing phytochemical stabilized gold nanoparticles as optical contrast enhancement agents to detect circulating tumor cells (CTC) in cancer patients.[34] The high-z value of gold, coupled with their nontoxic nature, makes them excellent candidates as contrast agents in X-ray CT imaging in cancer diagnostics and also in the detection of various disorders.[34] The detailed pharmacokinetics studies of Cin-AuNPs have shown that these nanoparticles predominantly accumulate in lungs and can be used as a lung imaging agent. Cum-AuNPs and T-AuNPs are also expected to exhibit similar diagnostic potential. Inherently therapeutic EGCg-AuNPs have been synthesized for potential applications in cancer therapy.[39] The therapeutic potential of EGCg-AuNPs has been investigated in detail in tumor-bearing mice are outlined in the following section.

6.4.1   Thereapeutic Efficacy of EGCG-[198]AuNPs

[198]Au possesses a desirable beta energy emission and half-life for effective destruction of tumor cells/tissue (beta max = 0.96 MeV; half-life of 2.7 days). The range of the [198]Au β-particle (up to 11 mm in tissue or up to 1100 cell diameters (depends on energy)) is sufficiently long to provide crossfire radiation dose to cells within the prostate gland and short enough to minimize significant radiation dose to critical tissues near the periphery of the capsule.[40] Formulation of these inherently therapeutic and biocompatible [198]AuNPs utilizes the redox chemistry of the prostate tumor-specific phytochemical of green tea, EGCg, without the intervention of any other toxic chemical. Radioactive gold nanoparticles ([198]AuNPs) coated with nontoxic phytochemical EGCg, results in the production of EGCg-[198]AuNPs with uniform size distribution (40–55 nm). EGCg-[198]AuNPs has high stability in biological media and is suitable for use in cancer therapy applications. EGCG-[198]AuNPs specifically targets prostate tumors through the high affinity of EGCg for laminin receptors

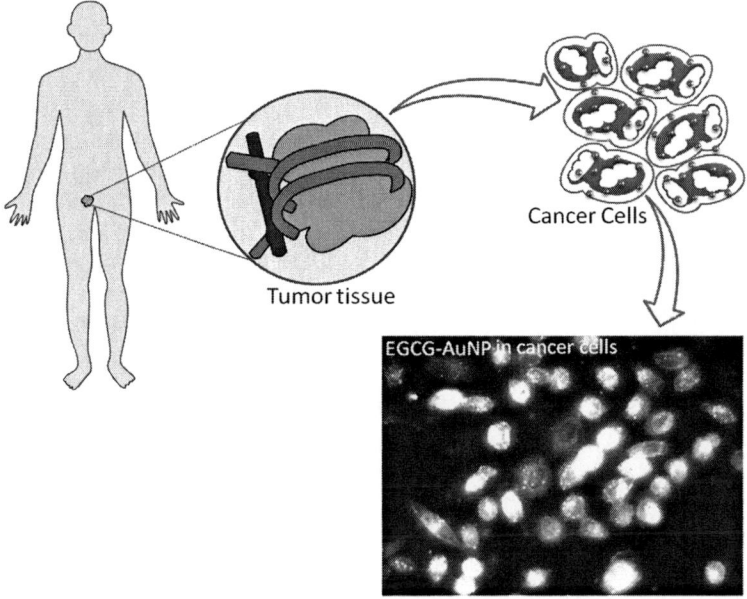

Figure 6.5   Selective internalization of EGCG-AuNPs in tumor cells, *inset* shows dark field microscope image of the EGCG-AuNPs in prostate cancer cells.

(Lam-67R) that are overexpressed on human prostate cancer cells (Figure 6.5).[39,41] The homogeneous size distribution allows penetration and retention within tumor vasculature with uniform tumor dosimetry. Their natural internalization within prostate tumor cells through receptor-mediated endocytosis resulted in improvement of therapeutic efficacy by optimal dose delivery and distribution to the primary prostate tumor while minimizing damage to neighboring tissue, as corroborated with *in vivo* therapeutic efficacy data obtained from mice models.[39]

The diagnostic and therapeutic potential of gold nanoparticles synthesized by green processes provide clear evidence that phytochemicals from readily available herbs (tea, cinnamon and cumin) serve as important building blocks in the development of biocompatible and tumor-specific nanoceuticals. Therefore, green nanotechnology presents realistic futuristic potential for the design and development of sophisticated molecular imaging and therapeutic agents and will contribute immensely to the field of nano-oncology.

6.5   Sustainability

Sustainability in the production of nanoparticles to meet the growing demands of the nanorevolution is a question of immense technological and societal importance. Production of nanoparticles should be considered from both the economic viability and environmental adaptability points of view. In order to keep the prices of the final finished nanotechnology-based products affordable to the consumers, industries will always look for a delicate balance between environmentally sound green processes and their sustainability. In addition, what may appear to be a green process may not turn out to be a fundamentally "game changing" process if the entire lifecycle of the process is not carefully addressed. For example, electric cars provide for zero emission automotive technologies. However, if the sources of electricity that power the electric cars result in negative environmental impacts, the overall approach of zero-emission automotive technology would fail. Likewise, if a specific nanotechnology product is used in cleaning air and purifying water, it is absolutely imperative to recognize that the nanoparticle production processes, from which such products are derived, create no negative environmental impacts through near- and long-term toxicity with severe irreversible consequences. Therefore, the development of environmentally benign and economically viable nanotechnological processes will drive major progresses in this emerging area of science and technology. Green nanotechnology processes, as described in this chapter, provide a strong foundation for the production of a variety of phytochemical or functionalized nanoparticles that can serve as building blocks in the development of new products that can be applicable in medical, energy, agricultural and environmental restoration sectors. The green nanotechnology-based production processes, as described herein, operate under 100% "green" conditions without the intervention of toxic chemicals. These production processes can be carried out without significant environmental pollution—thereby setting new standards in highly sustainable and economically viable "clean and green" technologies.

Acknowledgements

This work has been supported by the generous support from the National Institutes of Health/National Cancer Institute under the Cancer Nanotechnology Platform program (grant number: 5R01CA119412-01).

References

1. L. Y. Chou and W. C. Chan, *Nature Nanotechnol.*, 2012, **7**, 416.
2. J. Ge, J. Lei and R. N. Zare, *Nature Nanotechnol.*, 2012, **7**, 428.
3. C. Li, J. Adamcik and R. Mezzenga, *Nature Nanotechnol.*, 2012, **7**, 421.
4. M. I. Solar and M. J. Buehler, *Nature Nanotechnol.*, 2012, **7**, 417.
5. R. N. Kostoff, R. G. Koytcheff and C. G. Lau, *J Nanosci. Nanotechnol.*, 2009, **9**, 6239.
6. J. Zeng and Y. Xia, *Nature Nanotechnol.*, 2012, **7**, 415.
7. Y. Dang, Y. Zhang, L. Fan, H. Chen and M. C. Roco, *J. Nanopart. Res.*, 2010, **12**, 687.

8. M. D. Yang, Y. K. Liu, J. L. Shen, C. H. Wu, C. A. Lin, W. H. Chang, H. H. Wang, H. I. Yeh, W. H. Chan and W. J. Parak, *Opt. Exp.*, 2008, **16**, 15754.
9. P. K. Jain, X. Huang, I. H. El-Sayed and M. A. El-Sayed, *Acc. Chem. Res.*, 2008, **41**, 1578.
10. C. J. Murphy, A. M. Gole, J. W. Stone, P. N. Sisco, A. M. Alkilany, E. C. Goldsmith and S. C. Baxter, *Acc. Chem. Res.*, 2008, **41**, 1721.
11. Y. Wang, C. A. Mirkin and S. J. Park, *ACS Nano*, 2009, **3**, 1049.
12. S. Sonkaria, S. H. Ahn and V. Khare, *Recent Pat. Food Nutr. Agric.*, 2012, **4**, 8.
13. S. B. Nair, A. Dileep and G. K. Rajanikant, *Curr. Med. Chem.*, 2012, **19**, 744.
14. M. C. Roco, C. A. Mirkin and M. C. Hersam, *Nanotechnology Research Directions for Societal Needs in 2020*, 1st edn, Springer, 2011.
15. J. Curtis, M. Greenberg, J. Kester, S. Phillips and G. Krieger, *Toxicol. Rev.*, 2006, **25**, 245.
16. H. Nishino, Y. S., H. Tokuda and M. Masuda, *Curr. Pharmac. Des.*, 2007, **13**, 3394.
17. I. T. Johnson, *Proc. Nutrit. Soc.*, 2007, **66**, 207.
18. M. V. R. Dekker, *Eur. J. Nutr.*, 2003, **42**, 67.
19. W. G. B. Holst, *Curr. Opin. Biotechnol.*, 2008, **19**, 73.
20. G. M. Fent, S. W. Casteel, D. Y. Kim, R. Kannan, K. Katti and N. Chanda, *Nanomedicine*, 2009, **5**, 128.
21. K. C. N. Katti, R. Shukla, A. Zambre, T. Suibramanian, R. Kulkarni, R. R. Kannan and K. V. Katti, *InterNatureional J. Nanotechnol.: Biomed.*, 2009, **1**, B39.
22. K. K. Katti, V. Kattamuri, S. Bhaskaran, R. Kannan and K. V. Katti, *International J. Nanotechnol.: Biomed.*, 2009, **1**, B53.
23. V. Kattumuri, K. Katti, S. Bhaskaran, E. J. Boote, S. W. Casteel, G. M. Fent, D. J. Robertson, M. Chandrasekhar, R. Kannan and K. V. Katti, *Small*, 2007, **3**, 333.
24. V. C. M. Kattumuri, S. Guha, R. Kannan, K. V. Katti, K. Ghosh and R. J. Patel, *Appl. Phys. Lett.*, 2006, **88**, 153114.
25. S. C. N. Nune, R. Shukla, K. Katti, R. R. Kulkarni, S. Thilakavathy, S. Mekapothula, R. Kannan and K. V. Katti, *J. Mater. Chem.*, 2009, **19**, 2912.
26. R. Shukla, S. K. Nune, N. Chanda, K. Katti, S. Mekapothula, R. R. Kulkarni, W. V. Welshons, R. Kannan and K. V. Katti, *Small*, 2008, **4**, 1425.
27. R. N. S. Shukla, N. Chanda, K. Katti, S. Mekapothula, R. R. Kulkarni, W. V. Welshons, R. Kannan and R. V. Katti, *Science Editors' Choice*, 2008, **322**, 167.
28. R. T. Gow, D. L. G. W. Sypert and R. S. Alberte, *US 2007/0292540 A1* 2007.
29. S. M. a. T. E. Abraham, *Food Chem.*, 2006, **94**, 520.
30. P. Lopez, C. Sanchez, R. Batlle and C. Nerin, *J. Agric. Food Chem.*, 2007, **55**, 4348.

31. B. Shan, Y. Z. Cai, J. D. Brooks and H. Corke, *J. Agric. Food Chem.*, 2007, **55**, 5484.

32. B. Shan, Y. Z. Cai, M. Sun and H. Corke, *J. Agric. Food Chem.*, 2005, **53**, 7749.

33. J. Usta, S. Kreydiyyeh, P. Barnabe, Y. Bou-Moughlabay and H. Nakkash-Chmaisse, *Hum. Exp. Toxicol.*, 2003, **22**, 355.

34. N. Chanda, R. Shukla, A. Zambre, S. Mekapothula, R. R. Kulkarni, K. Katti, K. Bhattacharyya, G. M. Fent, S. W. Casteel, E. J. Boote, J. A. Viator, A. Upendran, R. Kannan and K. V. Katti, *Pharm. Res.*, 2011, **28**, 279.

35. N. R. Armstrong and R. K. Q. N. E. Vanderborgh, *Analyt. Chem.*, 1974, **46**, 1759.

36. L. Au, X. Lu and Y. Xia, *Adv. Mater. Deerfield*, 2008, **20**, 2517.

37. K. Katti, N. Chanda, R. Shukla, A. Zambre, T. Suibramanian, R. R. Kulkarni, R. Kannan and K. V. Katti, *Int. J. Green Nanotechnol. Biomed.*, 2009, **1**, B39.

38. S. K. Nune, N. Chanda, R. Shukla, K. Katti, R. R. Kulkarni, S. Thilakavathi, S. Mekapothula, R. Kannan and K. V. Katti, *J. Mater. Chem.*, 2009, **19**, 2912.

39. R. Shukla, N. Chanda, A. Zambre, A. Upendran, K. K. Katti R. K. Kulkarni, S. K. Nune, S. W. Casteel, C. J. Smith, J. Vimal, E. Boote, D. Robertson, P. Kan, H. Engelbrecht, L. D. Watkinson, T. J. Carmack, J. L. Lever, C. J. Cutler, C. Caldwell, R. Kannan, K. V. Katti *Proc. Natl. Acad. Sci. U.S.A.*, 2012, in press.

40. N. Chanda, P. Kan, L. D. Watkinson, R. Shukla, A. Zambre, T. L. Carmack, H. Engelbrecht, J. R. Lever, K. Katti, G. M. Fent, S. W. Casteel, C. J. Smith, W. H. Miller, S. Jurisson, E. Boote, J. D. Robertson, C. Cutler, M. Dobrovolskaia, R. Kannan and K. V. Katti, *Nanomed.: Nanotechnol. Biol. Med.*, 2010, **6**, 201.

41. R. Kannan, A. Zambre, N. Chanda, R. Kulkarni, R. Shukla, K. Katti, A. Upendran, C. Cutler, E. Boote and K. V. Katti, *Wiley Interdiscip. Rev. Nanomed. Nanobiotechnol.*, 2012, **4**, 42.

CHAPTER 7

Biofuels and High Value Added Chemicals from Biomass Using Sustainably Prepared Metallic and Bimetallic Nanoparticles

JARED T. WABEKE,[a] CLARA P. ADAMS,[a]
SETARE TAHMASEBI NICK,[a] LIYANA A. WAJIRA
ARIYADASA,[a] ALI BOLANDI,[a] DARRYL W. CORLEY,[a]
ROBERT Y. OFOLI[b] AND SHERINE O. OBARE*[a]

[a] Department of Chemistry, Western Michigan University, Kalamazoo, Michigan 49008, USA; [b] Department of Chemical Engineering an Materials Science, Michigan State University, East Lansing, Michigan 45224, USA
*Email: sherine.obare@wmich.edu

7.1 Introduction

One of the major concerns of the 21st century is the need to find effective means to replace crude oil for meeting global energy needs.[1-6] There is also general agreement that the world has access to a variety of nonfossil resources including biomass, solar, wind, and geothermal sources, that can collectively meet global energy needs. Central to achieving this goal is the availability of functional and comprehensive "biomass refineries" (*biorefineries*) with a suite of transformations capable of handling the large diversity of materials that fall under the general classification of biomass. While this suite of transformations

RSC Green Chemistry No. 19
Sustainable Preparation of Metal Nanoparticles: Methods and Applications
Edited by Rafael Luque and Rajender S Varma
© The Royal Society of Chemistry 2013
Published by the Royal Society of Chemistry, www.rsc.org

is well developed for petroleum processing, core chemical differences between crude oil and biomass prevent its direct application in the biorefinery.[7] For example, hydrocarbons are nonpolar, volatile, thermally tough, and unfunctionalized; on the other hand, biorenewables are typically polar, nonvolatile, and acid- and heat-sensitive. As a result, the traditional high-temperature vapor-phase conversions commonly used in petroleum processing must optimally be replaced by liquid-phase catalytic reactions that operate at low temperature and pressure.

The body of work on transforming renewable materials to fuels and chemicals in the last few years has been considerable, providing clear evidence of its increasing importance in science, engineering and technology. Indeed, there are many reports of successful conversion of biomaterials to hydrocarbons and biofuels. The list of materials converted include carbohydrates, sugars, alcohols and polyols[8–12] oils and acids[13–17] and solid biomass by fast pyrolysis followed by catalytic upgrading.[18–22] These studies, some of them pioneering, are important because they demonstrate technical feasibility for exhaustive reduction of biomaterials to high-density liquid fuels. This chapter provides an overview of some of the most recent advances in the use of nanoscale materials for converting biomass-derived materials into biofuels and high value added chemicals. Nanomaterials have recently emerged as important materials in biomass conversion, mainly due to their high surface to volume ratio, their high catalytic activity, robustness and ability to tune their morphology through sustainable methods. The chapter addresses some of the important roles that nanomaterials have played in the areas of oxidation of small molecules, cellulose conversion, production of liquid fuels, fatty acid decarboxylation and design of fuel cells.

7.2   Synthetic Procedures

7.2.1   Synthesis of Metallic Nanoparticles

The chemical and physical properties and reactivity of nanoparticles are dependent on their size, shape and morphology. Thus, synthetic procedures that produce well-defined nanoparticles with controlled dispersity are significant toward understanding the properties and applications of nanoscale materials. In this section, we provide an overview of the common methods for synthesis of metallic and bimetallic nanoparticles that are relevant toward degradation of organohalide pollutants. The formation of nanoparticles involves two important processes: (1) nucleation and (2) growth.[23] The initial nucleation process occurs through homogeneous, heterogeneous or secondary nucleation in a supersaturated solvent. The nucleation process is followed by growth of the nuclei *via* molecular addition. Molecular addition causes the concentration of the target element to drop below the critical point, resulting in an end to the nucleation process. However, the growth process continues until the concentration of the precipitates reaches equilibrium. The unity of the size of nanoparticles takes place due to the difference in free energy (ΔG) of the driving force within nanoparticles of different size. Smaller particles have larger free energy relative to larger particles. This difference in free energies results in faster growth for the smaller particles. It should be noted that under these conditions, the reaction should be stopped or more reactant should be added to the system in order to retrieve the depletion of its concentration due to particle growth. Alternatively, Ostwald ripening occurs, whereby the larger particles continue growing, as a result of generation of smaller particles. Formation of smaller particles is kinetically favored while the formation of larger particles is thermodynamically favored. The process continues until the smaller particles completely dissolve in the solvent.

In 1992 Bönneman *et al.*[24] expressed that the synthesis of metal nanoparticles requires at least three parameters: a metal precursor, a stabilizer and a reducing agent. Choosing a good stabilizer is also important for synthesis of metallic and bimetallic nanoparticles. Stabilizers are important in the synthesis of nanoparticles as they control the particle size and shape, while reducing the tendency to aggregate. In general, desired metal nanoparticles can be obtained by carefully choosing and manipulating these three substrates. Equation (7.1) demonstrates a general reaction for synthesizing metal nanoparticles.

$$MX_n(NR_4)_m + (n - m)Red \rightarrow M_{NP} + (n - m)(Red^+X^-) + m(NR_4^+X^-) \quad (7.1)$$

where M represents a Group VII, IX or X transition metal, X is a leaving group and *Red* is an effective reducing agent.[25]

Elemental metals, their inorganic salts and inorganic complexes are examples of metal precursors that are used for metallic nanoparticle synthesis. As an example, eqn (7.2) shows the chemical equation for the synthesis of zero-valent metal nanoparticles, in this case, iron (Fe), using ferric chloride as the precursor, sodium borohydride as the reducing agent and cetyltrimethyl-ammonium bromide as the surfactant:

$$2FeCl_2 + 6NaBH_4 + 18H_2O \rightarrow 2Fe^0 + 21H_2 + 6H_2BO_2 + 6NaCl \quad (7.2)$$

The synthesis method of metal nanoparticles can also be categorized by the media in which the nanoparticles are formed. According to this classification, there are two major methods for synthesis of metallic nanoparticles solution synthesis or a water-in-oil emulsion synthesis.[26–28]

Solution synthesis involves the synthesis of metal nanoparticles by chemical reduction of the metal salt in solution.[29] Generally, stirring the solution after addition of the reducing agent is needed in order to make sure that the reducing agent is distributed uniformly throughout the solution. Furthermore, to prevent possible oxidation of the synthesized nanoparticles, usually an excess amount of a reducing agent is added into the reaction solution. Although this method is straightforward, the limitation is that control of particle size and size distribution remains a challenge.

The *water-in-oil emulsion* method provides nanoparticles with improved size control.[30] In this method, an aqueous solution of metal ions is added into an oil phase, and a water-soluble cosolvent is added into the formed emulsion. The reducing agent is usually added into the system as the last step in the reaction. Metal nanoparticles synthesized by this method are more uniform in size and also have fewer tendencies to aggregate.

Metallic or bimetallic nanoparticles can be synthesized by an *in situ* method. In this case, metal precursors deposit on the surface of the stabilizer. The stabilizer is then soaked in a solution of reducing agent, resulting in reduction of the precursor and formation of nanoparticles on the polymer.[31,32]

7.2.2 Synthesis of Bimetallic Nanoparticles

Bimetallic alloy nanoparticles (*i.e.* metallic entities comprising atoms of two different metallic elements) offer improved catalytic properties relative to their single-metal counterparts.[33,34] Studies have shown that the alloy structure of bimetallic nanoparticles affects their catalytic properties.[35–37] The crystal structure of bimetallic alloy nanoparticles may vary from the bulk alloy, depending on the synthetic procedures and distribution of each metal.[38] Bimetallic nanoparticle alloys are usually synthesized by one of three methods, depending on the desired composition of the nanoparticles: (1) coreduction, (2) successive reduction, or (3) thermal decomposition.

Coreduction: Two corresponding metal ions are simultaneously reduced by refluxing in alcohol or other suitable solvent, in the presence of a stabilizing ligand. Most bimetallic nanoparticles are susceptible to oxidation, and there-fore Schlenk line techniques are usually employed during the synthesis procedure and once prepared, if needed, the nanoparticles are stored under nitrogen to avoid oxidation by air.

As an example and shown in eqn (7.3), addition of sodium borohydride to a solution of ferrous sulfate ($FeSO_4$) and nickel (II) chloride ($NiCl_2$) with a 4:1 Fe:Ni ratio resulted in the coreduction of the metal ions leading to the formation of Fe/Ni bimetallic nanoparticles.[39]

$$2Fe^{2+}(Ni^{2+}) + BH_4^- + 2H_2O \rightarrow 2Fe^0(Ni^0) + 2H_2 + BO_2^- + 4H^+ \qquad (7.3)$$

Successive Reduction: Successive reduction of metal ions is a suitable procedure for bimetallic alloy nanoparticles where the metal ions have a large difference in reduction potential. In this case, the metal ion with the more negative reduction potential is reduced first, followed by the second metal that will be reduced either chemically or by reflux in an alcohol solution.

Successive reduction was used to synthesize iron/palladium (Fe/Pd) nanoparticles because of their positions in the electrochemical series. In this case, iron nanoparticles were first synthesized by solution synthesis using sodium borohydride as the reducing agent. This was followed by addition of the palladium salt to the solution containing the iron nanoparticles. Equation (7.4) shows formation of Fe/Pd nanoparticles by the successive reduction method.[40–43]

$$Fe^0 + Pd^{2+} \rightarrow Fe^{2+} + Pd^0 \qquad (7.4)$$

<u>Thermal Decomposition</u>: Increased temperature can have significant effects on nucleation and growth during nanoparticle formation. Thermal decomposition usually takes place by heating the corresponding metal precursors with varied ratios in the presence of the stabilizing ligands.

7.3 Applications of Metallic and Bimetallic Nanoparticles for Biomass Conversion

7.3.1 Oxidation of Alcohols and Sugars

Liang and coworkers[44] studied the catalytic oxidation of glycerol in a base-free aqueous solution using a Pt-Cu/C catalyst. The Pt-Cu/C was prepared by pretreating active carbon (AC) with hydrogen peroxide (H_2O_2) for 5 h followed by dropwise addition of a platinum (II) chloride solution into the AC. Pt ions were reduced to zero-valent Pt using formaldehyde. By varying the amount of Pt ions added, the reaction resulted in a Pt catalyst consisting of 1Pt/C (1 wt% Pt), 3Pt/C (3 wt% Pt), and 5Pt/C (5 wt% Pt). The 3Pt/C was impregnated with a copper (II) chloride ($CuCl_2$) solution, dried and reduced by H_2 at 450 °C, followed by alloying with Cu ions at 750 °C under a nitrogen atmosphere. Using this process, and by varying mole ratios of Cu and Pt, the developed catalysts consisted of 1Pt-Cu/C, 3Pt-Cu/C, and 5Pt-Cu/C.

Characterization studies were conducted using X-ray diffraction (XRD) and transmission electron microscopy (TEM). XRD results illustrated that the Pt-Cu alloy was formed in Pt-Cu/C catalysts rather than simply a mixture of the two metal nanoparticles. TEM images depicted nanoparticle sizes in the range of 2.4 to 3.6 nm.

Analyses were conducted using the carbon-supported monometallic Pt, Cu and bimetallic Pt-Cu catalysts for the oxidation of glycerol. Results indicated that the 5Pt/C and 5Pt-Cu/C yielded the highest conversions ($\sim 70\%$ and $\sim 86\%$, respectively) of glycerol with the major product being glyceric acid. Increasing amounts of Pt showed an increase in selectivity for the glyceric acid. The Cu/C catalyst was analyzed and shown to be inactive towards the oxidation of glycerol in a base-free environment. Overall, the Pt-Cu/C catalyst was effective towards the selective oxidation of glycerol in a base-free solution.

Taarning *et al.*[45] investigated the oxidation of glycerol, 1,2-propanediol and 1,3-propanediol in methanol over titanium dioxide (TiO_2) and iron(III) oxide (Fe_2O_3) supported Au catalyst. No synthesis of gold catalyst, Au/C, Au/TiO_2, and Au/Fe_2O_3, was reported as they were supplied by the World Gold Council (WGC).[46] There was no conversion of glycerol in methanol when the Au/C catalyst was used. This was most likely due to the larger size (10.5 nm) of the Au NPs on Au/C in comparison with the 3.5 nm Au NPs on Au/TiO_2 and Au/Fe_2O_3. Thus, Au/TiO_2 and Au/Fe_2O_3 were found to be able to oxidize glycerol in methanol. When either Au/TiO_2 or Au/Fe_2O_3 was used several products were formed before reaching the fully oxidized dimethyl mesoxalate with 89% selectivity. Methyl glycolate was formed in very small quantities as a byproduct that was due to the C–C bond cleavage of a C_3. The C–C bond cleavage is thought to originate from a retro aldol reaction of methyl tartronate to produce methyl 2-hydroxyacetate and dimethyl carbonate that is in close relation between the formation of H_2O_2 and the C–C bond cleavage observed when the oxidation of glycerol takes place in water with Au-Pt NPs.[47,48] Thus, Taarning and coworkers came to the conclusion that it is likely that the same correlation was found in methanol, and that H_2O_2 cleaves a C–C bond in either dihydroxyacetone, methyl glycerate or dimethyl tartronate to form methyl glycolate.

According to the study, the oxidation of glycerol occurs at a higher rate and selectivity when over Au/Fe_2O_3 than over Au/TiO_2 (89% *vs.* 79%, respectively) due to the fact that Au/Fe_2O_3 is better at decomposing H_2O_2 that in turn decreases the amount of C–C cleavage that is induced by H_2O_2. Taarning *et al.* observed the same trend as others[49–52] when 1,2-propanediol was oxidized over gold catalyst. The oxidation was slower than glycerol and so by increasing the amount of catalyst they were able to obtain full conversion in approximately 1 h with 87% yield and with the major products being methyl lactate and 1-hydroxyactetone. Conversion of 1,2-propanediol increased to >98% after 21 h and had a selectivity of 72% to methyl lactate.

Hayashi *et al.*[53] had reported their investigations of the oxidation of 1,3-propanediol in methanol over Au catalyst previously. Nevertheless, Taarning *et al.* wanted to explore this oxidative esterification and compare it to the 1,2-propanediol results. They found that the oxidation of 1,3-propanediol was highly selective towards methyl 3-hydroxypropionate with 91% yield initially and 94% conversion after 21 h. Methyl acrylate was formed in small amounts (5%) after 21 h from the conversion of methyl 3-hydroxypropionate, which in turn is a building block for green polymers.

Dimitratos *et al.*[54] developed Au, Pd and Au-Pd supported catalysts using a wet impregnation and sol-immobilization method. Characterization of the catalysts was conducted using transmission electron microscopy (TEM), STEM-high-angle annular dark-field (HAADF), and X-ray photoelectron spectroscopy (XPS). Ultimately, STEM-HAADF revealed that the Au-Pd NPs prepared by sol-immobilization are most likely homogeneous alloys.

The oxidation of glycerol using the monometallic (Au, Pd) and bimetallic (Au-Pd) supported catalyst using the impregnation method was first investigated. The catalysts were supported on carbon (C) or titania (TiO_2). The Au/C catalyst consisting of 5 wt% showed a higher glycerol conversion ($\sim 60\%$) when compared to Pd/C catalyst (5 wt%, $\sim 44\%$ conversion of glycerol), the main product being glyceric acid. The Au/TiO_2 and Pd/TiO_2 (5 wt% for both) catalysts displayed similar glycerol conversion, $\sim 23\%$. TiO_2 resulted in being a less effective support for the monometallic catalysts when compared to carbon. A noticeable increase in catalytic activity was observed when the bimetallic supported catalysts (Au-Pd/C and Au-Pd/TiO_2, 2.5 wt% for each metal) were used, yielding glycerol conversions of 87% and 95%, respectively. In this case, TiO_2 resulted in being a better support for the bimetallic catalysts, which opposed the results indicated for the monometallic catalysts. Due to the acidic nature of the surface sites, the carbon-supported catalysts were found to be more selective towards glycolic acid and formic acid, whereas TiO_2-supported catalysts were selective toward glyceric acid.

Next, the group investigated the oxidation of glycerol using the Au, Pd, Au-Pd supported catalysts prepared by the sol-immobilization method. Lower concentrations of the metals were used (1%) because of the restrictions of the sol concentrations. When compared to samples prepared by wet impregnation, the catalysts illustrated an increase in activity for the oxidation of glycerol (Au/C-100%, Pd/C-21%, Au-Pd/C-100%, Au/TiO_2-95%, Pd/ TiO_2-40%, Au-Pd/ TiO_2-100%) when prepared by sol-immobilization. The catalysts also showed increased selectivity toward production of glyceric acid.

In other work, Yan *et al.*[55] developed an electrochemical method for glucose oxidation using the redox enzyme glucose oxidase (GOx). A three-electrode cell was used which consisted of Au as the working electrode, a saturated calomel reference electrode (SCE), and glassy carbon as the counter electrode. Figure 7.1 is a representation of the preparation of GOx functionalized electrode. 1.4 nm Au nanoparticles (NPs) were covalently linked to GOx leading to a surface coverage area of 2.5×10^{-12} mol cm^{-2}. The growth of the Au NPs in association with the GOx was performed by the GOx-catalyzed oxidation of glucose with O_2 to gluconic acid and H_2O_2, followed by the reduction of

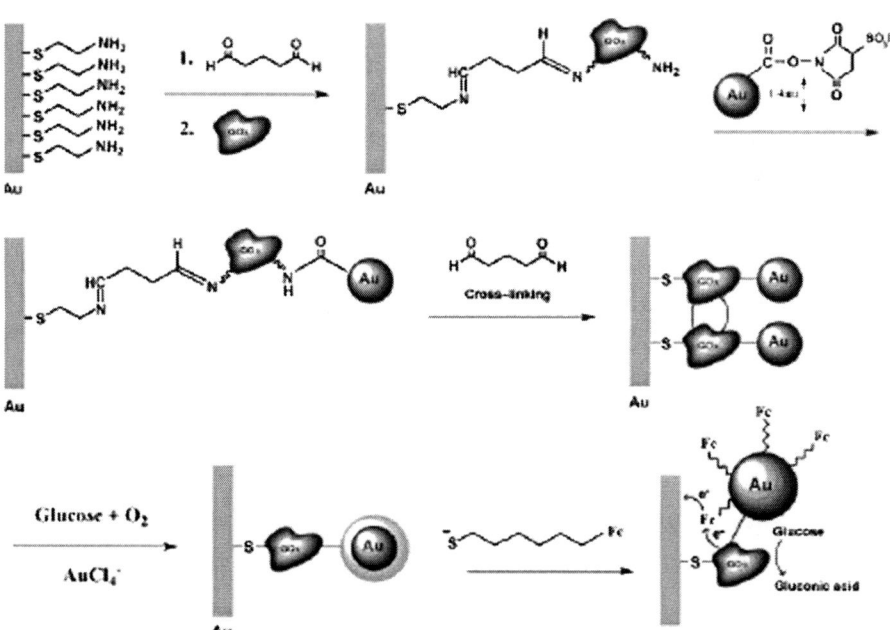

Figure 7.1   Construction of the Au NPs-GOx electrode and its growth by the enzyme-catalyzed reduction of HAuCl$_4$.[55]

$AuCl_4^-$ on the Au NPs by the H_2O_2 (Figure 7.1). The prolonged growth (30 min) of the Au NPs-functionalized GOx generously increased the bioelectrocatalytic current upon the oxidation of glucose. Cyclic voltammograms depicted that in the presence of Au NPs-functionalized GOx there was a 35% increase in the bioelectrocatalytic current. The bioelectrocatalytic current decreased as the time of growth was extended beyond 30 min, which is in reference to the blocking of the GOx towards the penetration of the substrate. Impedance spectra showed that immobilization of GOx on the electrode increased the interfacial electron resistance.[56–58] In turn, this is consistent with the formation of a nonconductive hydrophobic layer of protein, revealing that the covalent linking of Au NPs to the protein decreases the interfacial electron transfer resistance and as the Au NPs grow the values of resistance continue to decrease. The study further revealed that the main effect of the bioelectrocatalytic current is due to the accelerated charge transport between the electrode and the biocatalyst by the larger Au NPs. Some of the advantages of this electrode are its stability under 24 h of continuous operation and a degradation of < 5% upon storage at 4 °C for one week.

7.3.2   Conversion of Sugars

Zhao *et al.*[59] developed a Ru catalyst supported by carbon nanofibers (CNF) for the hydrogenolysis of sorbitol to ethylene glycol and propylene glycol. The Ru/CNF catalyst was prepared by wetness impregnation, calcination and reduction.[60] This study was conducted to observe the effect of calcinations on catalyst properties. The as-prepared Ru/CNF catalyst was calcined at 180, 240, and 300 °C for 5 h leading to catalysts designated Ru/CNF-180 °C, Ru/CNF-240 °C, Ru/CNF-300 °C, respectively. TEM images illustrated 1.0 nm as-prepared Ru/CNF, Ru/CNF-180 °C and Ru/CNF-240 °C particles, while Ru/CNF-300 °C showed a particle size of 10 nm.

The catalyst was characterized by thermogravimetric analysis (TGA), temperature-programmed reduction (TPR), X-ray diffraction (XRD), transmission electron microscopy (TEM) and X-ray photoelectron spectroscopy (XPS). TGA revealed the catalyst weight loss and Ru loading of the Ru/CNF catalysts calcined at different temperatures. Only slight weight losses for the Ru/CNF-180 °C and Ru/CNF-240 °C were detected (2.74 wt%, 3.04 wt%, respectively), however, there was a 13.33 wt% loss for Ru/CNF-300 °C, which was most likely due to catalytic gasification of CNFs and/or phase transformation of Ru.[61–63] TPR profiles revealed that the peak at lower temperatures became predominant as calcination temperatures increased. The one peak observed for Ru/CNF-300 °C reflected the reduction of RuO_2 particles, which was in accord with the XRD patterns. Thus, the phase transformation took place during calcinations and the TPR profiles indicated that $RuCl_3$ could be oxidized and transformed into $RuCl_xO_y$ (where, $0 < x < 3$, $0 < y < 2$) or RuO_2. XPS investigations disclosed the amount of surface oxygen-containing groups (SOCGs) produced from calcinations on

Ru/CNF catalyst surface. It is noted that calcinations increased the amounts of SOCGs.

Overall, calcination treatment changed the catalytic performance of as-prepared Ru/CNF catalysts. The calcination performance on Ru/CNF-180 °C and Ru/CNF-240 °C showed a decrease in sorbitol conversion but an increase in glycol selectivities when compared to previous studies.[64,65] Ru/CNF-240 °C appeared to be the best catalyst for sorbitol conversion because it displayed the highest selectivity for ethylene glycol and propylene glycol with logical yields. Sorbitol conversion and glycol selectivities decreased for the Ru/CNF-300 °C catalyst. This was most likely due to the larger-size nanoparticles (10 nm) or lower Ru dispersion, which meant fewer Ru active sites for sorbitol hydrogenolysis.

In further work, Zhao *et al.*[60] developed a Ru catalyst supported over carbon nanofiber (CNF) prepared by incipient wetness impregnation, and utilized the catalyst for sorbitol hydrogenolysis. Incipient wetness impregnation is a process by which CNF is added to an aqueous solution of ruthenium chloride trihydrate ($RuCl_3 \cdot 3H_2O$). The catalyst is first allowed to dry at room temperature and then it is dried in an oven at 120 °C for 12 h. Later, it is reduced in a H_2/Ar atmosphere at 300 °C for 5 h before use in sorbitol decomposition.

A comparison study was done using the Ru/CNF catalyst and a commercial activated carbon supported ruthenium catalyst (Ru/AC). Sorbitol hydrogenolysis was conducted using both catalysts and the results displayed not only successful conversion of sorbitol but also selectivity. The Ru/CNF catalyst showed a higher activity (85.73% of sorbitol) and a higher selectivity to the glycols (glycerol, ethylene glycol, and propylene glycol) than the Ru/AC catalyst (71.39% of sorbitol). N_2 physisorption and CO chemisorption studies were done using Ru/CNF and Ru/AC. The results indicated that Ru/CNF had a larger surface area and pore volume, and that the Ru nanoparticles were highly dispersed on Ru/CNF relative to Ru/AC. Studies also justified that the interaction between the CNF surface and metal nanoparticles could be somewhat accountable for the enhanced performance of Ru/CNF catalyst due to the electronic perturbation of the active metal particles. Varying amounts (6.0, 7.0, 8.0, 9.0, 10.0 MPa) of hydrogen partial pressure also had an effect on the sorbitol conversion. As the hydrogen partial pressure increased so did the sorbitol conversion yield, however, once the pressure reached 9.0 MPa the sorbitol conversion yield began to decrease. Thus, the preferred hydrogen pressure was 8.0 MPa. The hydrogen pressure also had a complex effect on the selectivity of the glycols.

Calcium oxide and sodium hydroxide were used as a base promoter in sorbitol hydrogenolysis. A base promoter is of good use for sorbitol conversion, however, it may reduce the product selectivity. According to the results, when sodium hydroxide was used the sorbitol conversion yield was higher but the selectivity was lower than that of calcium hydroxide. Others[66-69] suggested that the primary use of a base promoter was to prevent leaching of metal particles from the catalyst. Zhao *et al.* confirmed that the base promoter could improve the sorbitol hydrogenolysis by neutralizing the organic acids derived from side reactions. The very opposite occurred when calcium oxide was used as a basic promoter. In this case, there was a lower sorbitol conversion but higher product selectivity. Increasing the catalyst amount also increased the sorbitol conversion but decreased the selectivity for glycerol and ethylene glycol.

7.3.3 Production of Hydrocarbons

Kunkes and coworkers[1,70] were able to develop a Pt-Re catalyst that was used to convert sugars and polyols to monofunctional hydrocarbons (such as hydrophobic alcohols, ketones, carboxylic acids, and heterocyclic compounds) and liquid fuels. In this catalytic pathway the group sought to control the rates of C–C and C–O cleavage because this would help lead to the formation of monofunctional hydrocarbons. Most, if not all, oxygen atoms in the starting materials must be removed in order to convert biomass-derived carbohydrates (glucose and sorbitol in this case) into liquid fuels. The process of C–C coupling reactions and/or isomerization must take place in accord with removal of the oxygen atoms in order to produce branched hydrocarbons for gasoline and increase the molecular weight for diesel and jet fuels.

In the construction of the hydrocarbons, a portion of the polyol/sugar feed is reformed over the supported Pt-Re catalyst; this in turn supplies enough hydrogen to partially deoxygenate the remainder of the feed and more than 90% of the energy content of the polyol/sugar feed is retained in the reaction products. This reforming reaction can lead to the water gas shift reaction where CO is adsorbed onto the catalyst surface that then reacts with water to produce H_2 and CO_2. CO_2 production is important for the conversion of polyols to monofunctional hydrocarbons because it helps generate the hydrogen required for the deoxygenation reactions. In regards to cleavage of C–C bonds on Pt for an oxygenated species, studies suggest that C-O cleavage should be favored versus C–C cleavage because the surface is highly covered by strongly adsorbed species, such as adsorbed CO and highly oxygenated reaction intermediates.[71] In the case of Re, the rate of C–O in hydrogenolysis reactions is promoted for oxygenated hydrocarbons.[72] The effects of Re in the Pt-Re/C catalyst may be mediated by the presence of oxygen and/or hydroxyl groups because the binding energies of oxygen atoms and hydroxyl groups are stronger on Re than Pt.[73]

While conducting pyrolysis at temperatures greater than 770 K, biomass is converted to fuels and chemicals that form a liquid product called "bio-oil."[74,75] "Bio-oil" is a complex mixture with more than 300 highly oxygenated compounds, however, the liquid product from the catalytic conversion of glucose or sorbitol using the Pt-Re/C contains a well-defined mixture of liquid alkanes, olefins, and/or aromatics that have molecular weights that are suitable for transportation fuels.

The Pt-Re/C showed excellent stability for longer than 1 month time-on-stream (Sorb_503_18) and was used for successive catalytic processing. Studies showed that the maximum amount of carbon that could be converted to Sorb_503_18 would be 75% if all the H_2 produced was utilized by deoxygenated reactions. The actual amount of carbon in sorbitol that was converted to Sorb_503_18 was 52% (1: 3.5 kg, Sorb_503_18: sorbitol).

Additional studies were conducted using Sorb_503_18 in the presence of H_2 with $CuMg_{10}Al_7O_x$ catalyst. They wanted to achieve C–C coupling of the C_4 to C_6 ketones and secondary alcohols by aldol condensation but the catalyst ($CuMg_{10}Al_7O_x$) was deactivated due to the small amounts of organic acids and esters present in Sorb_503_18. An advantage of this catalytic approach is that it is simple and can be used in a limited number of flow reactors that ultimately has low capital cost, and it is flexible to where it can be employed to produce a variety of liquid-fuel components.

7.3.4 Conversion of Cellulose

Fukuoka *et al.*[76] successfully demonstrated the conversion of cellulose into sugar alcohols, sorbitol and mannitol, using supported Ru (Ru/HUSY (H form of ultrastable Y zeolite)) and Pt (Pt/γ-Al_2O_3) catalysts. In this study a variety of support materials and metal precursors that include Rh, Pd, Ir, and Ni were used, but only Ru and Pt illustrated high yields of the sugar alcohols. Figure 7.2 displays the production of sorbitol and mannitol by the hydrogenation of glucose.

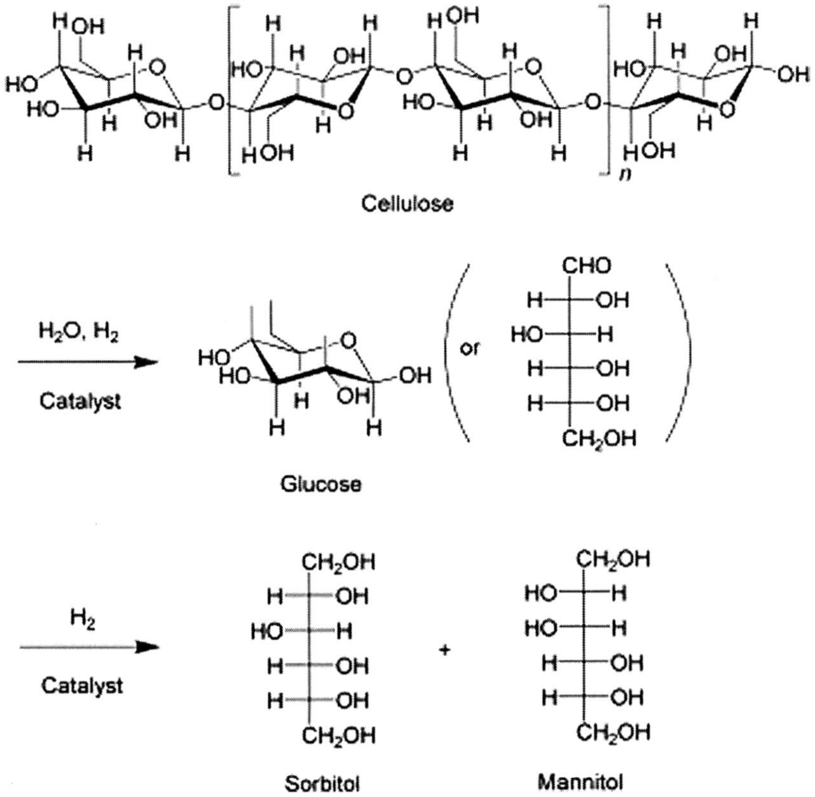

Figure 7.2 Hydrolysis of cellulose to glucose and the reduction of glucose to sorbitol and mannitol.[76]

The support material (γ-Al$_2$O$_3$, HUSY(40), SiO$_2$-Al$_2$O$_3$, and HUSY(20)) of these catalysts were chosen because of their high yields relative to other supports. Using a new "green" catalytic process it was found that Pt/γ-Al$_2$O$_3$ gave sugar alcohols in 31% yield with sorbitol being the main product (sorbitol: 25%, mannitol: 6%) with a molar ratio of 4:1 (sorbitol: mannitol).

The Ru/HUSY(20) catalyst for the conversion of cellulose displayed ∼26% yield. The reaction temperature range was 443–473 K with the highest yields observed over the Pt/γ-Al$_2$O$_3$ catalyst at 463 K. Catalytic runs (up to three cycles with similar yields) were able to be repeated with the supported metal catalysts, showing their environmentally friendly properties. This implied that there was no deactivation of the catalyst.

The study also included a reaction that consisted of the hydrolysis of cellulose into glucose by the support materials under H$_2$ pressure. According to the mechanistic pathway, H$_2$ is adsorbed onto the metal surface in a dissociative behavior and the hydrogen species is reversibly deposited onto the support surface. The results yielded less than 4% of glucose, which indicated that the metal drives the hydrolysis of cellulose. This study is in accordance with Zhang *et al.*,[77] which illustrates that crystallinity, degree of polymerization, availability of chain ends, and fraction of accessible bonds are important factors that contribute to the conversion of cellulose.

In other work, Deng *et al.*[78] developed a Ru carbon nanotube (CNT) catalyst to convert cellulose with a crystallinity of 33% into sorbitol in an aqueous medium in the presence of hydrogen. Ru catalysts loaded on various supports (SiO$_2$, CeO$_2$, MgO, Al$_2$O$_3$, and CNT) were prepared by impregnating the supports with an aqueous solution of RuCl$_3$, followed by drying, calcination in air at 350 °C and reduction by H$_2$ at 350 °C. The group prepared cellulose samples that consisted of treating the commercial grade cellulose (Alfa Aesar, crystallinity of 85%) with H$_3$PO$_4$ under varying time and temperature conditions.

Various catalysts (Fe, Co, Ni, Pd, Pt, Rh, Ru, Ir, Ag, and Au) supported on CNTs for the conversion of the treated cellulose with a crystallinity of 33% were studied. While results indicated that Pd, Pt, Rh, and Ir showed an active formation of sorbitol, Ru/CNT gave the highest yield of sorbitol amongst all the catalysts that were investigated. Conversion of cellulose (crystallinity 33%) over various supported (SiO$_2$, CeO$_2$, MgO, Al$_2$O$_3$, and CNT) Ru catalysts to yield sorbitol, mannitol, erythritol, and glycerol was also analyzed. Studies illustrated that SiO$_2$, CeO$_2$, MgO were not adequate supports (giving yields of 7%, 5%, and 0%, respectively) when considering sorbitol. The acidic Al$_2$O$_3$-supported Ru displayed a better catalytic performance with a 22% yield of sorbitol, but the CNT-supported Ru provided the highest yield of 69% of sorbitol. The analysis again showed that CNT was the most effective support for the Ru catalyst for the conversion of cellulose into sorbitol. The Ru/CNT catalyst was reused to illustrate its stability. There were four recycles with only an ∼8% decrease in the sorbitol yield.

Solid acids can catalyze the hydrolysis of cellulose to glucose,[79–85] which is considered a possible intermediate for sorbitol formation and one of the more important factors for the conversion of cellulose into sorbitol. Analyses were conducted to pretreat the CNT with concentrated HNO_3 to generate acidic groups, such as carboxylic groups, on their surfaces. It was found that acidic functional groups generated on CNT surfaces contributed to the conversion of cellulose into sorbitol.

The effects of percent crystallinity also played an important role in the sorbitol yield from the Ru/CNT catalyst. The lower the crystallinity, the higher the sorbitol yields. It was found that greater efficiency was attained using cellulose with a crystallinity of 33% that yielded 69% sorbitol relative to using cellulose with a crystallinity of 85% that only yielded 11% of sorbitol.

Fujita *et al.*[86] investigated the bioelectrocatalytic process for cellulosic materials (fructose and cellobiose) in an ionic liquid (IL) medium such as choline dihydrogen phosphate ([ch][dhp]). ILs are unique materials with physicochemical properties such as thermal and chemical stability. It has been anticipated that ILs will act as a matrix for applications within protein preservation, drug delivery, biofuel cells and biosensor development.

Three steps are necessary for energy conversion in an IL medium: (1) extraction of cellulose from biomass, (2) depolymerization of cellulose to produce mono-, di-, or oligosaccharides, and (3) collection of electrons from the saccharides. In this study, cellobiose dehydrogenase (CDH) and fructose dehydrogenase (FDH) were analyzed in hydrated [ch][dhp]. Gold nanoparticles (Au NPs) were prepared according to the literature[87] and used to develop the FDH-immobilized AuNP electrode. Catalytic oxidation currents based on the direct electron transfer (DET) reaction of FDH in hydrated [ch][dhp] were observed in the presence of D-fructose but not in its absence. Also, no current was observed with the FDH-unmodified electrode even in the presence of D-fructose which indicates that the bioelectrocatalysis of FDH successfully took place in the hydrated [ch][dhp]. Hence, the FDH-AuNP electrode was effective in hydrated [ch][dhp] due to the large surface area-to-volume ratio of the Au NPs which enabled a high level of protein loading. Using a carbon nanoparticle (CNP) modified carbon paper electrode, Fujita *et al.* investigated the DET reaction of CDH immobilized on the electrode without a mediator in an aqueous solution. Catalytic currents were observed when D-cellobiose was added with the hydrated [ch][dhp]. However, in the absence of D-cellobiose, no current of CDH was detected nor was a current detected with the CDH-unmodified electrode, even in the presence of D-cellobiose. The results show that CDH catalyzes the oxidation of cellobiose through DET in hydrated [ch][dhp]. The CDH-immobilized electrode stored in hydrated [ch][dhp] at room temperature was found to be stable for more than 3 weeks.

7.3.5 Decarboxylation of Fatty Acids

Among the various fields of growing interest is the development of fuels from renewable sources, for example, the utilization of plant and animal fats and oils.[88] Many synthesized oxygenated fuels or biodiesel consist of alkyl esters of long-chain fatty acids. Unfortunately, the oxygenated fuels have disadvantages such as high pour point, high cloud point, low energy density and potential for energy incompatibility.[89] To overcome the above-mentioned limitations, deoxygenation of biorenewable feedstock presents a viable alternative.

Hydrodecarboxygenation has commonly been accomplished using cobalt/molybdenum- or nickel/molybdenum-based catalysts. However, such catalysts are not cost effective and generally require expensive operational conditions such as high temperature and pressure.[90] In 2007, Murzin and coworkers[91] reported that catalytic decarboxylation is an alternative low energy technique for deoxygenation where they demonstrated that zero-valent palladium (Pd) exhibited the highest activity.

In further studies, Jones and coworkers[92] reported well-defined nanostructured catalysts for decarboxylation of free fatty acids. The catalysts were composed of Pd supported on siliceous mesocellular foam (MCF) functionalized with different silanes containing amine (N), urea (U), mercapto (S) and were designated Pd-MCF-X (where X is N, U or S). The support minimized the influence of internal mass-transfer limitations and prevented the pore blockage.[93] Porous silica was modified to promote even distribution of the Pd precursor and the reduction of Pd ions to zero-valent Pd was controlled to obtain evenly distributed 2 nm nanoparticles within the support.

Low-cost feedstock was obtained from a waste stream of a rendering facility and the free fatty acid (FFA) content was analyzed *via* gas chromatography. Carboxylic acids with a C_{18} chain length were found to be the main constitutes of the feedstock. Batch decarboxylation of stearic acid was performed over Pd-MCF-X at 300 °C for 6 h. 89.1%, 84.8% and 37% of stearic acid was converted to *n*-heptadecane using Pd-MCF-U, Pd-MCF-N and Pd-MCF-S, respectively (Figure 7.3). Pd-MCF-U and Pd-MCF-N were identified as the best catalyst candidates for overall Pd availability and low internal mass-transfer limitations. Pd-MCF-S showed less catalytic activity, due to formation of PdS during the high-temperature treatments that led to a limited amount of accessible Pd surface area.[94]

Han *et al.*[95] demonstrated a one-step method for the preparation of Pd supported on SBA-15 (a highly ordered mesoporous silicate material).[96] Murzin and coworkers[97] made use of Pd nanoparticles supported on SBA-15 for catalyzing the deoxygenation of stearic acid. A semibatch reactor, consisting of a condenser and heating mantle with a liquid-phase volume of 100 ml,

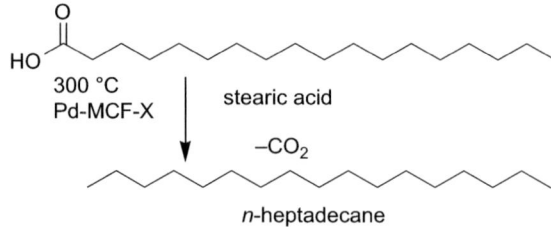

Figure 7.3 Decarboxylation of stearic acid to *n*-heptadecane.[94]

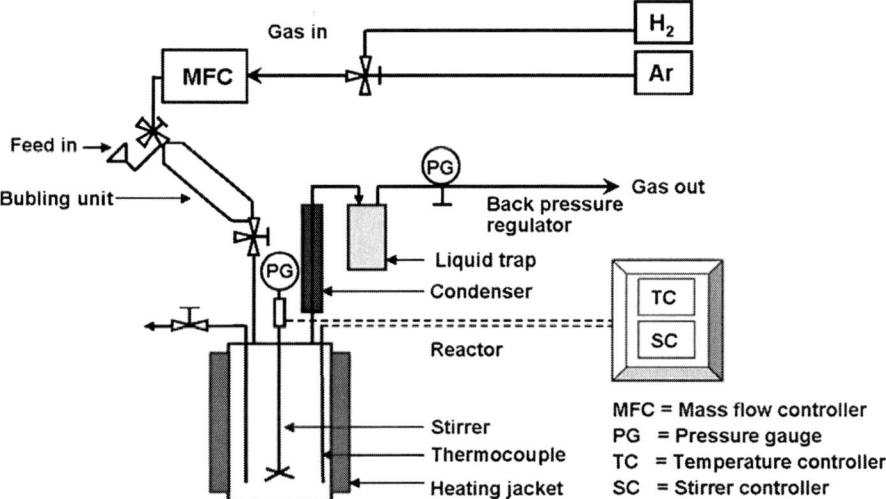

Figure 7.4 Semibatch reactor used to perform the deoxygenation of stearic acid.[95]

was used for the deoxygenation process, Figure 7.4. The reactor allowed for control of the flow rate of the carrier gas and the inlet and the outlet pressure. To prevent the external mass-transfer limitation, the reaction mixture was stirred at a high speed of 1100 rpm. To avoid catalysts oxidation, the reaction 4.8 bar of hydrogen gas was introduced to the reaction mixture at a temperature of 60 °C before the actual experiment. During the conversion process, 0.5 g of the catalyst was added to 1.4 g of stearic acid in dodecane in the presence of an inert gas. After the reaction, the catalyst was filtered and washed with acetone. The recovered catalysts were characterized for palladium content, metal dispersion and average crystalline size. It was found that the metal dispersion of the catalysts decreased as the particle size increased. Furthermore, increasing the metal loading resulted in an increase in particle size. A comparison of the physical properties of SBA-15 to Pd-SBA-15 by N_2 adsorption-desorption isotherms showed that the mesoporous channels of the support were not affected by the deposition of Pd.

In 2008, Murzin and coworkers[98] reported a carbon-supported Pd catalyst for deoxygenation of stearic acid. High metal dispersion (37%) of the new catalyst was achieved by oxidizing the support prior to the deposition. This method provides a smoother surface and high reactivity. The deoxygenation reaction of stearic acid was performed at three different temperatures (270, 300, and 330 °C) and was found to be most effective at 330 °C. The main observation of this experiment was the product distribution depending on the type of support and nature of surface groups. The analysis of the liquid phase of the reaction showed the formation of two main products, *n*-pentadecane and *n*-heptadecane in equal amounts. This was not the case when Pd nanoparticles catalysts were supported on SBA-15 or silica using the above process. Analysis of the gas mixtures after complete conversion of stearic acid showed the presence of CO, CO_2, methane, propane and compounds containing C1 and C3 in a ratio of 2.7. This ratio indicates the presence of high amounts of C1 compounds. The wide distribution of products was achieved due to the increasing number of acidic groups on the catalyst surface, in addition to having high mesoporous volume and narrow pore distribution from the supporting material. This property enables the efficient transportation of the reactant to the catalyst and removal of product from the catalyst.

7.3.6   Biodiesel Production

Deng *et al.*[99] synthesized Mg-Al (hydrotalcite) nanoparticles (which were used as a catalyst) in a 3:1 ratio by a coprecipitation method using urea as the precipitating agent with microwave-hydrothermal treatment (MHT) for biodiesel production from *Jatropha* oil. *Jatropha curcas L.* trees grow abundantly in Central and South America, Africa, India and South China. This nonedible oil is similar to edible oils (rapeseed, sunflower, palm and soybean)[100] that are used to produce biodiesel. The hydrotalcite nanoparticles were used in the transesterfication of *Jatropha* oil and several experiments were conducted under various reaction conditions that included the methanol/*Jatropha* oil molar ratio, catalyst concentration, temperature and ultrasonic power to assure biodiesel yield.

Characterization studies (XRD, AFM, XRF, SEM) were performed on the hydrotalcite before and after calcinations. XRD patterns of hydrotalcite particles with Mg-Al (3:1) revealed peaks that were typical of a hydrotalcite structure.[101] SEM images revealed that before calcination hydrotalcite nanoparticles had a well-developed platelet structure, whereas, after calcination the nanoparticles had a hexagonal platelet-like structure, which was due to the calcination temperature of 773 K. AFM studies revealed higher dimensions of the nanoparticle width and thickness, which was a result of agglomeration that was formed by the particles and hard to separate for AFM analysis. XRF spectroscopy verified the atomic ratio of Mg/Al that was accurately determined to be 2.78:1 versus 3:1.

The transesterfication reaction to produce biodiesel was a two-step process. The first step involved acid esterfication of the *Jatropha* oil that was previously investigated.[102] The second step was the solid base catalytic transesterfication. *Jatropha* oil was converted into diglyceride, monoglyceride, glycerin and fatty acid methyl esters (FAMEs).[103,104] The FAMEs consisted of palmitic acid, palmitoleic acid, stearic acid, oleic acid, linoleic acid and linolenic acid with a biodiesel yield of 13.79%, 0.95%, 6.33%, 42.61%, 26.34%, and 0.06%, respectively. As the molar ratio increased from 3:1 to 4:1, an increase in methyl esters was observed. Any ratio beyond 4:1 had no effect on the methyl ester yield. Due to the strong basic sites and large surface area of the calcined nanoparticles high activity was detected. As the catalyst concentration increased from 0.5 to 1.0 wt% the ester yield increased from ~53% to ~94%. Anything above 1.0 wt% would cause the biodiesel yield to decrease. Another important factor was the reaction temperature. At low temperatures the biodiesel yield was only ~52% at 303 K, however, at 318 K the biodiesel yield was ~94%. Any temperature beyond 318 K resulted in a decrease in biodiesel yield. The best ultrasonic power was 210 W, which resulted in a biodiesel yield of 95.2%. Ultrasonic radiation along with catalytically active nanoparticles synthesized by using urea and MHT facilitates the mixing of reactants with catalysts and is the possible reason for the high conversion rate. Deactivation studies were conducted on the catalyst. Overall, the Mg-Al catalyst was able to be used 8 times, yielding 89.1% biodiesel.

Pd nanohybrids (5 wt% Pd/SWNT/SiO$_2$) were used to determine the catalytic reactions at the liquid/liquid interface and were found to stabilize water–oil emulsions. Oxide nanoparticles are also of interest because they have been previously studied[105–107] to stabilize oil-in-water emulsions due to their hydrophilicity and carbon nanotubes were attractive because they produce emulsions of the water-in-oil nature due to their hydrophobicity.[108] The hybrid nanoparticles were prepared using a previous method[109] and the group was able to tune the composition so as to modify the hydrophilic–hydrophobic balance and assemble water-in-oil or oil-in-water emulsions.[110] The objectives of Crossley *et al.*'s study was to (1) extend the utility of the nanohybrids by incorporating a transition metal (Pd) to make them catalytically active for hydrogenation and (2) add a solid base to catalyze condensation reactions. Aqueous reactions would be catalyzed by depositing Pd on the hydrophilic face, thus deposition of Pd on the hydrophobic face would support organic solvent reactions. Two types of preparations with nanotubes were conducted: (1) incipient wetness impregnation to deposit Pd on single-walled carbon nanotubes (SWNT) and (2) MgO is used as a support instead of SiO$_2$. Vanillin (4-hydroxy-3-methoxybenzaldehyde) was an ideal compound of choice to investigate the catalytic activity of Pd-nanohybrids to phenolic hydrodeoxygenation in a water-in-oil (decalin) emulsion because it contains three different types of oxygenated functional groups (hydroxyl, aldehyde, ether) and it is partially soluble in both the aqueous and organic phase. The reaction took place in a batch reactor at three different temperatures (100, 200, and 250 °C). Filtering of the nanohybrid particles allowed the emulsion to be broken and the two liquid phases were separated by GC-MS and GS-FID in order to monitor the migration of the products when going from aqueous to organic phase.

As the temperature increased, changes in the chemoselectivity were observed, and at 100 °C vanillin alcohol was produced by hydrogenation of the aldehyde and further converted into 2-methoxy-4-methylphenol (*p*-creosol) by means of hydrogenolysis (this is the dominant pathway at 200 °C). Decarbonylation of the aldehyde group became the leading path when the temperature was increased to 250 °C (conversion to *o*-methoxyphenol (guaiacol)).

Octanal (soluble in oil) and glutaraldehyde (soluble in water) were chosen to explore the reactivity of molecules that are soluble in either the organic or aqueous phase. Reactions were carried out where octanal and glutaraldehyde were (1) separately placed in equimolar solutions of water and decalin, (2) separately dissolved in pure decalin, and (3) separately dissolved in pure water. The group expected to see a considerable effect of the conversion of glutaraldehyde since the Pd-nanohybrids had been placed along the hydrophilic side. While only 58% conversion of glutaraldehyde was obtained in the aqueous case, 98% was converted in the emulsion under the same conditions (3 h at 100 °C). The major products from this conversion were cyclic hemiacetal and valerolactol. The results illustrated that by controlling the rates of reaction and migration out of a phase where the catalyst is located then you could modify the selectivity. In the case of the octanal reaction where Pd was found on the hydrophobic side, the very opposite of results were observed where only 9.1% of 1-octanol was yielded as the dominant product.

Finally, a tandem reaction catalyzed by MgO instead of SiO_2, where Pd-catalyzed hydrogenation followed by aldol condensation of 5-methylfurfural and acetone, was investigated. MgO imparts basicity to the nanohybrids that in turn helped stabilize the Pd particles on both the hydrophilic and hydrophobic sides. This ultimately allowed reactions to place at the liquid/liquid interface by using a bifunctional catalyst that contains both metal and basic sites.

7.3.7 Design of Fuel Cells

Biofuel cells are becoming an increasingly important alternative energy source. Typically, biofuel cells can be classified either as a microbial-based or enzymatic fuel cells depending on the location of the enzymes, *i.e.* inside of microorganisms or outside of living cells.[111–120] Enzyme-based fuel cells offer the advantage of cathodes that reduce O_2 without generating hydrogen peroxide as a possible byproduct. Below, we demonstrate the latest advances of enzyme-based fuel cells that utilize nanoscale materials for their operation.

Basu *et al.*[121] developed platinum-based bimetallic nanoparticles in the design of a direct glucose alkaline fuel cell. Carbon-supported Pt-Au and Pt-Bi bimetallic catalysts were prepared in the liquid phase using $NaBH_4$ as a reducing agent. The catalysts were characterized by scanning electron microscopy (SEM), transmission electron microscopy (TEM), and X-ray diffraction (XRD). The bimetallic nanoparticles constituted the anode material in a batch direct glucose fuel cell (DGFC). Activated charcoal was used as the cathode in different glucose and electrolyte concentrations at ambient temperature. The development of DGFCs in which metallic catalysts are used can be approached through two different pathways. The first attempts to minimize the poisoning of the catalysts and the second looks to enhance the kinetic of electro-oxidation of glucose using more active catalysts.[122–124] By releasing two electrons, glucose is chemically oxidized to gluconic acid at the metal catalyst surface. The reactions involved are:

$$\text{Anode reaction:} \quad C_6H_{12}O_6 + 2OH^- \rightarrow C_6H_{12}O_7 + H_2O + 2e^-$$

$$\text{Cathode reaction:} \quad 0.5O_2 + H_2O + 2e^- \rightarrow 2OH^-$$

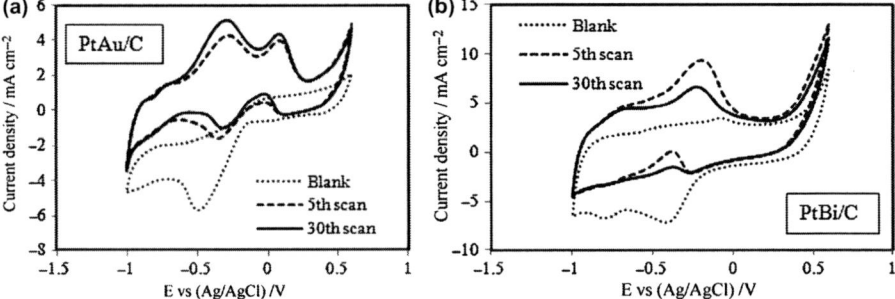

Figure 7.5 a) PtAu/C and b) PtBi/C catalysts in 0.5 M KOH without and with 0.05 M glucose.

The electro-oxidation of glucose in the presence of carbon-supported PtAu and PtBi was studied by cyclic voltammetry (CV) and chronoamperometry (CA). Figure 7.5 shows the comparison of voltammograms between the 5th scan and the 30th scan for Pt-Au/C and Pt-Bi/C catalysts in glucose- KOH solution and in KOH solution.

The CV data revealed that the reduction of the anodic peak height in the case of Pt-Bi/C indicates that the Pt-Bi/C catalysts undergo poisoning due to the strong adsorption of intermediate products formed. It was also found that the Pt-Au/C is less susceptible to poisoning and is thus more active than the Pt-Bi/C catalyst for glucose electro-oxidation in an alkaline medium.

Chronoamperometric measurements conducted at constant potential indicated that the long-term poisoning rate was much higher for the Pt-Bi/C relative to Pt-Au/C catalysts for glucose electro oxidation. Results from the batch DGFC experiments revealed that in 0.2 M glucose and 0.1 M KOH, the open circuit voltage (OCV) obtained using Pt-Bi/C anode was 0.8 V and 0.9 V for Pt-Au/C. In comparison, commercial PtRu/C had an OCV value of 0.9 V. In the case of Pt-Bi/C and commercial Pt-Ru/C, the power density increased up to 0.2 M glucose concentration and then decreased, while in the case of Pt-Au/C the cell performance increased with the increase in glucose concentration up to 0.3 M and then it decreased on increasing the concentration of glucose. Overall, it was found that bimetallic nanoparticles consisting of Pt-Au and supported on C were an effective catalysts for glucose oxidation.

Liu Deng et al.[125] reported the construction of a biofuel cell that was powered from ethanol and an alcoholic beverage. The oxygen biocathode consisted of laccase and Au nanoparticles immobilized on a composite membrane composed of partially sulfonated (3-mercaptopropyl)-trimethoxysilane sol-gel (PSSG)-Chitosan (CHI). The ethanol bioanode consisted of a film in which Au nanoparticles were coimmobilized with alcohol dehydrogenase (ADH) and Meldola's blue (MDB). Figure 7.6 shows the schematic illustration of this cell as well as some electrochemical properties. NAD^+ dependent ADH is a useful enzyme for the construction of biofuel cells but one limitation is its high overpotential caused by oxidation of NADH at the electrode surface. The

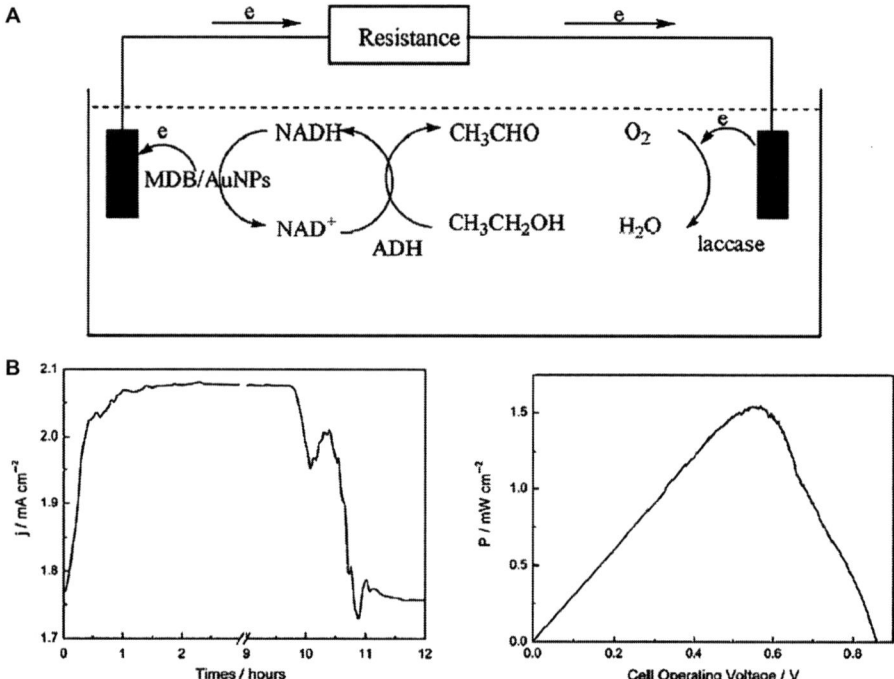

Figure 7.6   (A) Schematic illustration of the one-compartment biofuel cell. (B) Current generation under constant load in the biofuel cell in ethanol solution and polarization curve of the biofuel cell.

combination of Au nanoparticles and MDB improves the biofuel cell performance by decreasing the high potential caused by NADH oxidation. The power density obtained using the ethanol/oxygen biofuel cell is five times higher than that of most fuel cells reported in the literature.[126] The reasons for this improvement are the 3D structure of carbon increased the surface area of the electrode that resulted in a higher loading degree of the enzyme, the combination of ion-exchange capacity sol-gel and biopolymer chitosan that improved the protein environment, and the combination of Au nanoparticles and MDB that improved the biofuel performance by lowering the overpotential. Alternatively, wine was used in the biofuel cell instead of ethanol. The use of wine improved the overall performance of the ethanol/O_2 biofuel cell due to the presence of chemicals such as ethyl acid, glycol, phenolic components and vitamins of group B. Laccase is capable of oxidizing these chemicals, which results in improved power output.

Pt-Au bimetallic alloy nanoparticles were synthesized and used as the anode material for glucose oxidation.[127] Cyclic voltammetry was used to investigate the reactivity of various alloy nanoparticles including Pt, Au, Pt_{50}-Au_{50} alloy and Pt_{20}-Au_{80} in the presence of 0.1 M D-glucose. It was found that the glucose electro-oxidation happens at three potential regions: the potential range for pure Pt, the potential range for pure Au, and the potential range between

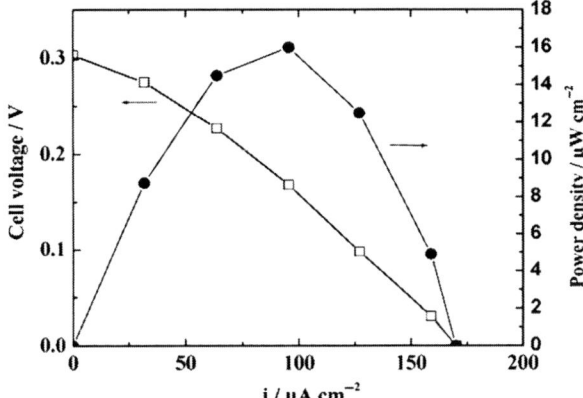

Figure 7.7 Performance of glucose/O_2 fuel cell using $Pt_{20}Au_{80}$ anode and biocathode (ABTS + laccase 27 mU cm^{-2}) at pH 5, 37 °C. a) cell voltage *vs.* current density; b) power density *vs.* current density.

Pt-OH and Au-OH formation. Previous work by Moller and Pistorius[128] showed that the presence of Au increased the current densities while the presence of Pt was useful for the dehydrogenation of glucose. Pt_{20}-Au_{80} alloy nanoparticles were applied at the anode in glucose oxidation while in the cathode laccase coimmobilized with 2,2′-azinobis (3-ethylbenzothiazoline-6-sulfonate) diammonium salt (ABTS 2-) played the role of a reducing agent for O_2. The power density and the cell voltage as a function of current density for this biofuel cell is shown in Figure 7.7.

To enhance the performance of the biofuel cell, two main parameters must be taken into consideration: the optimization of the nominal surface composition of Pt-Au, and maintaining the glucose concentration below 10 mM, which is advantageous for medical purposes.

Habrioux and coworkers[129] were able to electrochemically characterize adsorbed bilirubin oxidase (BOD) on Vulcan XC 72R to prepare a biocathode and apply it toward a glucose/O_2 biofuel cell that utilized $Au_{70}Pt_{30}$ nanoparticles as the anode material. The biocathode consisted of BOD adsorbed on Vulcan XC 72R and was immobilized onto a Nafion matrix. It was demonstrated that direct electron transfer was possible between BOD and Vulcan XC 72R. In the presence of an electrochemical mediator, the kinetic parameters of biocathode were enhanced. So in the designed biofuel cell, ABTS2- was used in the composition of biocathode. On the other hand, the application of $Au_{70}Pt_{30}$ nanoparticles as anode catalyst explained previously.[127,130] The activity of this material arises from the synergistic effect between Au and Pt. The dehydrogenation of the anomeric carbon of glucose happens on the Pt part at lower potential while Au increases the current density after the oxidation process. The performance of this biofuel cell was found to be higher relative to the membrane-less concentric system.[131]

Hoa *et al.*[132] designed a biohydrogen fuel cell in which the anode composed of conductive polymer nanocomposite. The fuel cell consisted of three main components: (1) a bioreactor that was responsible for producing hydrogen, (2) a filter to remove unwanted gases, and (3) a fuel cell to produce electricity from the pure humidified hydrogen. An *Escherichia coli* strain MC13-4 was used to produce hydrogen. PANI was mainly selected due to its ability to support the catalyst and enhance electron and proton transfer. The composition of platinum nanoparticles decorated on functionalized multiwalled carbon nanotubes (Pt/fMWCNTs) and polyaniline (PANI) were used as the anode for the biohydrogen fuel cells in two different configurations. In the first configuration thin films of PANI nanofibers were deposited on Pt/fMWCNTs, while in the second configuration, the cell consisted of a core–shell structure of PANI/Pt/fMWCNTs and Pt/fMWCNTs without PANI-based anodes. The results showed that the biohydrogen fuel cell consisting of the thin film of PANI nanofibers deposited on Pt/fMWCNTs/carbon paper as the anode showed much higher power density in comparison to the cell using a core–shell structure.

Ryu *et al.*[133] demonstrated the use of carbon nanotubes (CNTs) with Pt nanoislands for the design of a glucose biofuel cell. The goal of the work was to maximize glucose oxidation while maintaining a consistent amount of Pt. For this purpose an intense pulsed light (IPL) was used to modify the morphology of the catalyst layer. Using cyclic voltammetry the effect of modifying the Pt morphology on the glucose biofuel cell efficiency was monitored. The results indicate that enhancement in the glucose oxidation occurred with IPL induced on modified Pt-CNT electrode rather than the Pt/CNT electrodes which were not irradiated by IPL. As shown in Figure 7.8, from the *I–V* curve data, it was found that glucose oxidation on the Pt-CNT mat electrode was enhanced by the IPL treatment method. The nanoisland modification improved the cell power density 4.3 times compared to the as-sputtered Pt/CNT mat.

Thus, IPL-induced Pt nanoislands increased the activity platinum surface area, which resulted in an improved efficiency. On the other hand, higher

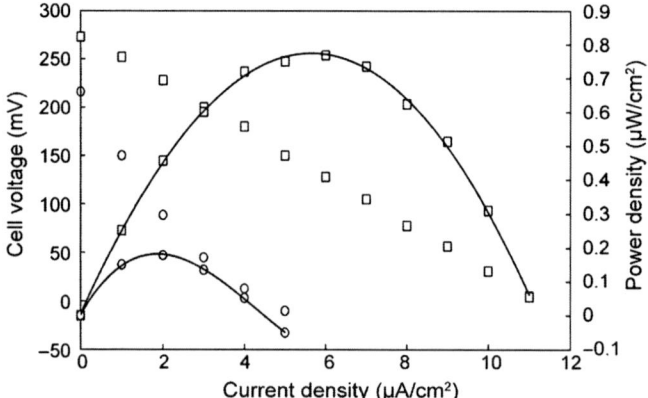

Figure 7.8 Performance of glucose/air biofuel cell using Pt–CNT mat electrodes: 3 min/IPL/7 min (□) and 10 min sputter (○).

stability output in long-term performance was reported for the biofuel cell based on modified Pt-CNT electrodes.

Murata *et al.*[134] studied the application of the direct electron transfer (DET) reaction of bilirubin oxidase (BOD) on three-dimensional Au nanoparticle electrodes in a biofuel cell. BOD can be used in biofuel cells because it catalyzes the O_2 reduction with low over potential at neutral pH. Murata *et al.* studied the DET reaction of BOD at thiol-modified and unmodified 3D Au nanoparticles surfaces. Remarkable stability of the catalytic current was observed from the electrodes. For construction of the biofuel cell carbon paper was used as the substrate electrode due to its large surface area that increased the current density. The performance of this biofuel cell was limited by the mass transfer of O_2 because O_2 reduction of BOD cathode is limited by the mass transfer of O_2. However, it was found that stirring was of crucial importance for practical applications in biofuel cells. It may be possible to improve the performance of the cell by applying the air-diffusion biocathode, because it can use gas-phase O_2. The use of BOD on carbon nanotubes[135] or BOD functionalized on various metals[136] is also an important area of investigation.

Kadirgan *et al.*[137] investigated the effect of carbon-supported Pt-Pd and Pt-Co nanoparticles as electrocatalysts for proton-exchange membrane fuel cells. They explored the advantages of using bimetallic nanoparticles as catalysts instead of common carbon-supported Pt nanoparticles. The Pt-Co/C catalyst was found to have better activity and stability (up to 60 °C) relative to commercially available Pt-C. On the other hand, synthesized Pt-Pd/C catalyst was found to have superior CO tolerance in comparison with commercial catalysts. This can result in an inhibition of the poisoning effect of the catalyst by the CO.

7.4 Future Perspectives

Literature reports have extensively demonstrated a suite of catalysts capable of converting biomass-derived materials into a variety of products ranging from chemicals to biofuels. In the case of biomass transformation to chemicals, the catalysts used include enzymes or bulk heterogeneous catalysts. The use of heterogeneous metallic catalysts requires high reaction temperatures and/or pressures for complete transformation. In addition, many of the reactions suffer low selectivities and/or low conversions, which then necessitate several intermediate separations to yield the final product. As a result, economic analyses of these systems have generally shown that the capital investment required for implementation is likely prohibitive. Since the conversion pathways are thermodynamically feasible, a reasonable conclusion is that the critical technological barrier to less energy- and capital-intensive biomass processing is the lack of catalysts capable of fulfilling the two essential requirements of **high selectivity** and **high reactivity** in executing the required cascade of elementary reactions. Nanoscale materials have demonstrated a significant role toward advancing biomass conversion processes, as showcased by select examples described in this chapter. Success in this area will require progress toward the rational design nanoscale catalysts that offer **highly selectivity, optimal efficiency, robustness** in aqueous and organic media, and the ability of the catalysts to **preselect and control reaction pathways** to ensure minimal or no separation to obtain the final product or intermediary of interest.

References

1. E. L. Kunkes, D. A. Simonetti, R. M. West, J. C. Serrano-Ruiz, C. A. Gartner and J. A. Dumesic, Catalytic conversion of biomass to mono-functional hydrocarbons and targeted liquid-fuel classes, *Science*, 2008, **322**, 417–421.
2. A. J. Ragauskas, C. K. Williams, B. H. Davison, G. Britovsek, C. A. Eckert, W. J. Frederick, J. P. Hallett, D. J. Leak, C. L. Liotta, J. R. Mielenz, R. Murphy, R. Templer and T. Tschaplinski, The path forward for biofuels and biomaterials, *Science*, 2006, **311**, 484–489.
3. J. N. Chheda, G. W. Huber and J. A. Dumesic, Liquid-phase catalytic processing of biomass-derived oxygenated hydrocarbons to fuels and chemicals, *Angewandte Chemie-International Edition*, 2007, **46**, 7164–7183.
4. J. C. Serrano-Ruiz and J. A. Dumesic, Catalytic routes for the conversion of biomass into liquid hydrocarbon transportation fuels, *Energy & Environmental Science*, 2011, **4**, 83–99.
5. J. C. Serrano-Ruiz, R. M. West and J. A. Durnesic, Catalytic Conversion of Renewable Biomass Resources to Fuels and Chemicals. In: J. M. Prausnitz, M. F. Doherty and M. A. Segalman, Eds. *Annual Review of Chemical and Biomolecular Engineering*, 2010, **1**, 79–100.
6. D. A. Simonetti and J. A. Dumesic, Catalytic Strategies for Changing the Energy Content and Achieving C–C Coupling in Biomass-Derived Oxygenated Hydrocarbons, *ChemSusChem*, 2008, **1**, 725–733.
7. D. G. Vlachos, J. G. G. Chen, R. J. Gorte, G. W. Huber and M. Tsapatsis, Catalysis Center for Energy Innovation for Biomass Processing: Research Strategies and Goals, *Catalysis Letters*, 2010, **140**, 77–84.
8. E. I. Gurbuz, E. L. Kunkes and J. A. Dumesic, Dual-bed catalyst system for C-C coupling of biomass-derived oxygenated hydrocarbons to fuel-grade compounds, *Green Chemistry*, 2010, **12**, 223–227.
9. B. G. Harvey and R. L. Quintana, Synthesis of renewable jet and diesel fuels from 2-ethyl-1-hexene, *Energy & Environmental Science*, 2010, **3**, 352–357.
10. U. V. Mentzel and M. S. Holm, Utilization of biomass: Conversion of model compounds to hydrocarbons over zeolite H-ZSM-5. *Applied Catalysis A-General*, 2011, **396**, 59–67.
11. J. O. Metzger, Production of liquid hydrocarbons from biomass, *Angewandte Chemie-International Edition*, 2006, **45**, 696–698.
12. R. M. West, E. L. Kunkes, D. A. Simonetti and J. A. Dumesic, Catalytic conversion of biomass-derived carbohydrates to fuels and chemicals by

formation and upgrading of mono-functional hydrocarbon intermediates, *Catalysis Today*, 2009, **147**, 115–125.

13. D. J. Braden, C. A. Henao, J. Heltzel, C. T. Maravelias and J. A. Dumesic, Production of liquid hydrocarbon fuels by catalytic conversion of biomass-derived levulinic acid, *Green Chemistry*, 2011, **13**, 1755–1765.

14. D. M. Alonso, J. Q. Bond, J. C. Serrano-Ruiz and J. A. Dumesic, Production of liquid hydrocarbon transportation fuels by oligomerization of biomass-derived C(9) alkenes, *Green Chemistry*, 2010, **12**, 992–999.

15. Y. G. Chen, C. Wang, W. P. Lu and Z. Y. Yang, Study of the co-deoxyliquefaction of biomass and vegetable oil for hydrocarbon oil production, *Bioresource Technology*, 2010, **101**, 4600–4607.

16. Y. G. Chen, F. Yang, L. B. Wu, C. Wang and Z. Y. Yang, Co-deoxyliquefaction of biomass and vegetable oil to hydrocarbon oil: Influence of temperature, residence time, and catalyst, *Bioresource Technology*, 2011, **102**, 1933–1941.

17. G. Fogassy, N. Thegarid, Y. Schuurman and C. Mirodatos, From biomass to bio-gasoline by FCC co-processing: effect of feed composition and catalyst structure on product quality, *Energy & Environmental Science*, 2011, **4**, 5068–5076.

18. T. R. Carlson, T. R. Vispute and G. W. Huber, Green gasoline by catalytic fast pyrolysis of solid biomass derived compounds, *ChemSusChem*, 2008, **1**, 397–400.

19. D. C. Elliott, T. R. Hart, G. G. Neuenschwander, L. J. Rotness and A. H. Zacher, Catalytic Hydroprocessing of Biomass Fast Pyrolysis Bio-oil to Produce Hydrocarbon Products, *Environmental Progress & Sustainable Energy*, 2009, **28**, 441–449.

20. A. G. Gayubo, A. T. Aguayo, A. Atutxa, R. Prieto and J. Bilba, Deactivation of a HZSM-5 zeolite catalyst in the transformation of the aqueous fraction of biomass pyrolysis oil into hydrocarbons, *Energy & Fuels*, 2004, **18**, 1640–1647.

21. S. Thangalazhy-Gopakumar, S. Adhikari, R. B. Gupta, M. B. Tu and S. Taylor, Production of hydrocarbon fuels from biomass using catalytic pyrolysis under helium and hydrogen environments, *Bioresource Technology*, 2011, **102**, 6742–6749.

22. C. Wang, Q. L. Hao, D. Q. Lu, Q. Z. Ja, G. J. Li and B. Xu, Production of light aromatic hydrocarbons from biomass by catalytic pyrolysis, *Chinese Journal of Catalysis*, 2008, **29**, 907–912.

23. C. B. Murray, D. J. Norris and M. G. Bawendi, Synthesis and characterization of nearly monodisperse CdE (E = sulfur, selenium, tellurium) semiconductor nanocrystallites, *Journal of the American Chemical Society*, 1993, **115**, 8706–8715.

24. H. Bönneman, W. Brijoux, R. Brinkmann, E. Dinjus, T. Fretzen, B. Joussen and J. Korall, Highly dispersive metal-clusters and colloids for the preparation of active liquid-phase hydrogenation catalysts, *Angwandte Chemie-International Edition in English*, 1992, **31**, 323.

25. R. G. Freemantle, M. Liu, W. Guo and S. O. Obare, 'Approaches to Synthesis and Characterization of Spherical & Anisotropic Palladium Nanomaterials,' in *Metallic Nanomaterials for Life Sciences*. C. S. S. R. Kumar, ed., Wiley-VCH, Weinheim, 2009, Vol. 1, pp. 305–355.
26. C. B. Wang and W. X. Zhang, Synthesizing nanoscale iron particles for rapid and complete dechlorination of TCE and PCBs, *Environmental Science and Technology*, 1997, **31**, 2154–2156.
27. A. Martino, M. Stoker, M. Hicks, C. H. Bartholomew, A. G. Sault and J. S. Kawola, The synthesis and characterization of iron colloids catalysts in inverse micelle solutions, *Applied Catalysis A General*, 1997, **161**, 235–248.
28. F. Li, C. Vipulanandan and K. K. Mohanty, Microemulsion and solution approaches to nanoparticle iron production for degradation of tri-chloroethylene, *Colloids and Surfaces A: Physicochemical and Engineering Aspects*, 2003, **223**, 103–112.
29. L. Wu, M. Shamsuzzoha and S. M. C. Ritchie, Preparation of cellulose acetate supported zero-valent iron nanoparticles for the dechlorination of trichloroethylene in water, *Journal of Nanoparticle Research*, 2005, **7**, 469–476.
30. L. Wu and S. M. C. Ritchie, Removal of trichloroethylene from water by cellulose acetate supported bimetallic Ni/Fe nanoparticles, *Chemosphere*, 2006, **63**, 258–292.
31. Z. H. Meng, H. L. Liu, Y. Liu, J. Zhang, S. Yu, F. Y. Cui, N. Q. Ren and J. Ma, Preparation and characterization of Pd/Fe bimetallic nanoparticles immobilized in PVDF. Al_2O_3 membrane for dechlorination of mono-chloroacetic acid, *Journal of Membrane Science*, 2011, **372**, 165–171.
32. J. Xu and D. Bhattacharyya, Fe/Pd nanoparticle immobilization in microfiltration membrane pores: Synthesis, characterization, and appli-cation in the dechlorination of polychlorinated biphenyls, *Industrial Engineering Chemical Research*, 2007, **46**, 2348–2359.
33. J. H. Sinfelt, Supported bimetallic-cluster catalysts, *Journal of Catalysis*, 1973, **29**, 308–315.
34. J. H. Sinfelt, Structure of bimetallic clusters, *Accounts of Chemical Research*, 1987, **20**, 134–139.
35. G. Meitzner, G. H. Via, F. W. Lytle and J. H. Sinfelt, Structure of bimetallic clusters – extended X-ray absorption fine-structure (EXAFS) studies of Ir-Rh clusters, *Journal of Chemical Physics*, 1983, **78**, 2533–2541.
36. G. Meitzner, G. H. Via, F. W. Lytle and J. H. Sinfelt, Structure of bimetallic clusters – extended X-ray absorption fine-structure (EXAFS) studies of Rh-Cu clusters, *Journal of Chemical Physics*, 1983, **78**, 882–889.
37. G. Meitzner, G. H. Via, F. W. Lytle and J. H. Sinfelt, Structure of bimetallic clusters – extended X-ray absorption fine-structure (EXAFS) studies of Ag-Cu and Au-Cu clusters, *Journal of Chemical Physics*, 1985, **83**, 4793–4799.
38. N. Toshima and T. Yonezawa, Bimetallic nanoparticles – Novel materials for chemical and physical applications, *New Journal of Chemistry*, 1998, **22**, 1179–1201.

39. T. T. Lim, J. Feng and B. W. Zhu, Kinetic and mechanistic examination of reductive transformation pathways of brominated methanes with nano-scale Fe and Ni/Fe particles, *Water Research*, 2007, **41**, 875–883.
40. T. T. Dong, H. Luo, Y. Wang, B. Hu and H. Chen, Stabilization of Fe-Pd bimetallic nanoparticles with sodium carboxymethyl cellulose for catalytic reduction of *para*-nitrochlorobenzene in water, *Desalination*, 2011, **271**, 11–19.
41. Z. Zhang, Q. Shen, N. Cissoko, J. Wo and X. Xu, Catalytic dechlorination of 2,4-dichlorophenol by Pd/Fe bimetallic nanoparticles in the presence of humic acid, *Journal of Hazardous Materials*, 2010, **182**, 252–258.
42. G. Zhang, Y. Kuang, J. Liu, Y. Cui, J. Chen and H. Zhou, Fabrication of Ag/Au bimetallic nanoparticles by UPD-redox replacement: Application in the electrochemical reduction of benzyl chloride, *Electrochemistry Communications*, 2010, **12**, 1233–1236.
43. C. J. Lin, Y. H. Liou and S. L. Lo, Supported Pd/Sn bimetallic nanoparticles for reductive dechlorination of aqueous trichloroethylene, *Chemosphere*, 2009, **74**, 314–319.
44. D. Liang, J. Gao, J. Wang, P. Chen, Y. Wei and Z. Hou, Bimetallic Pt–Cu catalysts for glycerol oxidation with oxygen in a base-free aqueous solution, *Catalysis Communications*, 2011, **12**, 1059–1062.
45. E. Taarning, A. T. Madsen, J. M. Marchetti, K. Egeblad and C. H. Christensen, Oxidation of glycerol and propanediols in methanol over heterogeneous gold catalysts, *Green Chemistry*, 2008, **10**, 408–414.
46. http://www.gold.org/discover/sci_indu/gold_catalysts/refcat.html [Accessed March 30, 2012].
47. W. C. Ketchie, Y. Fang, M. S. Wong, M. Murayama and R. J. Davis, Influence of gold particle size on the aqueous-phase oxidation of carbon monoxide and glycerol, *Journal of Catalysis*, 2007, **250**, 94–101.
48. W. C. Ketchie, M. Murayama and R. J. Davis, Selective oxidation of glycerol over carbon-supported AuPd catalysts, *Journal of Catalysis*, 2007, **250**, 264–273.
49. C. Bianchi, F. Porta, L. Prati and M. Rossi, Selective liquid phase oxidation using gold catalysts, *Topics in Catalysis*, 2000, **13**, 231–236.
50. F. Porta, L. Prati, M. Rossi, S. Coluccia and G. Marta, Metal sols as a useful tool for heterogeneous gold catalyst preparation: reinvestigation of a liquid phase oxidation, *Catalysis Today*, 2000, **61**, 165–172.
51. S. Demirel, P. Kern, M. Lucas and P. Claus, Oxidation of mono- and polyalcohols with gold: Comparison of carbon and ceria supported catalysts, *Catalysis Today*, 2007, **122**, 292–300.
52. S. Demirel, K. Lehnert, M. Lucas and P. Claus, Use of renewables for the production of chemicals: Glycerol oxidation over carbon supported gold catalysts, *Applied Catalysis B*, 2007, **70**, 637–643.
53. T. Hayashi, T. Inagaki, N. Itayama and H. Baba, Selective oxidation of alcohol over supported gold catalysts: methyl glycolate formation from ethylene glycol and methanol, *Catalysis Today*, 2006, **117**, 210–213.

54. N. Dimitratos, J. A. Lopez-Sanchez, J. M. Anthonykutty, G. Brett, A. F. Carley, R. C. Tiruvalam, A. A. Herzing, C. J. Kiely, D. W. Knight and G. J. Hutchings, Oxidation of glycerol using gold–palladium alloy-supported nanocrystals, *Physical Chemistry Chemical Physics*, 2009, **11**, 4952–4961.
55. Y. M. Yan, R. Tel-Vered, O. Yehezheli, Z. Cheglakov and I. Willner, Biocatalytic growth of au nanoparticles immobilized on glucose oxidase enhances the ferrocene-mediated bioelectrocatalytic oxidation of glucose, *Advanced Materials*, 2008, **20**, 2365–2370.
56. E. Katz and I. Willner, Probing biomolecular interactions at conductive and semiconductive surfaces by impedance spectroscopy: Routes to impedimetric immunosensors, DNA-sensors, and enzyme biosensors, *Electroanalysis*, 2003, **15**, 913–947.
57. L. Alfonta, A. Bardea, O. Khersonsky, E. Katz and I. Willner, Chrono-potentiometry and Faradaic impedance spectroscopy as signal transduction methods for the biocatalytic precipitation of an insoluble product on electrode supports: routes for enzyme sensors, immunosensors and DNA sensors, *Biosensors and Bioelectronics*, 2001, **16**, 675–687.
58. E. Katz, L. Alfonta and I. Willner, Chronopotentiometry and Faradaic impedance spectroscopy as methods for signal transduction in immuno-sensors, *Sensors and Actuators B-Chemistry*, 2001, **76**, 134–141.
59. L. Zhao, J. Zhou, H. Chen, M. Zhang, Z. Sui and X. Zhou, Carbon nanofibers supported Ru catalyst for sorbitol hydrogenolysis to glycols: Effect of calcinations, *Korean Journal of Chemical Engineering*, 2010, **27**, 1412–1418.
60. L. Zhao, J. H. Zhou, Z. J. Sui and X. G. Zhou, Hydrogenolysis of sorbitol to glycols over carbon nanofiber supported ruthenium catalyst, *Chemical Engineering Science*, 2010, **65**, 30–35.
61. V. Mazzieri, F. Coloma-Pascual, A. Arcoya and F. L'Argentière, XPS, FTIR and TPR characterization of Ru/Al_2O_3 catalysts, *Applied Surface Science*, 2003, **210**, 222–230.
62. P. G. J. Koopman, A. P. G. Kieboom and H. Van Bekkum, Character-ization of ruthenium catalysts as studied by temperature programmed reduction, *Journal of Catalysis*, 1981, **69**, 172–179.
63. A. Infantes-Molina, J. Mérida-Robles, J. L. G. Rodríguez-Castellón Fierro and A. Jiménez-López, Synthesis, characterization and catalytic activity of ruthenium-doped cobalt catalysts, *Applied Catalysis A-General*, 2008, **341**, 35–42.
64. K. Y. Wang, M. C. Hawley and T. D. Furney, Mechanism study of sugar and sugar alcohol hydrogenolysis using 1,3-diol model compounds, *Industrial & Engineering Chemistry Research*, 1995, **34**, 3766–3770.
65. M. A. Andrews and S. A. Klaeren, Selective hydrocracking of mono-saccharide carbon-carbon single bonds under mild conditions. Ruthe-nium hydride-catalyzed formation of glycols, *Journal of the American Chemical Society*, 1989, **111**, 4131–4133.
66. L. M. Ye, X. P. Duan, H. Q. Lin and Y. Z. Yuan, mproved performance of magnetically recoverable Ce-promoted Ni/Al_2O_3 catalysts

for aqueous-phase hydrogenolysis of sorbitol to glycols, *Catalysis Today*, 2012, **183**, 65–71.

67. J. C. Chao and D. T. A. Huibers, *Catalytic hydrogenolysis of alditols to product glycerol and polyols*, 1982, *US Patent, 4366332*.
68. S. P. Chopade, D. J. Miller, J. E. Jackson, T. A. Werpy, J. G. Frye and A. H. Zacher, *Catalysts and process for hydrogenolysis of sugar alcohols to polyols*. 2001, *US Patent 6291725*.
69. A. K. Sirkar. *Catalytic hydrogenolysis of alditols to produce polyols*. 1982, *US Patent 4338472*.
70. D. A. Simonetti, J. Rass-Hansen, E. L. Kunkes, R. R. Soares and J. A. Dumesic, Coupling of glycerol processing with Fischer-Tropsch synthesis for production of liquid fuels, *Green Chemistry*, 2007, **9**, 1073–1083.
71. R. Alcala, M. Mavrikakis and J. A. Dumesic, DFT studies for cleavage of C=C and C=O bonds in surface species derived from ethanol on Pt(111), *Journal of Catalysis*, 2003, **218**, 178–190.
72. V. Pallassana and M. Neurock, Reaction paths in the hydrogenolysis of acetic acid to ethanol over Pd(111), Re(0001), and PdRe alloys, *Journal of Catalysis*, 2002, **209**, 289–305.
73. J. Zhang, M. B. Vukmirovic, K. Sasaki, A. U. Nilekar, M. Mavrikakis and R. R. Adzic, Mixed-metal Pt monolayer electrocatalysts for enhanced oxygen reduction kinetics, *Journal of the American Chemical Society*, 2005, **127**, 12480–12481.
74. S. Czernik and A. V. Bridgwater, Overview of applications of biomass fast pyrolysis oil, *Energy Fuels*, 2004, **18**, 590–598.
75. A. V. Bridgwater, Catalysis in thermal biomass conversion, *Applied Catalysis A. General*, 1994, **116**, 5–47.
76. A. Fukuoka and P. L. Dhepe, Catalytic conversion of cellulose into sugar alcohols, *Angewandte Chemie International Edition*, 2006, **45**, 5161–5163.
77. Y. P. Zhang and L. R. Lynd, Toward an aggregated understanding of enzymatic hydrolysis of cellulose: noncomplexed cellulase systems, *Biotechnology & Bioengineering*, 2004, **88**, 797–824.
78. W. Deng, X. Tan, W. Fang, Q. Zhang and Y. Wang, Conversion of cellulose into sorbitol over carbon nanotube-supported ruthenium catalyst, *Catalysis Letters*, 2009, **133**, 167–174.
79. D. Klemm, B. Heublein, H. P. Fink and A. Bohn, Cellulose: Fascinating biopolymer and sustainable raw material, *Angewandte Chemie International Edition*, 2005, **44**, 3358–3393.
80. G. Pierre, Conversion of biomass to selected chemical products, *Chemical Society Reviews*, 2012, **41**, 1538–1558.
81. G. W. Huber, S. Iborra and A. Corma, Synthesis of transportation fuels from biomass: chemistry, catalysts, and engineering, *Chemical Reviews*, 2006, **106**, 4044–4098.
82. P. L. Dhepe and A. Fukuoka, Cracking of cellulose of supported metal catalysts, *Catalysis Surveys from Asia*, 2007, **11**, 186–191.
83. P. L. Dhepe and A. Fukuoka, Cellulose conversion under heterogeneous catalysis, *ChemSusChem*, 2008, **1**, 969–975.

84. N. Yan, C. Zhao, C. Luo, P. J. Dyson, H. Liu and Y. Kou, One-step conversion of cellobiose to C_6-alcohols using a ruthenium nanocluster catalyst, *Journal of the American Chemical Society*, 2006, **128**, 8714–8715.

85. S. Suganuma, K. Nakajima, M. Kitano, D. Yamguchi, H. Kato, S. Hayahsi and M. Hara, Hydrolysis of cellulose by amorphous carbon bearing SO_3H, COOH, and OH groups, *Journal of the American Chemical Society*, 2008, **130**, 12787–12793.

86. K. Fujita, N. Nakamura, K. Murata, K. Igarashi, M. Samejima and H. Ohno, Electrochemical analysis of electrode-immobilized dehydrogenases in hydrated choline dihydrogen phosphate type ionic liquid, *Electrochim. Acta*, 2011, **56**, 7224–7227.

87. G. Frens, Controlled nucleation for the regulation of the particle size in monodisperse gold suspensions, *Nature Physical Science*, 1973, **241**, 20–22.

88. M. Ash and E. Dohlman, Oil Crops Situation and Outlook Year Book: Electronic Outlook Report from the Economic Research Service, United States Department of Agriculture, 2007, 1–83.

89. E. Lotero, Y. Liu, D. E. Lopez, K. Suwannakarn, D. A. Bruce and J. G. Goodwin, *Industrial & Engineering Chemical Research*, 2005, **44**, 5353–5363.

90. A. Centeno, E. Laurent and B. Delmon, Influence of the support of CoMo sulfide catalysts and of the addition of potassium and platinum on the catalytic performances for the hydrodeoxygenation of carbonyl, carboxyl, and guaiacol-type molecules, *Journal of Catalysis*, 1995, **154**, 288–298.

91. P. Maki-Arvela, I. Kubickova, M. Snare, K. Eranen and D. Y. Murzin, Catalytic deoxygenation of fatty acids and their derivatives, *Energy Fuels*, 2007, **21**, 30–41.

92. E. W. Ping, R. Wallace, J. Pierson, T. F. Fuller and C. W. Jones, Highly dispersed palladium nanoparticles on ultra-porous silica mesocellular foam for the catalytic decarboxylation of stearic acid, *Microporous and Mesoporous Materials*, 2010, **132**, 174–180.

93. P. Schmidt-Winkel, W. W. Lukens, D. Y. Zhao, P. D. Yang, B. F. Chmelka and G. D. Stucky, *Journal of the American Chemical Society*, 1999, **121**, 254–255.

94. E. W. Ping, J. Pierson, R. Wallace, J. T. Miller, T. F. Fuller and C. W. Jones, On the nature of the deactivation of supported palladium nanoparticle catalysts in the decarboxylation of fatty acids, *Applied catalysis A: General*, 2011, **396**, 85–90.

95. P. Han, X. Wang, X. Qui, X. Ji and L. Gao, One-step synthesis of palladium/SBA-15 nanocomposites and its catalytic application, *Journal of Molecular Catalysis A. Chemistry*, 2007, **272**, 136–141.

96. A. Katiyar, S. Yadev, P. G. Smirniotis and N. G. Pinto, Synthesis of ordered large pore SBA-15 spherical particles for adsorption of biomolecules, *Journal of Chromatography A*, 2006, **1122**, 13–20.

97. S. Lestari, P. Maki-Arvela, K. Eranen, J. Beltramini, G. Q. Max-Lu and D. Y. Murzin, Hydrocarbons from catalytic deoxygenation of stearic acid over supported Pd nanoparticles on SBA-15 catalysts, *Catalysis Letters*, 2010, **134**, 250–257.

98. S. Lestari, I. Simakova, A. Tokarev, P. Maki-Arvela, K. Eranen and D. Y. Murzin, Synthesis of biodiesel via deoxygenation of stearic acid over supported Pd/C catalyst, *Catalysis Letters*, 2008, **122**, 247–251.

99. X. Deng, Z. Fang, Y. H. Liu and C. L. Yu, Production of biodiesel from jatropha oil catalyzed by nanosized solid basic catalyst, *Energy*, 2011, **36**, 777–784.

100. M. M. Gui, K. T. Lee and S. Bhatia, Feasibility of edible oil *vs.* non-edible oil *vs.* waste edible oil as biodiesel feedstock, *Energy*, 2008, **33**, 1646–1653.

101. M. A. Aramendía, V. Borau, C. Jiménez, J. M. Marinas, J. R. Ruiz and F. J. Urbano, Catalytic hydrogen transfer from 2-propanol to cyclohexanone over basic Mg–Al oxides, *Applied Catalysis A. General*, 2003, **255**, 301–308.

102. X. Deng, Z. Fang and Y. H. Liu, Ultrasonic transterification of Jatropha curcas L. oil to biodiesel by a two-step process, *Energy Conversion and Management*, 2010, **51**, 2802–2807.

103. W. L. Xie and H. Li, Alumina-supported potassium iodide as a heterogeneous catalyst for biodiesel production from soybean oil, *Journal of Molecular Catalysis A. Chemistry*, 2006, **255**, 1–9.

104. N. Barakos, S. Pasias and N. Papyannakos, Transesterification of triglycerides in high and low quality oil feeds over an HT2 hydrotalcite catalyst, *Bioresource Technology*, 2008, **99**, 5037–5042.

105. B. P. Binks, Particles as surfactants-similarities and differences, *Current Opinion in Colloid & Interface Science*, 2002, **7**, 21–41.

106. B. P. Binks and C. P. Whitby, Silica particle-stabilized emulsions of silicone oil and water: Aspects of emulsification, *Langmuir*, 2004, **20**, 1130–1137.

107. B. P. Binks, J. Philip and J. A. Rodrigues, Inversion of silica-stabilized emulsions induced by particle concentration, *Langmuir*, 2005, **21**, 3296–3302.

108. R. K. Wang, H. O. Park, W. C. Chen, C. Silvera-Batista, R. D. Reeves, J. E. Butler and K. J. Ziegler, *Journal of the American Chemical Society*, 2008, **130**, 14721–14728.

109. D. E. Resasco, W. E. Alvarez, F. Pompeo, L. Balzano, J. E. Herrera, B. Kitiyanan and A. Borgna, *Journal of Nanoparticle Research*, 2002, **4**, 131–136.

110. M. Shen and D. E. Resasco, Emulsions stabilized by carbon nanotube-silica nanohybrids, *Langmuir*, 2009, **25**, 10843–10851.

111. A. Heller, Miniature biofuel cells, *Physical Chemistry Chemical Physics*, 2004, **6**, 209–216.

112. E. Katz, I. Willner and A. B. Kotlyar, A non-compartmentalized glucose/O_2 biofuel cell by bioengineered electrode surfaces, *Journal of Electroanalytical Chemistry*, 1999, **479**, 64–68.

113. G. Tayhas, R. Palmore and H.-H. Kim, Electro-enzymatic reduction of dioxygen to water in the cathode compartment of a biofuel cell, *Journal of Electroanalytical Chemistry*, 1999, **464**, 110–117.

114. T. Ikeda and K. Kano, An electrochemical approach to the studies of biological redox reactions and their applications to biosensors, bio-reactors, and biofuel cells, *Journal of Bioscience and Bioengineering*, 2001, **92**, 9–18.

115. T. Chen, S. C. Barton, G. Binyamin, Z. Gao, Y. Zhang, H.-H. Kim and A. Heller, A miniature biofuel cell, *Journal of the American Chemical Society*, 2001, **123**, 8630–8631.

116. S. Tsujimura, H. Tatsumi, J. Ogawa, S. Shimizu, K. Kano and T. Ikeda, Bioelectrocatalytic reduction of dioxygen to water at neutral pH using bilirubin oxidase as an enzyme and 2,2′-azinobis (3-ethylbenzothiazolin-6-sulfonate) as an electron transfer mediator, *Journal of Electroanalytical Chemistry*, 2001, **496**, 69–75.

117. H.-H. Kim, N. Mano, Y. Zhang and A. Heller, A miniature membrane-less biofuel cell operating under physiological conditions at 0.5 V, *Journal of the Electrochemical Society*, 2003, **150**, A209–A213.

118. N. Mano, F. Mao and A. Heller, Characteristics of a miniature compartment-less glucose–O_2 biofuel cell and its operation in a living plant, *Journal of the American Chemical Society*, 2003, **125**, 6588–6594.

119. N. Mano, F. Mao and A. Heller, On the parameters affecting the characteristics of the "wired" glucose oxidase anode, *Journal of Electroanalytical Chemistry*, 2005, **574**, 347–357.

120. E. Farneth and M. B. D'Amore, Encapsulated laccase electrodes for fuel cell cathodes, *Journal of Electroanalytical Chemistry*, 2005, **581**, 197–205.

121. D. Basu and S. Basu, Synthesis, characterization and application of platinum based bi-metallic catalysts for direct glucose alkaline fuel cell, *Electrochimica Acta*, 2011, **56**, 6106–6113.

122. C. Jin and I. Taniguchi, Electrocatalytic activity of silver modified gold film for glucose oxidation and its potential application to fuel cells, *Materials Letters*, 2007, **61**, 2365–2367.

123. L. H. Li, W. D. Zhang and J. S. Ye, Electrocatalytic oxidation of glucose at carbon nanotubes supported PtRu nanoparticles and its detection, *Electroanalysis*, 2008, **20**, 2212–2216.

124. I. Becerik, S. Suzer and F. Kadırgan, Platinum–palladium loaded poly-pyrrole film electrodes for the electrooxidation of d-glucose in neutral media, *Journal of Electroanalytical Chemistry*, 1999, **476**, 171–176.

125. L. Deng, L. Shang, D. Wen, J. Zhai and S. Dong, A membraneless biofuel cell powered by ethanol and alcoholic beverage, *Biosensors and Bioelectronics*, 2010, **26**, 70–73.

126. L. Deng, F. A. Wang, H. J. Chen, L. Shang, L. Wang, T. Wang and S. J. Dong, A biofuel cell with enhanced performance by multilayer biocatalyst immobilized on highly ordered macroporous electrode, *Biosensors and Bioelectronics*, 2008, **24**, 329–333.

127. A. Habrioux, E. Sibert, K. Servat, W. Vogel, K. B. Kokoh and N. Alonso-Vante, Activity of platinum-gold alloys for glucose electro-oxidation in biofuel cells, *Journal of Physical Chemistry B*, 2007, **111**, 10329–10333.
128. H. Möller and P. C. J. Pistorius, The electrochemistry of gold–platinum alloys, *Electroanalytical Chemistry*, 2004, **570**, 243–255.
129. A. Habriouxa, T. Napporna, K. Servata, S. Tingryb and K. B. Kokoh, Electrochemical characterization of adsorbed bilirubin oxidase on Vulcan XC 72R for the biocathode preparation in a glucose/O_2 biofuel cell, *Electrochimica Acta*, 2010, **55**, 7701–7705.
130. A. Habrioux, W. Vogel, M. Guinel, L. Guetaz, K. Servat, B. Kokoh and N. Alonso-Vante, Structural and electrochemical studies of Au–Pt nanoalloys, *Physical Chemistry Chemical Physics*, 2009, **11**, 3573–3579.
131. A. Habrioux, K. Servat, S. Tingry and K. B. Kokoh, Enhancement of the performances of a single concentric glucose/O_2 biofuel cell by combination of bilirubin oxidase/Nafion cathode and Au–Pt anode, *Electrochemical Communications*, 2009, **11**, 111–113.
132. L. Q. Hoa, Y. Sugano, H. Yoshikawa, M. Saito and E. Tamiya, A bio-hydrogen fuel cell using a conductive polymer nanocomposite based anode, *Biosensors and Bioelectronics*, 2010, **25**, 2509–2514.
133. J. Ryu, H. S. Kim, H. T. Hahn and D. Lashmore, Carbon nanotubes with platinum nano-islands as glucose biofuel cell electrodes, *Biosensors and Bioelectronics*, 2010, **25**, 1603–1608.
134. K. Murata, K. Kajiya, N. Nakamura and H. Ohno, Direct electro-chemistry of bilirubin oxidase on three-dimensional gold nanoparticle electrodes and its application in a biofuel cell, *Energy & Environmental Science*, 2009, **2**, 1280–1285.
135. L. Hussein, Y. J. Feng, N. Alonso-Vante, G. Urban and M. Krüge, Functionalized-carbon nanotube supported electrocatalysts and bucky-paper-based biocathodes for glucose fuel cell applications, *Electrochimica Acta*, 2011, **56**, 7659–7665.
136. Y.-M. Yan, I. Baravik, R. Tel-Vered and I. Willner, An ethanol/O_2 biofuel cell based on an electropolymerized bilirubin oxidase/Pt nano-particle bioelectrocatalytic O_2-reduction cathode, *Advanced Materials*, 2009, **21**, 4275–4279.
137. F. Kadirgan, A. M. Kannan, T. Atilan, S. Beyhan, S. Ozenler, S. Suzer and A. Yörür, Carbon supported nano-sized Pt–Pd and Pt–Co electro-catalysts for proton exchange membrane fuel cells, *International Journal of Hydrogen Energy*, 2009, **34**, 9450–9460.

CHAPTER 8

Toxicology of Designer/ Engineered Metallic Nanoparticles

H.-M. HWANG,*[a] P. C. RAY,[b] H. YU[b] AND X. HE[a]

[a] Jackson State University, Department of Biology, Box 18540, 1400 Lynch Street, Jackson, MS 39217, USA; [b] Jackson State University, Department of Chemistry and Biochemistry, Box 17910, 1400 Lynch Street, Jackson, MS 39217, USA
*Email: huey-min.hwang@jsums.edu

8.1 Introduction

In the last few years, nanotechnology has been rapidly evolving from the discovery phase to the application phase; and as a result, nanomaterials in consumer products will be released to environmental media.[1] The recent report on "An Inventory of Nanotechnology-based Consumer Products Currently on the Market" indicates that as of March 10, 2011, the nanotechnology consumer products inventory contains 1317 products.[2] Metallic nanoparticles include both elemental metallic nanoparticles (e.g., Ag, Au) and metal-oxide nanoparticles (e.g., TiO_2, ZnO). Metallic nanoparticles are believed to be the basis of many of the future technological and biomedical innovations of this century.[3–5] The ability to integrate metallic nanoparticles into biological systems will have the greatest impact in biology and medicine.[6–20] Engineered metal-oxide nanoparticles have been widely used in consumer products such as cosmetics and sunscreens, self-cleaning coatings and textiles. Additionally, they are also

RSC Green Chemistry No. 19
Sustainable Preparation of Metal Nanoparticles: Methods and Applications
Edited by Rafael Luque and Rajender S Varma
© The Royal Society of Chemistry 2013
Published by the Royal Society of Chemistry, www.rsc.org

used as water-treatment agents, ingredients of solar batteries and the newer automobile catalytic converters.[21] However, it has been shown recently that some nanosized particles, not their macro- or microcounterparts, are toxic to some organisms.[22] It is therefore possible that sunscreens that contain these particles may be more hazardous than UV-radiation itself and that the use of some of these solar batteries may introduce higher environmental risk than carbon dioxide emission from conventional energy sources. Lack of toxicological data on nanomaterials makes it difficult to determine if there is a risk associated with nanomaterials exposure. Thus, there is an urgent need to develop rapid, accurate and efficient testing strategies to assess the potential hazard of these emerging materials.

Prevention is always the best strategy to minimize human or environmental exposure to hazardous nanomaterials. Early identification of the potentially hazardous properties of nanomaterials could enable us to design or redesign these materials with less health impact, while retaining the main desirable properties. Much of the existing research on nanotoxicity has concentrated on empirical evaluation of the toxicity of various nanoparticles with less attention given to the relationship between nanoparticle physicochemical properties (including chemical composition, shape, size and size distribution, dispersion, aggregation state, surface area, surface chemistry, surface charge, and porosity) and toxicity.[23] This approach gives limited information, and should not be considered adequate for developing predictions of toxicity of seemingly similar nanomaterials. Instead, if condition permits, we should adopt a systematic approach for studying the toxicity of engineered nanoparticles by focusing on a group of test nanomaterials of similar physicochemical properties with variation in one particular parameter.[24] It is also worth noting that characterization of supplied nanoparticles may not actually represent the physicochemical properties of the particles during or following administration. Therefore, wherever possible, independent characterization of test nanoparticles should be conducted before and after administration.[23]

Induction of intracellular oxidative stress by reactive oxygen species (ROS) seems to be a common key event of the biological effects of many metallic nanoparticles. For example, TiO_2 exposure led to observable alteration in various intracellular structures in the study cyanobacteria cells and induced a series of stress responses, including production of ROS, appearance and increase in the abundance of membrane crystalline inclusions, and internal plasma membrane disruption.[25] In most cases of metal-oxide nanoparticles exposure, either by entering into the cells or attaching to the cell membrane, particles may also induce intracellular oxidative stress by disturbing the balance between oxidant and antioxidant processes.[26–29] Therefore, assessing formation and extent of ROS is regarded as the benchmark for studying the mechanism of nanotoxicity *in vitro* and *in vivo*.

The objectives of this review are to provide an overview on the relationship between individual physicochemical parameters and biological response after exposure to engineered elemental metallic nanoparticles (with a focus on gold and silver) and metal-oxide nanoparticles. Possible mechanisms of observed toxicity, the influence of biotic and abiotic factors on particle toxicity, and different biological models (*in vitro* and *in vivo*) for studying nanotoxicity are discussed. Doped metal-oxide nanoparticles, new designer products in the recent development of safe nanotechnology application, are also reviewed. Finally, a brief discussion on research gaps and collaboration needed in nanoecotoxicity studies is provided. To provide a systematic review on nano-toxicity, we arrange discussion points based on individual physicochemical parameters. Such an organization may seem arbitrary or redundant throughout the text, as some original studies in the cited literature may concern multiple influential parameters.

8.2 Biophysicochemical Interactions (Nano/Bio Interface)

Interactions between nanomaterials and cellular materials typically take place at the particle surface; therefore, particle surface-cellular interactions may take precedence over the core component or particle size or surface area in determining the ultimate toxic responses in biological tissues and organs.[30] Readers are encouraged to retrieve detailed information on biophysicochemical interactions at the nano/bio interface from the report by Nel *et al.*[31]

8.2.1 Engineered Nanoparticles

8.2.1.1 Physicochemical Factors

Dose Metrics. Careful consideration should be given to the metric used to quantify dose in a nanotoxicity screening study. Among various metric units, mass, surface area or particle number are frequently used to enable quantitative interpretation of data. In their report, Oberdörster *et al.*[23] strongly recommended that sufficient information should be collected to allow dose against the three aforementioned primary physicochemical metrics to be derived. Without this information, the experimental results could be misinterpreted. One of our recent studies is provided below[32] for discussion on the relevant parameters.

Shapes. Engineered nanomaterials can be synthesized in highly homogeneous forms of desired shapes (*e.g.*, rings, spheres, rods, planes). Particle interactions will be influenced by particle shape, because diffusion rates of the material will change with the aspect ratio of the material and steric hindrance in the collisions. Thus, the shape may create some difficulty for particles to approach each other.[33] Limited research on the relationship between shape and dose of a material, and biological endpoint response of the test bioassay had been conducted. The shape of nanoparticles may have effects on the kinetics of deposition and adsorption within the biological tissues/organs. For example, Chithrani *et al.*[34] investigated the intracellular uptake of different sized and shaped colloidal gold nanoparticles by HeLa cells. The results showed that kinetics and saturation concentrations are highly dependent upon the physical dimensions of the nanoparticles.

Sizes, Surface Area, Surface Reactivity, and Surface Coating. Hypothetically, smaller-sized nanomaterials have lower thermal stability and are likely to be more toxic because of greater bioavailability *via* greater specific surface area. As shown in Figure 8.1,[35,36] the per cent of surface molecules changes as a function of particle size. Surface molecules increase exponentially when particle size decreases < 100 nm, reflecting the importance of surface area for increased chemical and biologic activity of engineered nanoparticles. Note the dramatic increase in the slopes at diameter < 10 nm.

Indeed, Ma *et al.*[37] reported particle-size-dependent phototoxicity of ZnO nanoparticles to nematode *Caenorhabditis elegans*. The observed phototoxicity was well correlated with photocatalytic ROS generation of ZnO particles. In addition, the nanosized ZnO showed greater phototoxicity than bulk ZnO, despite their similar size of aggregates, indicating that primary particle size is more important than aggregate size in determining ultimate phototoxicity. In addition, Choi and Hu[38] reported size-dependent inhibition by Ag nanoparticles and ROS-related nanosilver toxicity to nitrifying bacteria cultivated in a stirred tank reactor. By examining the correlation between nanoparticle-size distribution (between 9–21 nm), photocatalytic ROS generation, intracellular ROS accumulation, and nitrification inhibition, they observed that inhibition to nitrifying organisms correlated with the fraction of Ag nanoparticles less than 5 nm in the suspension. These nanosized nanoparticles appeared to be more toxic to bacteria than any other fractions of nanoparticles or their counterpart bulk species.

Several cytotoxicity and biocompatibility studies in the literature primarily focused on *in vitro* cell cultures. These reports indicated that gold nanoparticle

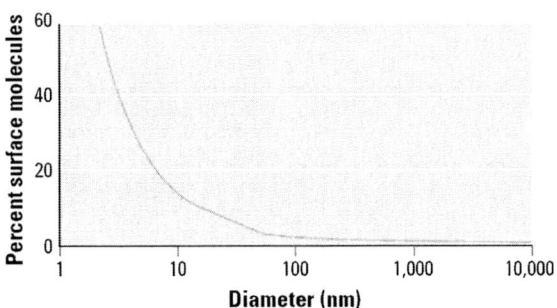

Figure 8.1 Surface molecules as a function of particle size. Surface molecules increase exponentially when particle size decreases < 100 nm, reflecting the importance of surface area for increased chemical and biologic activity of engineered nanoparticles. The increased biologic activity can be positive and desirable, negative and undesirable, or a mix of both. Figure courtesy of H. Fissan (personal communication).
(Reproduced with permission from refs. 35 and 36.)

are either not cytotoxic or they do not display size-dependent toxicity after coating with various ligands.[3,5–10,39–46] Spherical gold nanoparticles of different sizes are not inherently toxic to human skin cells, but gold nanorods are highly toxic due to the presence of Cetyltrimethyl ammonium bromide (CTAB) as coating material. Poly(4-styrene sulfonate) (PSS)-coated gold nanorods are not toxic. Therefore, it was concluded that coating with biocompatible and functionalization friendly stabilizing agents is essential for using of gold nanorods in living cells. Connor et al.[41] tested the toxicity of gold nanoparticles with different coating materials (citrate, cysteine, glucose, biotin) against K562 leukemia cell line, when leukemia cells were exposed to the nanoparticles for three days. They suggested none of the spherical gold nanoparticles were toxic to the human leukemia cells even up to ~100 µM in gold atom concentration.

Lu et al.[44] reported how surface coating plays an important role on the environmental fate, cellular uptake, and toxic effects of silver nanomaterials. Freshly prepared colloidal spherical silver nanoparticles and silver nanoprisms of 30 nm sizes are not cyto-, photo- or genotoxic to human skin HaCaT keratinocyte cells after 48 h of incubation, whereas silver nitrate is toxic at concentrations as low as 10 µg/ml and also phototoxicity increases in concentration dependent manner. Interestingly, Jiang et al.[47] reported that in their study gold and silver nanoparticles coated with antibodies can regulate the process of membrane receptor internalization. The binding and activation of membrane receptors and subsequent protein expression strongly depends on nanoparticle size. Although all nanoparticles within the size range (2–100 nm) were found to alter signaling processes essential for basic cell functions, 40- and 50-nm nanoparticles appeared to demonstrate the greatest effect. They speculated that the higher surface curvature of smaller nanoparticles actually restricted the relative orientation between molecules with certain degrees of conformational rigidity and their docking surface during the absorption process, thus resulting in large background areas without protein coverage. Consequently, larger nanoparticles (40–50 nm) eventually have a higher protein-to-nanoparticle ratio, which maximized protein loading. Apparently, biophysicochemical interactions should not be reasoned with particle size only.

When the particle size is decreased, the proportion of atoms available at the surface is increased relative to the proportion within its volume. Therefore, particles of nanoscale are likely to be more catalytically reactive.[48] From the perspective of human health, the reactive groups on the surface (or surface coating) of a particle are prone to modify the function of biological molecules and illicit toxicological effects. As indicated earlier, the size of nanoparticles alone may not be sufficient to determine their toxicity. Other factors such as overall number, total surface area and surface coating materials are also important. Ying and Hwang[32] investigated the dependence of cytotoxicity on particle size and surface coating of iron-oxide nanoparticles in an *in vitro* study using the A3 human T lymphocyte as a model system. Two different sizes (10 nm and 50 nm) and two different surface coatings (amine and carboxyl groups) of iron-oxide nanoparticles were tested with fluorescein diacetate (FDA) assay and Water Soluble Tetrazolium salt (WST)-1 assay. The results from both assays indicated size- and surface-coating-dependent toxicity to A3 cells in terms of mass concentration. Iron-oxide nanoparticles of the smaller size are more toxic than those of the larger size. Iron-oxide nanoparticles with the carboxyl group have higher toxicity than those with the amine group. Nevertheless, in terms of the number of particles per well and the resultant total surface area in the cell plates used for study, the 50-nm iron-oxide nanoparticles are more toxic than those of size 10 nm. This finding is in agreement with the report by Yin *et al.*[49] The speculation is that the effective interaction area for accessing the cell is greater for a larger particle than that for a smaller one. Within this specific area, there are more functional groups on the individual larger particles. Thus, each larger particle exerts a stronger stimulus on the cells. If an equal mass of particles with different sizes is considered, the significantly larger number of small particles increases the number of interaction points that are randomly distributed around the cell. Due to the likelihood of a smaller number of functional groups on individual small particles, each interaction point exerts a weak stimulus on the cells. The experimental results seem to suggest that the total sum of weak stimuli at different locations resulted in a reduced toxic effect than that by one localized strong stimulus from a bigger particle.

The findings in this study revealed that size and surface coating could affect the nanotoxicity in biological systems and interpretation of the cytotoxicity of nanoparticles could vary with the mass concentration, the total number of particles per well, and the total surface area of particles per well. In addition, modification of surface coating could overcome the cytotoxicity of metal-oxide nanoparticles. Based on the aforementioned findings by Ying and Hwang[32] and Jiang *et al.*,[47] with specific coating treatment nanoparticles could play an active role in mediating biological effects. Moreover, these types of finding may assist in the design of nanoscale delivery and clinical applications such as drug delivery and medical treatment.

Form (Liquid *vs.* Solid). The report "An Inventory of Nanotechnology-based Consumer Products Currently on the Market" indicates that as of March 10, 2011, the nanotechnology consumer products inventory contains 313 silver-nanoparticle-based products. This is about 30% of total nanotechnology-based products.[2] Asahi *et al.*[50] reported that a silver nanoparticle is capable of mitochondrial dysfunction, induction of ROS and chromosomal aberrations. Their cytokinesis-blocked micronucleus assay results indicated possible chromosomal breaks in Ag-nanoparticle-treated cells. Their result shows that the extent of DNA damage was much higher in cancer cells as compared to fibroblasts, and significant numbers of micronuclei were formed in cancer cells than fibroblasts. The toxicity of nanosilver can be attributed to the associated silver cation formation after oxidation of the nanosurface and also its ability to generate ROS.[4,10,46,51–67] A recent report shows that the majority of silver ions come from oxidation of the zero-valent metallic particle, typically by reaction with dissolved O_2 mediated by protons and other components in the surroundings.[4,56–64] Since it is well known that silver ions can bind to biological thiol groups in enzymes, they may disrupt the bacterial respiratory chain through oxidative stress and allow cell damage.

Recently, Gomes *et al.*[68] reported copper nanoparticles toxicity in the gills of mussels *Mytilus galloprovincialis*. Mussels accumulated copper in the gills; and it induced oxidative stress in mussels by overwhelming gills antioxidant defense system. Exposure to CuO nanoparticles resulted in lipid peroxidation in mussels. Griffit *et al.*[69] reported silver, copper, aluminum, nickel, and cobalt nanoparticles toxicity using zebrafish, daphnids, and an algal species model. Nanosilver and nanocopper cause toxicity in all organisms tested, with 48 h median lethal concentrations as low as 40 and 60 μg/L, respectively. Nanoparticulate forms of metals were less toxic than the soluble forms.

The role of released metal ions in liquid media was heavily taken into account for nanotoxicity studies on ZnO[70–72] and Ag.[73] Several investigators attributed nanoparticle toxicity to dissolved toxic ions, rather than the particles themselves. Thus the importance of particle solubility has received attention in the nanotoxicity research. In a recent study conducted by Jiang *et al.*,[27] the toxicity of nanoscaled aluminum, silicon, titanium and zinc oxides to several bacteria (*Bacillus subtilis*, *E. coli* and *Pseudomonas fluorescens*) was examined and compared to that of their respective bulk (microscaled) counterparts. All nanoparticles except titanium oxide showed higher toxicity (at 20 mg/L) than their bulk counterparts. The toxicity of released metal ions was differentiated from that of the oxide particles. TEM images showed attachment of nanoparticles to bacteria, suggesting that the toxicity was affected by the attachment. Based on this finding, they concluded that the toxicity of nanoparticles was not only from dissolved metal ions, but also from their greater tendency to attach to the cell walls.

Interestingly, Heinlaan *et al.*[72] noticed that bioavailable metal ions (detached from the surface) were responsible for the toxicity of ZnO and CuO to *Vibrio fisheri* bacteria. They also observed that, in the case of CuO, which is traditionally classified as insoluble, the bioavailability of metal ions from nanoparticles was much higher than bioavailability of ions from bulk. Thus, CuO metal oxides in their nanoforms were remarkably more "soluble". Moreover, the increased solubilization of metal ions and, in consequence, generation of ROS results in damaging of cell membranes. Because of that, metal-oxide nanoparticles do not necessarily have to even enter the cells to cause cellular toxicity.

Aggregation/Stability. One of the main challenges scientists are facing in understanding the toxicity of nanoparticle is that nanoparticles may aggregate in cell-culture media due to exposure to ions and proteins.[5,10,40,44–46] Aggregation refers to the collection of nanoparticles that are held together by various forces such as van der Waals and electrostatic interaction.[23] Under ambient conditions, some nanoparticles can form aggregates or agglomerates in various forms. Manufacturers thus frequently add coatings to prevent this process. Characterization of the agglomeration state should include primary (primary particles), secondary (primary particle agglomerates and self-assembled structures) and tertiary (assembles of secondary structures) scales.[23] This can be accomplished with measurement of the zeta potential. The zeta potential is an abbreviation for the electrokinetic potential in colloidal systems. From a theoretical viewpoint, the zeta potential is the electric potential in the interfacial double layer at the location of the slipping plane versus a point in the bulk fluid away from the interface. In other words, the zeta potential is the potential difference between the dispersion medium and the stationary layer of fluid attached to the dispersed particle. Zeta potential units are mV. The value of 25 mV can be taken as the boundary that separates low-charged surfaces from highly charged surfaces. The significance of the zeta potential is that its value is related to the stability of colloidal dispersions. Colloids with high zeta potential (negative or positive) are electrically stabilized. Recently, Albenese and Chan[40] examined the effect of aggregation on the cellular uptake and toxicity of nanoparticles. They found that aggregation of transferrin-coated Au nanoparticles reduce uptake *via* receptor-mediated endocytosis in HeLa and A549 cells. With the case of MDA-MB-435 cells, aggregates entered into the cells independently of transferrin receptor *via* unknown mechanisms, and this led to the accumulation of large structures inside cell vesicles. In addition, doping of metal-oxide nanoparticles could affect their agglomeration state in water and cell culture medium. For example, George *et al.*[74] reported that decrease of agglomerate size occurred when Fe content increased in the Fe-doping process of TiO_2 nanoparticles. Measurement of the zeta potential showed an increase in the negative surface charge in Fe-doped TiO_2. Thus, the net reduction in the agglomeration size of the particles was caused by the electrostatic repulsive force.

Assessing the degree of nanoparticles dispersion in the environment depends on the information of their size distribution. Size-distribution measurements are feasible in simulated biological fluids, but there are very few techniques available to directly measure agglomerate size in living cells or tissues.

The agglomeration state of a nanomaterial during and after administration may significantly influence the biological uptake of test nanoparticles and consequently the outcome of a nanotoxicity study. Many toxicologists envisage that uptake of nanoparticles is necessary for eliciting cytotoxicity intracellularly. However, cytotoxicity could also occur when the nanoparticles (including agglomerates) are associated with cell membrane or in near vicinity to cells.[74] Upon photoactivation, ROS could cause direct damage to cell-membrane components to disrupt the membrane integrity or some of the ROS could enter into the membrane to induce oxidative stress and cellular damage subsequently. Interestingly, according to Gorge *et al.*,[74] microscopic analyses indicated that agglomerated TiO_2 particles were taken up into the test mammalian cells through phagocytosis, irrespective of their Fe content *via* doping. This was in agreement with what was reported by Churg *et al.*[75] and Singh *et al.*[76] Beyond these, Geiser *et al.*[77] even suggested the possibility of direct interaction of TiO_2 nanoparticles with the nucleus. In light of what was reported, agglomerated particles could still enter into cell membranes and elicit oxidative stress intracellularly.

Cation Charge. In a study by Hu *et al.*[24] with *E. coli* bacteria, they adopted a systematic approach in assessing cytotoxicity of seven engineered metal-oxide nanoparticles of similar primary size range (30–40 nm) with variation in metal species. The objective of that study was to investigate the relationship between cytotoxicity of metal-oxide nanoparticles and their cation charges. The list of metal oxides includes ZnO, CuO, Al_2O_3, La_2O_3, Fe_2O_3, SnO_2 and TiO_2. The results indicated that the cytotoxicity decreased with the increase in the cation charge. It is well known that micro-organisms typically have an overall negative charge associated with their surface particles at neutral pH in the growth media solution.[78] The speculation was that these negatively charged *E. coli* cells associated with the positively charged metal-oxide particles, resulting in electrostatic condensation. Numerically, *E. coli* could attract more of the lower-charged cations to membrane surfaces on a per cell basis; hence, greater cytotoxicity was exerted *via* the lower valent cations. Nevertheless, additional information on the responsible physicochemical descriptor was obtained from our recent quantitative structure–activity relationship (QSAR) study of the nanotoxicology of metal oxides (see redox potential).

Redox Potential. Experimental evaluation of the safety of chemicals is expensive and time consuming. By comparison, computational toxicology is deemed as a green-chemistry approach by predicting the potential toxicity and environmental impact of toxicants before they are mass produced or released into consumer markets. In response to this awareness, QSAR models were developed recently in physical organic chemistry to aid in predicting the structure–property relationship of nanoparticles. Currently, a QSAR method was used to predict the toxicity of various metal oxides based on their physicochemical properties.[21] Besides conducting the experimental toxicity testing with *E. coli*, a set of 12 parameters was also calculated quantitatively describing the variability of the nanoparticles' structure. Based on these data, a model was developed to describe the cytotoxicity of 17 nano-sized metal oxides to bacteria *E. coli* (Figure 8.2).[79] The simple but statistically significant nano-QSAR equation developed, utilizing only one descriptor, was fund to successfully predict the cytotoxicity of the metal-oxide nanoparticles. It is worth noting that development of this QSAR model was based on a single set of toxicity data by using one bioassay that was conducted by the same laboratory. Hence, the extent of discrepancy and deviation among database caused by crossreference was minimized.

The predicted EC_{50} (the concentration of the metal oxides that causes the reduction of bacteria viability of 50% values) can be calculated by using descriptor ΔH_{Me+} that represents the enthalpy of formation of a gaseous cation having the same oxidation state as that in the metal oxide structure. According to Auffan *et al.*,[26] the most important parameter controlling *in vitro*

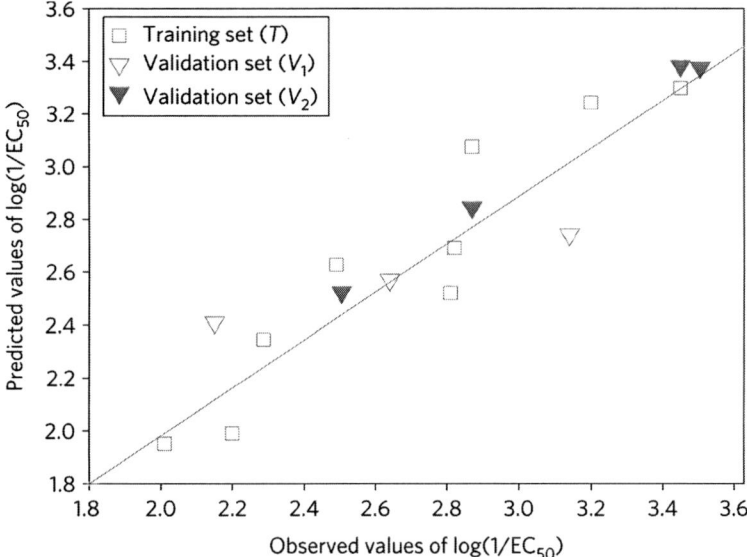

Figure 8.2 Plot of experimentally determined (observed) versus predicted log values of $1/EC_{50}$. The straight line represents perfect agreement between experimental and calculated values. Squares represent values predicted for the metal oxides from the training set; triangles represent data calculated for metal oxides from the validation sets. The distance of each symbol from the green line corresponds to its deviation from the related experimental value. (Reproduced with permission from ref. 79.)

cytotoxicity of metallic nanoparticles is their chemical stability. Chemically stable metallic nanoparticles in physiological redox conditions do not appear to exhibit cytotoxicity *in vitro*, whereas metallic nanoparticles with strong oxidation/reduction capability can be cytotoxic and genotoxic to biological targets *in vitro*. The stability is related to the dissolution of the metallic nanoparticles (release of cations) and catalytic properties and/or redox modifications of the surface. In the redox reactions with the molecules present in the biological media, release of ions is often accompanied by generation of ROS, such as superoxide ($^{\bullet}O_2{}^-$) and hydroxyl radicals (OH$^{\bullet}$). Many of the observed toxicity exerted by metal oxides can be induced by the released cations themselves, ROS or both.[26,80,81]

8.2.1.2 Biological Factors

***In Vitro* Testing.** More and more of the new toxicological studies have been conducted to better our understanding of the physicochemical properties of nanomaterials that may lead to adverse biological outcomes.[82] Nevertheless, information of their environmental fate and transport must be collected to conduct realistic risk assessment of human exposure to nanomaterials. There have been some cytotoxicity and genotoxicity studies of selected metal-oxide nanomaterials in the past decade. For example, ZnO and several other nanoparticles caused toxic responses in bacteria, copepods, crustaceans, fish, and mammalian cells.[24,72,83,84] However, most of them were derived from clinical toxicology. They were mainly conducted in the laboratory using theoretical concentrations. Information on ecotoxicity or environmental impacts of nanomaterials release is basically lacking. Moreover, intuitively toxic response caused by the same nanoparticle is expected to vary dramatically among the organisms across different trophic levels. To date, very few nanotoxicology studies had been conducted to address the effects of nanoparticles in a variety of organisms and environments. In order to evaluate realistic risk of nanomaterials release into natural environment and biota exposure, we should conduct nanotoxicology studies with a holistic approach by utilizing test specimens of different trophic levels. In addition, both *in vitro* and *in vivo* methodologies should be considered in the development of a screening strategy.

By simulating relevant routes for human exposure, *in vitro* testing is deemed a good surrogate for *in vivo* studies. Decisions need to be made regarding the most relevant and useful *in vitro* endpoints and their relative importance and suitability. For example, production of ROS (a common endpoint of nanotoxicity), cellular viability, and production of cations are frequently used for toxicity study of metal-oxide nanoparticles. In addition, a ranking of a suite of *in vitro* assays in order of appropriateness and importance should be provided. Testing with a prokaryote model,[24] human and animal cell lines of various organs[81,85] has been conducted to assess the potential toxicity of engineered nanoparticles as an adjunct to *in vivo* studies. Although this may lack validation against *in vivo* health studies, it does allow specific mechanisms of nanotoxicity to be elucidated under controlled conditions.

Prokaryotes such as bacteria serve as the foundation of many food chains in natural environments and are the most important agents that mediate biogeochemical cycling of naturally occurring and engineered materials; therefore, it is ecologically important to use a bacteria model to elucidate nanotoxicity mechanism, the severity of toxicity and the persistence of nanomaterials. As for eukaryotes, in general immortalized cell lines are an inexpensive alternative to *in vivo* model and may provide mechanistic information on nanparticle cytotoxicity. Nevertheless, if possible freshly derived primary cells and noncancer cell lines should be used.[23] In light of the difference between primary cell culture and transformed (immortalized) cell cultures, Wang *et al.*[81] conducted a study to investigate the mechanism of *in vitro* cytotoxicity of several metal-oxide nanoparticles using catfish primary hepatocytes and human HepG2 cells. The results demonstrated that HepG2 cells are more sensitive than catfish primary hepatocytes to the toxicity of metal-oxide nanoparticles. The overall ranking of the toxicity of metal oxides to the test cells is as follows: $TiO_2 < CoO < ZnO < CuO$. The stronger resistance to metal-oxide inhibition observed for catfish primary hepatocytes was attributed to their intact metabolic capability.

***In Vivo* Testing.**   The potential for bioaccumulation is an important aspect of hazard assessment of nanomaterials.[86] If condition permits, *in vitro* data generated should be critically tested or verified *in vivo*. Despite their limitation, results from *in vitro* assay could advance our knowledge in cellular toxicology pathways. They also provide a rapid indication of the potential toxicity of nanoparticles to humans or natural environments. These data frequently suffer the drawbacks of simplicity and overdosing in the experimental design. Consequently, *in vivo* studies should be conducted to evaluate specific endpoints (*e.g.*, bioaccumulation) at the intact organism level. Utilization of *in vivo* studies will be of great value in identifying hazard and assessing risk if exposure dose, route of exposure, and particle characteristics closely model those of human exposure.

Currently, rats and mice models are preferred animal models due to the existence of a large toxicity database of these systems. Nevertheless, recently there is growing interest in using piscine models for *in vivo* toxicological research. Indeed, this is deemed appropriate as long as exposures in the fish model are relevant to human exposures.[87,88] For example, zebrafish embryos were used for assessing phototoxicity of TiO_2 nanoparticles because they develop most of the major organ systems present in mammals, including the cardiovascular, nervous and digestive systems, in less than 1 week. Additional characteristics that make them advantageous for toxicant screening are their small size, transparency and ability to absorb those toxicants through the water.[46,89] By using zebrafish embryos, Asharini *et al.*[90] conducted toxicity tests of silver, gold, and platinum nanoparticles. In their experiment, toxicity was recorded in terms of mortality, hatching delay, phenotypic defects and metal accumulation. The data showed that addition of Ag nanoparticles resulted in a concentration-dependant increase in mortality rate. In addition, both Ag- and Pt-nanoparticles-induced hatching delays, as well as a concentration-dependant drop in heart rate.

8.2.1.3   Environmental Factors

Light Irradiation.   Many metal-oxide nanoparticles are known for their photocatalytic characteristics that enable their application to photodynamic therapy of certain cancers,[91,92] renewable energy[93,94] and remediation of inorganic and organic pollutants in wastewater.[95,96] However, the nanophotosensitizer used for cancer treatment could nonspecifically accumulate in the skin or eyes of the patient, which in turn induces skin cancer and then kills normal cells of the skin or eyes excited by exposure to light.[97] Meanwhile, when they are applied to wastewater treatment and as the waste of energy materials, metallic nanoparticles have been demonstrated to damage algae (including diatoms, chlorophytes, and prymnesiophytes),[98] *Pseudokirchneriella subcapitata*,[99] nitrate reducing bacteria,[100] and aquatic invertebrates such as *Daphnia magna*.[101]

In the photocatalysis application with metallic nanoparticles, production of one or more of the ROS such as hydroxyl radical ($^\bullet OH$),[102,103] hydrogen peroxide (H_2O_2),[104] superoxide radical ($^\bullet O_2^-$),[104,105] and singlet oxygen ($^1O_2$)[106,107] was reported. Ultraviolet irradiation can excite electrons (e^-) of

ZnO and TiO_2 nanoparticles into the conduction band and leave behind a hole (h^+) in the valence band. The generated charge could react with water molecule and oxygen (O_2) to produce superoxide radical anions and hydroxyl radicals, separately.[74,100] Increased ROS level has great hazard potential to cellular structure due to the presence of unpaired valence-shell electrons, which are capable of oxidizing many kinds of organic compounds.

In addition, in the photoactivation process of many metallic nanoparticles, cell death,[108] oxidative DNA damage,[109] RNA damage,[110] lipid peroxidation,[37,111] and micronuclei formation[112] were observed. Zhang *et al.*[113] suggested that cell apoptosis could be induced by the phototherapy with ZnO nanoparticles treatment. Phototoxicity of different shapes of silver nanomaterials to human skin HaCaT keratinocytes was also reported by Lu *et al.*[44] Up to now, the majority of reports on the phototoxicity of metal-oxide nanoparticles have focused on titanium dioxide nanoparticle *in vitro*. There is a lack of the systematic review on the phototoxicity of other metallic nanoparticles, especially case studies of the *in vivo* systems.

Ionic Strength/Composition, pH. The stability, and hence toxicity, of nanoparticles in environmental media could be affected by parameters such as pH, ionic strength, ionic composition, and dissolved organic matter (DOM). The solubility of metallic nanoparticles and thus bioavailability of cations are expected to increase when the pH decreases. High ionic strength promotes nanoparticle aggregation by screening electrostatic double-layer repulsion between like-charged particles, thereby decreasing metal-oxide dissolution, ROS production, and toxicity. Jin *et al.*[73] conducted a study to elucidate the influence of inorganic aquatic chemistry on silver nanoparticle stability (aggregation, dissolution, reprecipitation) and bacterial viability. In their study, a synthetic "freshwater" matrix was prepared comprising various combinations of cations and anions while maintaining a fixed ionic strength. The results indicated that Ca^{2+} and Mg^{2+} ions promoted aggregation of silver nanoparticles in simulated fresh-water matrices regardless of presence of the other ions. The antibacterial activity of silver nanoparticles was much lower than Ag^+ ions when compared on the basis of total mass added. Thus, this study reveals the inherent complexity associated with understanding metallic-nanoparticle antibacterial efficacy as well as potential environmental impacts of metallic nanoparticles. With this in mind, we should exercise extra caution when extrapolating research data generated from environmentally irrelevant buffer or media solutions in the lab assay to natural environments.[114,115] It is noteworthy that in the study of Fe-doped ZnO by Li *et al.*[114] doping did not significantly reduce Zn^{2+} dissolution and nanotoxicity of Zn particles to the test mammalian cells as that reported by George *et al.*[116] Difference in ZnO test concentration, solubility limit of Zn^{2+}, and dissolution kinetics were attributed to the discrepancy (see the section below on doping metal oxides).

Dissolved Organic Matter. Humic acid (HA), a ubiquitously compound, is known to play an important role in the geochemical cycles of many organic and inorganic compounds in aquatic systems because of its unique electro-chemical and ion-exchange behavior. For example, HA may reduce agglom-eration of engineered nanopartocles (ENPs) by coating onto their surface.[117] HA was also found to influence the stability and agglomeration state of metal-oxide nanoparticles.[118,119] The fate of ENPs in aquatic systems was reported to depend on the interactions between ENPs and natural organic matter.[120] Molecular composition of HA plays an important role in affecting inhibition of bacterial growth by ENPs. Recently, Fabrega *et al.*[121] reported that bacterial toxicity was mitigated by sorption of humic substances (HS) onto silver nanoparticles (AgNPs) surfaces. HS could also significantly influ-ence the effect of sunlight irradiation on biological activity.[122] For example, absorption of light by HA can initiate various photochemical processes and result in the production of peroxy radicals and hydroxyl radicals which can damage cellular materials.[123,124]

Interestingly, Fabrega *et al.*[121] reported that HA could act as a physical barrier to cell–nanoparticle interactions and act as an antioxidant by reacting with ROS. Indeed, the presence of HA was found to mitigate short-term bac-terial toxicity caused by AgNPs to *Pseudomonas fluorescens*.[121] The effect of a terrestrial HA and a river HA on the cytotoxicity of AgNPs (size distribution range: 15–25 nm; up to 5 µM) to a natural aquatic bacterial assemblage was investigated by Dasari and Hwang.[125] The effect of HA (20 and 40 ppm) on the cytotoxicity of AgNPs was tested in the presence and in the absence of natural sunlight. A significant effect on bacterial viability count was observed by all treatment interactions with both terrestrial and river HA treatments except for the exposure to river HA in light. The toxicity of AgNPs to natural aquatic bacteria assemblage was concentration dependent. The data indicates that light exposure inhibited the viable count more than the darkness exposure. The HA treatment groups in the presence of light showed a greater reduced viability count compared to darkness exposure groups. The inhibition of bacterial-viability counts by AgNPs exposure was less in the light-treatment groups containing a terrestrial HA compared to that with a river HA. The difference in the extent of reactive oxygen species formation and adsorption/binding of AgNPs by HA was speculated to account for the observed phenomenon.

Besides HA, other dissolved compounds may also significantly affect bio-availability of metal-oxide nanoparticles in certain aquatic environments. For example, Li *et al.*[114] reported that tannic acid complexes free Zn^{2+} effectively, thereby reducing their bioavailability at a larger extent than humic, fulvic, and aliginic acid. As a polyphenolic compound containing glucose linkages *via* ether bonds to 9–10 gallic acid molecules, it thus has higher metal-binding capacity than the other dissolved organic compounds used in their study.

Interactions with Biological Media or Test Reagents. Cytotoxicity endpoints are frequently measured with viability-metabolic markers such as mitochon-drial reduction of color-changing tetrazolium salts in MTT assay. However, it has been documented that the results of nanotoxicity test with MTT assay may be unreliable due to marker interactions with nanoparticles.[23,126] As for the interaction with biological media, Li *et al.*[115] indicated that the generation of precipitates and complexes between zinc ion (from ZnO nanoparticles) and organic moieties (*e.g.*, citrate or amino acids) in the growth media solution dramatically decreased the concentration of zinc ions, resulting in lower toxicity to test *E. coli* in these media. Moreover, isotonic and rich nutrient conditions enhanced the tolerance of *E. coli* to ZnO nanoparticles. Owing to the possible effects (*e.g.*, agglomeration) of buffered salts, George *et al.*[116] indicated that in their nanotoxicity study ZnO nanoparticles were dispersed in deionized water and then stabilized by protein addition before introduction into the tissue culture media. Indeed, the agglomeration size was significantly reduced when serum albumin (BSA) was used to stabilize the nanoparticles.

In light of the dramatic difference of the toxicity of ZnO nanoparticles in various aqueous media, caution is warranted against the effect of water chemistry on the physicochemical properties of metal-oxide nanoparticles in conducting their nanotoxicity evaluations.

8.3 Designer Metal-Oxide Nanoparticles (Doped Metal-Oxide Nanoparticles)

In the recent development of safe nanotechnology applications, doped metal-oxide nanoparticles have received much attention from the scientific communities.[127,128] Among the nanoparticles, new forms of doped-ZnO and -TiO_2 were synthesized at faster rates than others due to their wide application in the areas of consumer products, renewable energy resources, and environmental remediation. As these two metal-oxide nanoparticles possess bandgap energies 3.22–3.33 eV, doping with various elements could change their physicochemical properties to fit the desirable feature of a safe nanoproduct. A common strategy in "re-engineering" those nanoparticles is to conduct an indepth evaluation of the mechanisms of their potentially hazardous physicochemical properties in order to guide the safe-by-design process.[74,116] For example, dissolution of ZnO was recognized as the mechanism of causing oxidative stress and cytotoxicity *in vitro* and *in vivo*; therefore, Fe doping was utilized to slow Zn^{2+} release to improve nanosafety of ZnO particles.[129] Photoactivation of TiO_2 could lead to formation of biomolecule-damaging ROS (see the previous section). Consequently, doping TiO_2 nanoparticles with various transition metals was used to create impurity energy levels in the valence and conduction band, and thus facilitate photoactivation at lower energy wavelength and reduce the bandgap energy requirement for photocatalytic activation. Indeed, metals of group IV-VIII and nonmetals such as B, C, N, S, or I (and/or their combinations) were used to narrow the bandgap of TiO_2[50,74,127,130–133] and promote the photocatalytic activities in the near-UV and the visible-light ranges.[134]

Doping with iron for TiO_2 nanoparticles is sound nanodesigning as iron is environmentally safe.[32,74] However, doping with some other metal elements (*e.g.*, Ag) could suffer the drawback of metal leaching and possible toxicity; thus diminish the potential of using metal-doped TiO_2 nanoparticles for drinking and wastewater treatment applications.[127] Shifting photocatalytic activities of TiO_2 from UV to the visible-light range by doping will not only make the doped TiO_2 ecologically friendly but also widen the application of TiO_2 for *in situ* remediation, as visible light can penetrate the water column deeper than UV irradiation. For example, doping with sulfur[128] and codoping with nitrogen–fluorine[127] enables TiO_2 nanoparticles to photocatalytically degrade microcystin-LR in water *via* visible-light activation. In the meantime, formation of singlet oxygen ($^1O_2$) was found to occur upon visible-light irradiation of N, S codoped TiO_2.[135] Although doping may increase the safety margin of metallic nanoparticles, many of the doped products remain detrimental to the biota. Therefore, eventually some decisions have to be made while designers face the tradeoffs between the economic, environmental, and commercial ramifications of new developments in designer metal-oxide nanoparticles.

8.4 Research Gaps and Collaboration Needed

If feasible, characterization of test nanoparticles before and after administration should be conducted. Importantly, instead of total dependence on the suppliers, separate characterizations should be carried out. Ideally, many if not all physicochemical characteristics of nanoparticles should be measured or be derivable in toxicity screening tests. These include particle size and size distribution, particle shape, chemical composition, surface area, surface chemistry (coating), surface charge, porosity, zeta potential (agglomeration state), and crystal structure. Although many existing analytical instruments enable us to measure the nominal concentration of nanomaterials in the laboratory, the database of the environmental concentration of engineered metallic particles is scarce. Mostly, this problem arises from a lack of capability in measurement of their concentration in environmental matrix, thus making the levels of uncertainty high in risk estimation. Moreover, very few studies have quantified chronic toxic effects or the extent of bioaccumulation of engineered metallic particles.[136] As agreed by the scientific community, nanotoxicity is a global concern and interdisciplinary collaboration on this issue is needed. Nanotechnologists can begin by focusing on developing common global standards, definitions, test methods, and on providing databases on physical and chemical properties of nanomaterials that could impact human and environmental health.[136] For example, the input of toxicologists across different nations is crucial to apply analytical efforts to some challenging questions such as: What dose metrics should be chosen as the appropriate dosage unit in an effective study of nanomaterial? Which treatment method and solvent should be chosen for preparing test nanomaterials in a liquid solution? Before a universal approach is adopted, selection of dispersion medium, treatment of the stock and working solutions (*e.g.*, sonication) need to be well documented, as many reports indicated alteration of material characteristics could occur and artifacts would be generated after aggressive treatment during preparation procedures.[137] After release or spill into natural environments, some environmental factors, such as pH, salinity, and presence of natural organic matter, could alter the toxicity and influence the ultimate fate of engineered nanomaterials in the natural environments. Besides what was discussed in the previous sections, readers are encouraged to read more details in the report by Klaine *et al.*[1] In addition, a critical review on the fundamental aspects of test methods that are adopted or being developed for assessing the ecotoxicity of engineered nanomaterials has just become available.[86] Detailed reports on practical experiences and recommendations is also provided in that report.

8.5   Summary and Outlook

In conclusion, in this chapter we have critically reviewed the relationship between the individual physicochemical parameters and biological response in toxicological studies of engineered gold, silver, and metal-oxide nanoparticles. We have discussed the possible effect of biological and nonliving factors on the outcomes of nanotoxicity assays. Since nanotoxicology is a relatively new field, the concepts of dose metrics, exposure assessment, hazard identification, and risk characterization are still far from the state in which a conclusive picture emerges. Although significant development has been made in safe nanotechnology application, considerable research remains to be conducted for us to balance between promoting commercial profits and reducing health hazards of the newly designed nanomaterials. Despite the advantages in using *in vitro* systems to address a mechanistic study in a nanotoxicity research, their biological relevance to the intact organism frequently needs to be verified with *in vivo* studies.

Acknowledgements

This study was supported by a grant received from NSF-CREST program (National Science Foundation-Centers of Research Excellence in Science and Technology) with grant #HRD-0833178.

References

1. S. J. Klaine, P. J. J. Alvarez, G. E. Batley, T. E. Fernandes, R. D. Handy, D. Y. Lyon, S. Mahendra, M. J. McLaughlin and J. R. Lead, *Environ. Toxicol. Chem.*, 2008, **27**, 1825–1851.
2. *An Inventory of Nanotechnology-based Consumer Products Currently on the Market*, http://www.nanotechproject.org/inventories/consumer/analysis_draft/, Accessed April, 2012.
3. P. Rivera Gil, G. Oberdörster, A. Elder, V. Puntes and W. J. Parak, *ACS Nano*, 2010, **4**, 5527–5531.
4. P. V. AshaRani, G. Low Kah Mun, M. P. Hande and S. Valiyaveettil, *ACS Nano*, 2009, **3**, 279–290.

5. S. H. Lacerda, J. J. Park, C. Meuse, D. Pristinski, M. L. Becker, A. Karim and J. F. Douglas, *ACS Nano*, 2010, **4**, 365–379.
6. M. Tarantola, D. Schneider, E. Sunnick, H. Adam, S. Pierrat, C. Rosman, V. Breus, C. Sönnichsen, T. Basché and J. E. A. Wegener, *ACS Nano*, 2009, **3**, 213–222.
7. Y. Zhang, S. F. Ali, E. Dervishi, Y. Xu, Z. Li, D. Casciano and A. S. Biris, *ACS Nano*, 2010, **4**, 3181–3186.
8. M. R. Wiesner, G. V. Lowry, K. L. Jones, M. F. J. Hochella, R. T. Di Giulio, E. Casman and E. S. Bernhardt, *Environ. Sci. Technol.*, 2009, **43**, 6458–6462.
9. R. D. Glover, J. M. Miller and J. E. Hutchison, *ACS Nano*, 2011, **5**, 8950–8957.
10. C. Levard, B. C. Reinsch, F. M. Michel, C. Oumahi, G. V. Lowry and G. E. Brown Jr., *Environ. Sci. Technol.*, 2011, **45**, 5260–5266.
11. T. Phenrat, T. C. Long, G. V. Lowry and B. Veronesi, *Environ. Sci. Technol.*, 2009, **43**, 195–200.
12. T. L. Kirschling, P. L. Golas, J. M. Unrine, K. Matyjaszewski, K. B. Gregory, G. V. Lowry and R. D. Tilton, *Environ. Sci. Technol.*, 2011, **45**, 5253–5259.
13. P. C. Ray, H. Yu and P. P. Fu, *J. Environ. Sci. Health C Environ. Carcinog. Ecotoxicol. Rev.*, 2009, **27**, 1–35.
14. A. Sassolas, B. D. Leca-Bouvier and L. J. Blum, *Chem. Rev.*, 2008, **108**, 109–139.
15. S. M. Borisov and O. S. Wolfbeis, *Chem. Rev.*, 2008, **108**, 423–461.
16. M. Famulok, J. S. Hartig and G. Mayer, *Chem. Rev.*, 2007, **107**, 3715–3743.
17. D. Senapati, A. K. Singh and P. C. Ray, *Chem. Phys. Lett.*, 2010, **487**, 88–91.
18. W. Lu, A. K. Singh, S. A. Khan, D. Senapati, H. Yu and P. C. Ray, *J. Am. Chem. Soc.*, 2010, **132**, 18103–18114.
19. M. Ljungman, *Chem. Rev.*, 2009, **109**, 2929–2950.
20. M. Ferrari, *Nature Rev. Cancer*, 2005, **5**, 161–171.
21. T. Puzyn, D. Leszczynska and J. Leszczynski, *Small*, 2009, **5**, 2494–2509.
22. K. L. Dreher, *Toxicol. Sci.*, 2004, **77**, 3–5.
23. G. Oberdörster, A. Maynard, K. Donaldson, V. Castranova, J. Fitzpatrick, K. Ausman, J. Carter, B. Karn, W. Kreyling, D. Lai, S. Olin, N. Monteiro-Riviere, D. Warheit, H. Yang and ILSI Research Foundation/Risk Science Institute Nanomaterial Toxicity Screening Working Group, *Part. Fibre Toxicol.*, 2005, **2**, 8.
24. X. Hu, S. Cook, P. Wang and H.-M. Hwang, *Sci. Total Environ.*, 2009, **407**, 3070–3072.
25. C. Cherchi, T. Chernenko, M. Diem and A. Z. Gu, *Environ. Toxicol. Chem.*, 2011, **30**, 861–869.
26. M. Auffan, J. Rose, M. R. Wiesner and J. Y. Bottero, *Environ. Pollut.*, 2009, **157**, 1127.
27. W. Jiang, H. Mashayekhi and B. Xing, *Environ. Pollut.*, 2009, **157**, 1619–1625.

28. J. Wang, G. Zhou, C. Chen, H. Yu, T. Wang, Y. Ma, G. Jia, Y. Gao, B. Li, J. Sun, Y. Li, F. Jiao, Y. Zhao and Z. Chai, *Toxicol. Lett.*, 2007, **168**, 176–185.

29. R. Dunford, A. Salinaro, L. Cai, N. Serpone, S. Horikoshi, H. Hidaka and J. Knoland, *FEBS Lett.*, 1997, **418**, 87–90.

30. D. B. Warneit, T. R. Webb, K. L. Reed, S. Frerichs and C. M. Sayes, *Toxicology*, 2007, **230**, 90–104.

31. A. E. Nel, L. Mädler, D. Velegol, X. T., H. E.M.V., P. Somasundaran, F. Klaessig, V. Castranova and M. Thompson, *Nature Mater.*, 2009, **8**, 543–557.

32. E. Ying and H.-M. Hwang, *Sci. Total Environ.*, 2010, **408**, 4475–4481.

33. R. D. Handy, R. Owen and E. Valsami-Jones, *Ecotoxicology*, 2008, **17**, 315–325.

34. B. D. Chithrani, A. A. Ghazani and W. C. W. Chan, *Nano Lett.*, 2006, **6**, 662–668.

35. W. Krämer and H. Fissan, *Wer soll leben? - Rationierung im Gesundheitswesen aus Statistiker- und Ökonomensicht: Nachhaltige Nanotechnologie*, Ferdinand Schöningh, Paderborn, München, Wien, Zürich, Schöningh, 2008.

36. G. Oberdörster, E. Oberdörster and J. Oberdörster, *Environ. Health Perspect.*, 2005, **113**, 823–839.

37. H. Ma, N. J. Kabengi, P. M. Bertsch, J. M. Unrine, T. C. Glenn and P. L. Williams, *Environ. Pollut.*, 2011, **159**, 1473–1480.

38. O. Choi and Z. Hu, *Environ. Sci. Technol.*, 2008, **42**, 4583–4588.

39. J. Lee, S. Mahendra and P. J. J. Alvarez, *ACS Nano*, 2010, **4**, 3580–3590.

40. A. Albanese and W. C. Chan, *ACS Nano*, 2011, **5**, 5478–5489.

41. E. E. Connor, J. Mwamuka, A. Gole, C. J. Murphy and M. D. Wyatt, *Small*, 2005, **1**, 325–327.

42. H. Meng, T. Xia, S. George and A. E. Nel, *ACS Nano*, 2009, **3**, 1620–1627.

43. N. Lewinski, V. Colvin and R. Drezek, *Small*, 2008, **4**, 26–49.

44. W. Lu, D. Senapati, S. Wang, O. Tovmachenko, A. K. Singh, H. Yu and P. C. Ray, *Chem. Phys. Lett.*, 2010, **487**, 92–96.

45. S. Wang, W. Lu, O. Tovmachenko, U. S. Rai, H. Yu and P. C. Ray, *Chem. Phys. Lett.*, 2008, **463**, 145–149.

46. O. Bar-Ilan, R. M. Albrecht, V. E. Fako and D. Y. Furgeson, *Small*, 2009, **5**, 1897–1910.

47. W. Jiang, B. Y. S. Kim, J. T. Rutka and W. C. W. Chan, *Nature Nanotechnol*, 2008, **3**, 145–150.

48. D. B. Warheit, *Toxicol. Sci.*, 2008, **101**, 183–185.

49. H. Yin, H. P. Too and G. M. Chow, *Biomaterials*, 2005, **26**, 5818–5826.

50. R. Asahi, T. Morikawa, T. Ohwaki, K. Aoki and Y. Taga, *Science*, 2001, **293**, 269–271.

51. K. J. Lee, P. D. Nallathamby, L. M. Browning, C. J. Osgood and X. H. N. Xu, *ACS Nano*, 2007, **1**, 133–143.

52. T. C. King-Heiden, P. N. Wiecinski, A. N. Mangham, K. M. Metz, D. Nesbit, J. A. Pedersen, R. J. Hamers, W. Heideman and R. E. Peterson, *Environ. Sci. Technol.*, 2009, **43**, 1605–1611.
53. S. George, S. Lin, Y. Zhao, T. Xia, H. Meng, Z. Ji, R. Liu, S. George, S. Xiong, X. Wang, H. Zhang, S. Pokhrel, L. Mädler, R. Damoiseaux, S. Lin and A. E. Nel, *ACS Nano*, 2011, **5**, 1805–1817.
54. Y. S. Kim, J. S. Kim, H. S. Cho, D. S. Rha, J. M. Kim, J. D. Park, B. S. Choi, R. Lim, H. K. Chang and Y. H. Chung, *Inhal. Toxicol.*, 2008, **20**, 575–583.
55. J. H. Sung, J. H. Ji, J. U. Yun, D. S. Kim, M. Y. Song, J. Jeong, B. S. Han, J. H. Han, Y. H. Chung, J. M. Kim, H. K. Chang, E. J. Lee, J. H. Lee and I. J. Yu, *Inhal. Toxicol.*, 2008, **20**, 567–574.
56. C. Levard, E. M. Hotze, G. V. Lowry and G. E. Brown Jr., *Environ. Sci. Technol.*, 2012, **46**, 6900–6914.
57. S. M. Hussain and J. J. Schlager, *Toxicol. Sci.*, 2009, **108**, 223–224.
58. S. Kittler, C. Greulich, J. Diendorf, M. Koller and M. Epple, *Chem. Mater.*, 2010, **22**, 4548–4554.
59. O. Choi and Z. Hu, *Environ. Sci. Technol.*, 2008, **42**, 4583–4588.
60. J. Liu and R. H. Hurt, *Environ. Sci. Technol.*, 2010, **44**, 2169–2175.
61. L. Y. Yin, Y. W. Cheng, B. Espinasse, B. P. Colman, M. Auffan, M. Wiesner, J. Rose, J. Liu and E. S. Bernhardt, *Environ. Sci. Technol.*, 2011, **45**, 2360–2367.
62. J. Roh, S. Sim, J. Yi, K. Park, K. Chung, D. Ryu and J. Choi, *Environ. Sci. Technol.*, 2009, **43**, 3933–3940.
63. J. Fabrega, J. C. Renshaw and J. R. Lead, *Environ. Sci. Technol.*, 2009, **43**, 9004–9009.
64. Z.-M. Xiu, J. Ma and P. J. J. Alvarez, *Environ. Sci. Technol.*, 2011, **45**, 9003–9008.
65. J. Y. Liu, D. A. Sonshine, S. Shervani and R. H. Hurt, *Environ. Sci. Technol.*, 2010, **4**, 6903–6913.
66. K. A. Huynh and K. L. Chen, *Environ. Sci. Technol.*, 2011, **45**, 5564–5571.
67. M.-N. Croteau, S. K. Misra, S. N. Luoma and E. Valsami-Jones, *Environ. Sci. Technol.*, 2011, **45**, 6600–6607.
68. T. Gomes, J. P. Pinheiro, I. Cancio, C. G. Pereira, C. Cardoso and M. J. Bebianno, *Environ. Sci. Technol.*, 2011, **45**, 9356–9362.
69. R. J. Griffitt, J. Luo, J. Gao, J.-C. Bonzongo and D. S. Barber, *Toxicol. Chem.*, 2008, **27**, 1972–1978.
70. N. M. Franklin, N. J. Rogers, S. C. Apte, G. E. Batley, G. E. Gadd and P. S. Casey, *Environ. Sci. Technol.*, 2007, **41**, 8484–8490.
71. D. Lin and B. Xing, *Environ. Pollut.*, 2007, **150**, 243–250.
72. M. Heinlaan, A. Ivask, I. Blinova, H. Bubourguier and A. Kahru, *Chemosphere*, 2008, **71**, 1308–1316.
73. X. Jin, M. Li, J. Wang, C. Marambio-Jones, F. Peng, X. Huang, R. Damoiseaux and E. M. V. Hoek, *Environ. Sci. Technol.*, 2010, **44**, 7321–7328.
74. S. George, S. Pokhrel, Z. Ji, B. L. Henderson, T. Xia, L. Li, J. I. Zink, A. E. Nel and L. Mädler., *J. Am. Chem. Soc.*, 2011, **133**, 11270–11278.

75. A. Churg, B. Stevens and J. L. Wright, *Am. J. Physiol. Lung Cell Mol. Physiol.*, 1998, **274**, L81–L86.

76. S. Singh, T. Shi, R. Duffin, C. Albrecht, D. van Berlo, D. Höhr, B. Fubini, G. Martra, I. Fenoglio, P. J. A. Borm and R. P. F. Schins, *Toxicol. Appl. Pharmacol.*, 2007, **222**, 141–151.

77. M. Geiser, B. Rothen-Rutishauser, N. Kapp, S. Schürch, W. Kreyling, H. Schulz, M. Semmler, Im Hof V, J. Heyder and P. Gehr, *Environ. Health Perspect.*, 2005, **113**, 1555–1560.

78. R. M. Maier, I. L. Pepper and C. P. Gerba, *Environmental Microbiology*, 2 edn., Academic Press, San Diego, 2009.

79. T. Puzyn, B. Rasulev, A. Gajewicz, X. Hu, T. P. Dasari, A. Michalkova, H.-M. Hwang, A. Toropov, D. Leszczynska and J. Leszczynski, *Nature Nanotechnol.*, 2011, **6**, 175–178.

80. A. L. Neal, *Ecotoxicology*, 2008, **17**, 362–371.

81. Y. Wang, W. G. Aker, H.-M. Hwang, C. G. Yedjou, H. Yu and P. B. Tchounwou, *Sci. Total Environ.*, 2011, **409**, 4753–4762.

82. T. Xia, M. Kovochich, J. Brant, M. Hotze, J. Sempf, T. Oberley, C. Sioutas, J. I. Yeh, M. R. Wiesner and A. E. Nel, *Nano Lett.*, 2006, **6**, 1794–1807.

83. X. S. Zhu, L. Zhu, Y. S. Chen and S. Y. Tian, *J. Nanopart. Res.*, 2009, **11**, 67–75.

84. H. A. Jeng and J. Swanson, *J. Environ. Sci. Health A*, 2006, **41**, 2699–2711.

85. X. Hu, S. Cook, P. Wang, H.-M. Hwang, X. Liu and Q. L. Williams, *Sci. Total Environ.*, 2010, **408**, 1812–1817.

86. R. D. Handy, G. Cornelis, T. Fernandes, O. Tsyusko, A. Decho, T. Sabo-Attwood, C. Metcalfe, J. A. Steevens, S. J. Klaine, A. A. Koelmans and N. Horne, *Environ. Toxicol. Chem.*, 2012, **31**, 15–31.

87. L. I. Zon and R. T. Peterson, *Nature Rev. Drug Discov.*, 2005, **4**, 35–44.

88. V. E. Fako and D. Y. Furgeson, *Adv. Drug Delivery Rev.*, 2009, **61**, 478–486.

89. A. Rubinstein, *Expert Opin. Drug Metab. Toxicol.*, 2006, **2**, 231–240.

90. P. V. Asharani, Y. Lianwu, Z. Gong and S. Valiyaveettil, *Nanotoxicology*, 2011, **5**, 43–54.

91. D. Gao, R. R. Agayan, H. Xu, M. A. Philbert and R. Kopelman, *Nano Lett.*, 2006, **6**, 2383–2386.

92. M. E. Wieder, D. C. Hone, M. J. Cook, M. M. Handsley, J. Gavrilovic and D. A. Russell, *Photochem. Photobiol. Sci.*, 2006, **5**, 727–734.

93. S. S. Mao and X. Chen, *Int. J. Energ. Res., Special Issue: Recent Advances in Micro and Nano Energy Systems*, 2007, **31**, 619–636.

94. P. V. Kamat, *J. Phys. Chem. C*, 2007, **111**, 2834–2860.

95. J.-M. Herrmann, *Catal. Today*, 1999, **53**, 115–129.

96. C. Karunakaran, S. Senthilvelan, S. Karuthapandian and K. Balaraman, *Catal. Commun.*, 2004, **5**, 283–290.

97. Y.-D. Choi, B.-S. Jang, I.-H. Kim and J.-Y. Park, *US Pat.*, 2011, Application number: 12/665,013.
98. R. J. Miller, S. Bennett, A. A. Keller, S. Pease and H. S. Lenihan, *PLoS One*, 2012, **7**, e30321.
99. J. L. Bouldin, T. M. Ingle, A. Sengupta, R. Alexander, R. E. Hannigan and R. A. Buchanan., *Environ. Toxicol. Chem.*, 2008, **27**, 1958–1963.
100. J. Gao, Y. Wang, A. Hovsepyan and J.-C. J. Bonzongo, *J. Hazard Mater.*, 2011, **186**, 940–945.
101. J. Kim, Y. Park, T. H. Yoon, C. S. Yoon and K. Choi, *Aquat. Toxicol.*, 2010, **97**, 116–124.
102. J. F. Reeves, S. J. Davies, N. J. F. Dodd and A. N. Jha, *Mutat. Res.*, 2008, **640**, 113–122.
103. T. Uchino, H. Tokunaga, M. Ando and H. Utsumi, *Toxicol. In Vitro*, 2002, **16**, 629–635.
104. T. C. Long, N. Saleh, R. D. Tilton, G. V. Lowry and B. Veronesi, *Environ. Sci. Technol.*, 2006, **40**, 4346–4352.
105. A. L. Lisenbigler, G. Lu and J. T. J. Yates, *Chem. Rev.*, 1995, **95**, 735–758.
106. T. Daimon, T. Hirakawa, M. Kitazawa, J. Suetake and Y. Nosaka, *Appl. Catal. A: Gen.*, 2008, **340**, 169–175.
107. R. Konaka, E. Kasahara, W. C. Dunlap, Y. Yamamoto, K. C. Chien and M. Inoue, *Redox Rep.*, 2001, **6**, 319–325.
108. W. K. Boyes, K. Sanders, L. Degn, R. M. Zucker, W. R. Mundy, B. Zhao and J. Roberts, Society of Toxicology (SOT) Annual Meeting, Washington, DC, 2011.
109. W. F. Vevers and A. N. Jha, *Ecotoxicology*, 2008, **17**, 410–420.
110. W. G. Wamer, J. Yin and R. R. Wei, *Free Radic. Biol. Med.*, 1997, **23**, 851–858.
111. P.-C. Maness, S. Smolinski, D. M. Blake, Z. Huang, E. J. Wolfrum and W. A. Jacoby, *Appl. Environ. Microbiol.*, 1999, **65**, 4094–4098.
112. J. Zhang, M. Wages, S. B. Cox, J. D. Maul, Y. Li, M. Barnes, L. Hope-Weeks and G. P. Cobb, *Environ. Toxicol. Chem.*, 2012, **31**, 176–183.
113. Y. Zhang, W. Chen, S. Wang, Y. Liu and C. Pope, *J. Biomed. Nanotechnol.*, 2008, **4**, 432–438.
114. M. Li, S. Pokhrel, T. Xia, X. Jin, L. Mädler, R. Damoiseaux and E. M. V. Hoek, *Environ. Sci. Technol.*, 2011, **45**, 755–761.
115. M. Li, L. Zhu and D. Lin, *Environ. Sci. Technol.*, 2011, **45**, 1977–1983.
116. S. George, S. Pokhrel, T. Xia, B. Gillbert, Z. Ji, M. Schowalter, A. Rosenauer, R. Damoiseaux, K. A. Radley, L. Mädler and A. E. Nel, *ACS Nano*, 2010, **4**, 15–29.
117. H. Hyung, J. D. Fortner, J. B. Hughes and J. H. Kim, *Environ. Sci. Technol.*, 2007, **41**, 179–184.
118. K. Yang, D. Lin and B. Xing, *Langmuir*, 2009, **25**, 3571–3576.
119. B. Wu, W.-Q. Zhuang, M. Sahu, P. Biswas and Y. J. Tang, *Sci. Total Environ.*, 2011, **409**, 4635–4639.
120. E. Navarro, F. Piccapetra, B. Wagner, F. Marconi, R. Kaegi, N. Odzak, L. Sigg and R. Behra, *Environ. Sci. Technol.*, 2008, **42**, 8959–8964.

121. J. Fabrega, S. R. Fawcett, J. C. Rensha and J. R. Lead, *Environ. Sci. Technol.*, 2009, **43**, 7285–7290.
122. M. Bittner, K. Hilscherová and J. Giesy, *Environ. Int.*, 2007, **33**, 812–816.
123. A. L. Balarezo, V. N. Jones, H. Yu and H.-M. Hwang, *Int. J. Mol. Sci,* 2002, **3**, 1133–1144.
124. A. Paul, S. Hackbarth, R. D. Vogt, B. Roder, B. Burnison and C. E. W. Steinberg, *Photochem. Photobiol. Sci.*, 2004, **3**, 273–280.
125. T. P. Dasari and H.-M. Hwang, *Sci. Total. Environ.*, 2010, **408**, 5817–5823.
126. S. Wang, H. Yu and J. Wickliffe, *Toxicol. In Vitro*, 2011, **25**, 2147–2151.
127. M. Pelaez, A. A. de la Cruz, E. Stathatos, P. Ealaras and D. D. Dionysiou, *Catal. Today*, 2009, **144**, 19–25.
128. C. Han, M. Pelaez, V. Likodimos, A. G. Montos, P. Falaras, K. O'Shea and D. D. Dionysiou, *Appl. Catal. B-Environ.*, 2011, **107**, 77–87.
129. T. Xia, Y. Zhao, T. Sager, S. George, S. Pokhrel, N. Li, D. Schoenfeld, H. Meng, S. Lin, X. Wang, M. Wang, Z. Ji, J. I. Zink, L. Mädler, V. Castranova, S. Lin and A. E. Nel, *ACS Nano*, 2011, **5**, 1223–1235.
130. W. Choi, A. Termin and M. R. Hoffmann, *J. Phys. Chem.*, 1994, **98**, 13669–13679.
131. Y. Cong, J. L. Zhang, F. Chen, M. Anpo and D. N. He, *J. Phys. Chem. C*, 2007, **111**, 10618–10623.
132. X. Hong, Z. Wang, W. Cai, F. Lu, J. Zhang, Y. Yang, N. Ma and Y. Liu, *Chem. Mater.*, 2005, **17**, 1548–1552.
133. W. Zhao, W. Ma, C. Chen, J. Zhao and Z. Shuai, *J. Am. Chem. Soc.*, 2004, **126**, 4782–4783.
134. T. L. Thompson and J. T. Yates, *Chem. Rev.*, 2006, **106**, 4428–4453.
135. J. A. Rengifo-Herrera, K. Pierzchata, A. Sienkiewicz, L. Forr, J. Kiwi and C. Pulgarin, *Appl. Catal. B-Environ.*, 2009, **88**, 398–406.
136. S. J. Klaine, A. A. Koelmans, N. Horne, S. Carley, R. D. Handy, L. Kapuska, B. Nowack and F. von der Kammer, *Environ. Toxicol. Chem.*, 2012, **31**, 3–14.
137. T. B. Henry, F.-M. Menn, J. T. Fleming, J. Wilgus, R. N. Compton and G. S. Sayler, *Environ. Health Perspect.*, 2007, **115**, 1059–1065.

CHAPTER 9

Introduction to Nanosafety

FRANCISCO BALAS

Aragon Nanoscience Institute (INA) – Universidad de Zaragoza, c/ Mariano Esquillor s/n, 50018 Zaragoza (Spain); Networking Research Center of Bioengineering, Biomaterials and Nanomedicine (CIBER-BBN), c/ Mariano Esquillor s/n, 50018 Zaragoza (Spain)
Email: fbalas@unizar.es

9.1 Introduction: Safety and Nanoparticles

Nanotechnology is setting the scientific bases of a scientific and social revolution in recent years. The different materials at the nanoscale that have been produced show amazing properties and have applications in different areas as information and communication technologies, materials for energy or even medicine. Such materials are now leaving laboratories and becoming part of the commercial world, reaching almost all the industrial sectors with a subsequent economic impact. The outlook for the nanotechnological market is outstanding and Lux Research has estimated at $245 billions the sales of products with nanotechnological components in 2009 and forecasts a ten times increase by 2015.[1]

In addition to the academic and industrial interest in nanotechnology, some voices have started enquiring on the impact of nanomaterials for human health and the environment. Nanoparticles and nanowires of a variety of compositions and nature have a proven capacity of crossing biological membranes, reaching not only cytoplasm but also the nucleus of different cell lines. This feature, besides giving promising biomedical applications, brings up some questions regarding the possibility of undesired absorption of biomaterials in

RSC Green Chemistry No. 19
Sustainable Preparation of Metal Nanoparticles: Methods and Applications
Edited by Rafael Luque and Rajender S Varma
© The Royal Society of Chemistry 2013
Published by the Royal Society of Chemistry, www.rsc.org

the organisms, with adverse biological effects that have been demonstrated in cell cultures and with *in vivo* tests. There is a growing concern not only in the actual risks in handling nanomaterials, that can be minimized with the appropriate measures, but also in the public risk perception that can change the currently favorable image of nanotechnology.[2] Also, the industries are realizing the potential impact that unsafe nanotechnology could imply and several studies have been committed to this topic in recent years.[3]

The four critical steps in the risk assessment of nanomaterials have been recently identified[4] and are largely the same proposed for other chemicals. An adequate risk-assessment scheme should start with the appropriate hazard identification that means the proper knowledge about the properties of those nanomaterials that pose a particular risk to health. The next step comprises the characterization of the hazard and the definition of the dose–response levels for critical targets in the body, as well as the description of the toxicity mechanisms[5] and the ability to cross biological barriers.[6] The third step is the exposure assessment that requires validated analytical procedures for determining the exposure levels in both large body tissues as well as the breathing atmosphere. It is a matter of fact that in occupational environments there could be more nanoparticles flowing in the air and therefore workplaces are significant places for analyzing human exposure to nanoparticles.[7,8] The last step is the understanding of the transport processes between the source and the receptor,[9] which concerns the formation of nanoparticle-laden air streams and the appropriate measurements that have to be performed to determine their impact.

All these emerging risks should not lead to consider nanotechnology as something dangerous. However, the use of the precautionary principle is justified in the early stages of the development of synthetic procedures of nanomaterials or during their applications. The European Union emphasizes this approach through the Scientific Committee on Emerging and Newly Identified Health Risks (SCENIHR) or the Scientific Committee on Consumer Safety (SCCS), when dealing with the evaluation of engineered nanoparticles (ENP). This approach had been characterized in the context of risk assessment and management by the statement "no data, no market". When the amount and quality of data are improved, the chances to make more accurate risk assessments are also improved, and the importance of emphasizing uncertainties is consequently reduced.[10] In recent years, the understanding of factors that influence the interaction between biological systems and ENP has been increased in a substantial manner. The evidence of the underlying mechanisms of these interactions have been described by Nel *et al.*[11] as a dynamical relationship between the nanoparticle surface and the solid/liquid interface, where chemical reactions occur. Understanding these nano–bio interactions will provide the needed basis of a reliable risk assessment of ENM.

9.1.1   Risks in Handling Nanoparticles

The terms "hazard" and "risk" will be frequently used in this chapter and it is therefore convenient to distinguish between them for the sake of clarity.

The United States Environmental Protection Agency (EPA) defines the term "hazard" as the inherent toxicity of a compound. For example, a chemical substance that has the property of being toxic is hazardous. Similarly, "risk" is defined as a measure of the probability that damage to life, health, property, and/or the environment will occur as a result of a given hazard.[12] Health risks are considered to be high if the hazard is adverse health effects even though the probability of occurrence is low. The distinction between hazard and risk is important as we can identify a hazardous substance, but the risk that could cause any harm is the fact that determines what precautionary measures should be taken.

9.1.2 Factors that Influence Nanoparticle Toxicity

There are many concerns regarding the impact on human health and environment that can arise when matter is at the nanoscale. In most cases, the expected impact is similar to that induced by chemicals but it is difficult to forecast the effects of the exposure to nanomaterials exclusively based on the chemical composition. The presence of low-solubility ENPs in biological fluids could pose a serious risk and reduced sizes induce a unique behavior in ENPs, being able to pass through the physiological defense mechanisms and be transported along the body fluids in insoluble form.[13] Therefore the ENPs end up in the bloodstream after passing through all the respiratory or gastro-intestinal membranes, which distributes them to various organs and they then accumulate at specific locations.

The reduction in size of the ENPs results in a substantial increase in the specific surface area and Gibbs free energy, which reflects the rapid increase of chemical reactivity as particle size reduces. It is easy to calculate that the surface energy of nanoparticles will be increased by a factor of one million as the size is decreased from millimeters to nanometers. Therefore, there is a reconsideration of the traditional concepts of toxic risk assessment, which is normally based on the dose:response ratio, with the dose expressed in terms of mass or concentration. Indeed, previous studies have shown that factors such as surface area, surface modification and number of particles are factors that should be considered in nanotoxicity assessment. Furthermore, contrary to common toxicity studies with particles with larger sizes, the initial exposure dose can involve a degree of uncertainty because nanoparticles can agglomerate into larger particles during the emission process or during translocation within the organism.

Classically, the exposure to a certain mass of the toxic agent during a specific time period is considered the critical factor to determine the hazard and assessing the risks. Normally, toxic effects are correlated to the quantity of product to which individual animals or humans are exposed. The greater the mass absorbed, the greater the effect. In the case of nanoparticles, it has been clearly shown that measured effects are not linked to the mass of the product, which challenges our entire approach to the classical interpretation of toxicity measurement.[14] Although several studies find a good correlation between the specific surface and the toxic effects, a consensus seems to be emerging in the scientific community that several factors can contribute to the toxicity of these products and that it is currently impossible, with our limited knowledge, to weigh the significance of each of these factors or predict the precise toxicity of a new nanoparticle. It is still a matter of debate what is the most relevant parameter for the dose–response analysis in nanotoxicology. It has been reported that for the same dose in terms of mass, the toxic response increases as a function of the decreasing nanoparticle size. Under these investigations, the surface area has been proposed as the adequate dose metric for analyzing the impact of engineered nanoparticles in a linear dose-response relationship.[15,16]

9.1.3 Inhalation of Engineered Nanoparticles

The exposure to environmental particles has been a health hazard ever since humans began to use fire and we have been inhaling considerable amounts of particles in the form of smoke or dust since then. Since the industrial revolution in the 19th century the amount of potentially noxious nanoparticles that we have been breathing has increased in a significant manner. In the latest years novel nanotechnologies have also included the presence of engineered nano-particles to the total amount of breathable environmental particles with a deeper impact in human health.[17] Since engineered nanoparticles (ENPs) are increasingly being applied in many consumer products such as in cosmetics, textiles, and paints, the human and environmental exposure to ENPs is becoming probable.[18]

Furthermore, the use of nanoparticles as drug-delivery carriers has been known for a long time and a large literature is devoted to the development of ENPs with sizes, surface functionalities and compositions that enhance their medical applicability. The respiratory system is usually the main entrance for nanoparticles to the human body. The deposition of particles in the lungs varies considerably according to the granulometry of ultrafine dusts and, normally, the deposition of the coarse particles of work environments in the alveolar region increases as particle diameter decreases, reaching a maximum value of around 20% for 3 μm. This situation could lead to the misconception of a deep-lung deposition of small nanoparticles.

Nanoparticles with aerodynamic diameters smaller than 1 nm cannot reach the alveoli and 80% of them are deposited in the nose and pharynx. The remaining 20% are trapped in the tracheobronchial region. At this size, retention of inhaled nanoparticles is nearly 100%.[19] By increasing the particle size to 5 nm, 90% of all inhaled particles are retained in the lung and then are deposited in the three regions with relative uniformity. Total pulmonary absorption of 20 nm nanoparticles decreases to 80% but more than 50% of 20 nm nanoparticles are deposited in the alveolar region. This means that 20% of inhaled particles penetrate the lung but leave it during exhalation. Particle granulometry thus has a major impact on the pulmonary deposition site.[20] In several nanoparticle production processes, the granulometry can also vary considerably according to the stage of production. Thus, even though the mass of 20-nm ultrafine particles deposited in the alveolar region represents over 50% of the total mass, the deposited dust concentration, expressed in lung surface units, will still be over 100 times greater in the nasal region and more than 10 times greater in the tracheobronchial region.[21] The quantity of particles and the particle deposition site in the pulmonary system are also influenced significantly by the presence of pre-existing lung diseases.[22]

9.2   Risk-Reduction Strategies

The potential exposure to ENPs can be controlled in research centers and industries through an adaptive risk-management program. Such programs should provide the framework to anticipate the emergence of nanotechnologies in the design of laboratories, recognize the potential hazards, evaluate the exposure to ENPs, develop controls to prevent or reduce exposure and confirm the efficiency of those controls. Exposure assessment is therefore the basic element of an effective risk-management scheme. Those tasks contributing more to exposure and workers conducting them should be adequately identified and a register of tasks should be developed. This inventory includes information on the duration and frequency of tasks, the amount of the nanomaterial being handled and its physical state and dustiness.[23]

Occupational exposure limit values (OELs) are a set of recommendations given by competent national authorities or other relevant national institutions as limits for concentrations of hazardous compounds in workplace air. OELs for hazardous substances represent an important tool for risk assessment and management and valuable information for occupational safety and health activities concerning hazardous substances. These can apply both to marketed products and to waste and byproducts from production processes, setting limits to protect against health effects, but do not address safety issues such as flammable concentrations. Nevertheless, in the case of nanotechnology there is no clear indication of the adequate metric for determining the impact in health of a specific nanoparticle. In fact, the majority of chemical substances that can be found in both research and industry have no established OELs. In this case, employers and workers often lack the necessary guidance on the extent to which occupational exposures should be controlled. This is especially interesting in an emerging field such as nanotechnology, where materials and applications are rapidly moving forward.

9.2.1   Prevention through Design and Good Laboratory Practices

Anticipating potential safety and health hazards early in the development of the technology or process is a key point for a risk-reduction scheme. Moreover, the incorporation of those safe practices into all design, implementation and operation phases have to be considered. Prevention through design (PtD) is a management tool to protect workers from potentially unsafe work conditions.

The PtD scheme addresses occupational safety and health needs by eliminating hazards and minimizing risks to workers throughout the life cycle of the process.[24] Many nanotechnology research laboratories recognize PtD as a cost-effective means to enhance occupational safety and health and have incorporated PtD management practices within their facilities.[25] PtD strategies follow the standard hierarchy of controlling workplace hazards, which includes (1) eliminating, substituting, or modifying the nanomaterials; (2) engineering the process to minimize or eliminate exposure to the nanomaterials; (3) implementing administrative controls that limit the quantity or duration of exposure to the nanomaterials; and (4) providing use of adequate personal protective equipment.

A set of adequate regulations should be implemented in all laboratories and installations where chemical reactions are being performed. The most common as well as easy to implement are the good laboratory practices (GLP) defined as a worldwide regulatory requirement primarily used in studies that are undertaken to generate data by which the hazards and risks to users, consumers and the environment are assessed.[26] In fact, GLPs are a set of principles that provides a framework within which laboratory studies are planned and archived and by definition is referred to the testing of chemicals in an OECD member country in accordance with OECD Test Guidelines. GLP helps assure regulatory authorities that the data submitted are a true reflection of the results obtained during the study and can therefore be relied upon when making risk and safety assessments.

9.3 Safety and Prevention in the Nanotechnology Laboratory

Handling nanomaterials in the workspace is a complex subject that could imply the exposure to potentially hazardous matter. In nanotechnology research, it is furthermore complex to eliminate or substitute the nanomaterial. However, some aspects of the process could be modified in a way that reduces release of the nanomaterial to the working environment. For instance, working with ENPs suspended in a liquid is a significant improvement over working with them in dry powder form, because the potential for airborne release is reduced in most laboratory processes, although physical agitation of the liquid by sonication may aerosolize small droplets containing the ENPs.[27]

Whenever possible, the use of hazardous substances should be eliminated or those materials should be substituted for less-hazardous forms. Research in the synthesis and handling of ENPs often requires the use of solvents and other potentially hazardous chemicals and researchers should select those chemical processes that utilize innocuous or less-toxic alternatives, in order to minimize worker exposures and environmental releases when the process is scaled up. While nanotechnology continues to be an engine for economic growth through its use in innovative products, all parties need to find out more about the environmental, health and safety risks arising from nanomaterials during their life cycle. The industrial interest is to keep pace with technological developments in order to promote risk awareness and management. In this section, we will explore the most common approaches for safety and prevention in nanotechnology, namely control banding and nanoparticle emission assessment technique.

9.3.1 Control Banding

Since there are no relevant exposure limits that could be considered in the workplace, the strategy that is used in this situation is the control banding (CB). This is a qualitative strategy to assess and manage hazards associated with chemical exposures in workplaces. As a generic approach, the control measure (*e.g.* dilution ventilation, engineering controls, containment, *etc.*) is based on a range or "band" of hazards (skin/eye irritant, very toxic, carcinogenic, *etc.*) and exposures (small, medium, large exposure). This method for controlling workers' exposure is based on the fact that there is limited number of control approaches and that many problems have been met and solved before. This latter aspect, in the case of nanotechnology, requires a deep knowledge of the characteristics and properties of the nanomaterials considered. As the CB approaches use the solutions that experts have developed to control an earlier occupational exposure to specific chemicals that are closely related to the synthesized nanoparticles and suggesting them to other tasks with similar exposure situations, this is therefore a procedure that focuses resources on exposure controls and describes how strictly a risk needs to be managed.

CB tools and strategies are essentially based in the grouping of the exposure to specific nanomaterials according to similar physical and chemical characteristics, planned operations and foreseen situations (amount of ENPs and exposure way). Control strategies for risk management are then determined for each group. One of the most common forms of CB, which has been recently proposed for nanotechnology applications, is based in a four-level hierarchy of risk management options:[28,29] (1) Good occupational hygiene practices, which is enhanced using the appropriate personal protective equipment; (2) Engineering controls, including local exhaust ventilation; (3) Containment; and (4) Seek specialist advice.

The correct determination of the control strategy requires information on the characteristics of a specific ENP and parent chemical substances, together with the potential for exposure (quantity in use, dustiness and the relative hazard as described in what is known as a risk phrase, or R-phrase). Determining potential exposures for airborne particulates or vapors involves characterizing the process or activity in which the substance is used. CB tools must be used in conjunction with health and safety practices such as substitution. Substitution for a less-hazardous material or precursor for the synthesis of the ENPs is highly recommended to prevent exposure. It is important to note that CB does not replace the experts in occupational safety and health or eliminate the need for exposure monitoring. In fact, the application of a CB approach recommends the use of professionals to provide recommendations.

9.3.2 Nanoparticle Emission Assessment Technique

The lack of exposure limits specific to ENPs significantly reduces the applicability of CB tools in nanotechnology. In addition, there is also an incomplete international consensus on standards and measurement techniques that should be considered for nanomaterials in both occupational environments and consumer products. However, the interest of research centers and industrial facilities devoted to the production and use of nanomaterials in determining the potential for exposure is currently growing. The National Institute for Occupational Health and Safety (NIOSH) has developed the nanoparticle emission assessment technique (NEAT) to evaluate the concentration of airborne ENPs in the workplace.[30] The NEAT approach is based on the direct reading of the release ENPs using aerosol spectrometers, coupled with air filtering to perform chemical and microscopic analysis for particle identification and chemical speciation. The purpose of NEAT is listing target areas, processes or tasks that involve a higher concentration of airborne ENPs, identifying the source of nanomaterial emission depending on the processing method and occupational procedures. As was mentioned for CB tools, an adequate understanding of ENPs is needed through material safety data sheets, amount of nanomaterial synthesized or handled and the available literature on both precursors and final product. Once potential sources of emissions have been identified and the nature of the ENPs and reagents are known, the procedure for a NEAT analysis involves the following steps:[31] (1) Observational survey of the processing area to identify tasks that require air sampling; (2) analysis of the frequency and duration of each operation and the type of equipment used for handling and containment of the ENPs; (3) determination of the presence or absence of exhaust ventilation, both general and local; (4) determination of the points where containment is deliberately breached such as for product retrieval or for cleaning. In the latter steps the existence of potential system failure points that could result in emission from the containment/control system, for instance holes in ducts or damaged joints are also considered.

NEAT is a useful tool for health and safety professionals to define whether a release and potential exposure to ENPs occur in the workplace. Several direct-reading instruments are used in a parallel and differential manner to evaluate the total particle number concentrations relative to background and the relative size distributions of the particles. If this initial evaluation indicates an elevated number of small particles, which could potentially be the ENPs of interest, then instruments are used to identify the source of the emissions. Based on this identification step, filter-based air samples are collected for qualitative analysis of particle size, shape, and morphology using TEM or SEM analysis and for the determination of mass concentration by chemical analysis.[32]

9.4 Conclusions

Nanotechnology is paving the way of a new industrial revolution. Materials synthesized with controlled sizes and tailored properties are solving a myriad of concerns in both science and technology that just ten years ago seemed impenetrable. Also, the development of nanomaterials for daily-life products, together with novel procedures and techniques that give nanoparticles their properties, are producing an impact in a society every day more aware about science and technology.

However, those features of engineered nanoparticles that bestow remarkable capacities in several research fields also induce an enhanced toxicological response upon exposure by inhalation or several other routes. The design of risk-reduction strategies for engineered nanoparticles is therefore an urgent need for a safer nanotechnology. Although exposure reduction by elimination is the best way to reduce risks, most of the time such a procedure it is not possible or even desirable.

Currently, the most common approaches for risk reduction in the nano realm come from the adaptation of the control measures used for the chemical industry. However, the lack of specific regulations and normative regarding nanomaterials reduces their applicability. The design of novel assessment methods and procedures for engineered nanomaterials depends on the measurement of the appropriate parameters of nanoparticles using specific techniques.

References

1. Lux Research, Global Trends in Nanotech and Cleantech, 2010 (www.luxresearchinc.com).
2. T. Satterfield, M. Kandlikar, C. E. Beaudrie, J. Conti and B. Herr Harthorn, *Nature Nanotech.*, 2009, **4**, 752.
3. K. Schmid and M. Riediker, *Env. Sci. Technol.*, 2008, **42**, 2253.
4. K. Savolainen, H. Alenius, H. Norppa, L. Pylkkänen, T. Tuomi and G. Kasper, *Toxicology*, 2010, **269**, 92.
5. A. E. Nel, T. Xia, L. Mädler and N. Li, *Science*, 2006, **311**, 622.
6. T. Cedervall, I. Lynch, M. Foy, T. Berggård, S. C. Donnelly, G. Cagney, S. Linse and K. A. Dawson, *Angew. Chem. Int. Ed.*, 2007, **46**, 5754.
7. F. Balas, M. Arruebo, J. Urrutia and J. Santamaría, *Nature Nanotech.*, 2010, **5**, 93.
8. T. M. Peters, S. Elzey, R. Johnson, H. Park, V. H. Grassian, T. Maher and P. O'Shaughnessy, *J. Occup. Env. Hyg.*, 2009, **6**, 73.
9. M. Seipenbusch, A. Binder and G. Kasper, *Ann. Occup. Hyg.*, 2008, **52**, 707.
10. A. Elder, I. Lynch, K. Grieger, S. Chan-Remillard, A. Gatti, H. Gnewuch, E. Kenawy, R. Korenstein, T. Kuhlbusch, F. Linker, S. Matias, N. Monteiro-Riviere, V. R. S. Pinto, R. Rudnitsky, K. Savolainen and A. Shvedova. In *Nanomaterials: Risks and Benefits.* ed. I. Linkov and J. Steevens, Springer, Dordrecht, 2009, p. 3.
11. A. E. Nel, L. Mädler, D. Velegol, T. Xia, E. M. Hoek, P. Somasundaran, F. Klaessig, V. Castranova and M. Thompson, *Nature Mater.*, 2009, **8**, 543.

12. USEPA. *Terms of the Environment*. U.S. Environmental Protection Agency, 2004 http://www.epa.gov/OCEPAterms/.
13. M. Simkó and M. O. Mattsson, *Part. Fibre Toxicol.*, 2010, **7**, 42.
14. SCENIHR, *Risk Assessment of Products of Nanotechnologies, Scientific Committee on Emerging and Newly Identified Health Risks*, 2009. http://ec.europa.eu/health/ph_risk/committees/04_scenihr/docs/scenihr_o_023.pdf.
15. G. Oberdörster, E. Oberdörster and J. Oberdörster, *Env. Health Perspect.*, 2005, **113**, 823.
16. T. Stoeger, C. Reinhard, S. Takenaka, A. Schroeppel, E. Karg, B. Ritter, J. Heyder and H. Schulz, *Env. Health Perspect.*, 2006, **114**, 328.
17. T. Xia, N. Li and A. E. Nel, *Ann. Rev. Pub. Health*, 2009, **30**, 137.
18. B. Nowack and T. D. Bucheli, *Env. Pollut.*, 2007, **150**, 5.
19. O. Witschger and J. F. Fabriès, *Hyg. Sec. Trav.*, 2005, **199**, 21.
20. G. Oberdörster, *Int. Arch. Occup. Env. Health*, 2001, **74**, 1.
21. G. Oberdorster, A. Maynard, K. Donaldson, V. Castranova, J. Fitzpatrick, K. Ausman, J. Carter, B. Karn, W. Kreyling, D. Lai, S. Olin, N. Monteiro-Riviere, D. Warheit and H. Yang, *Part. Fiber Toxicol.*, 2005, **2**, 8.
22. A. D. Maynard and E. D. Kuempel, *J. Nanopart. Res.*, 2005, **7**, 587.
23. *General safe practices for working with engineered nanomaterials in research laboratories*, National Institute of Occupational Safety and Health, 2012. http://www.cdc.gov/niosh/docs/2012-147/pdfs/2012-147.pdf.
24. P. Schulte, R. Rinehart, A. Okun, C. Geraci and D. Heidel, *J. Safety Res.*, 2008, **39**, 115.
25. V. Murashov and J. Howard, *Nature Nanotechnol.*, 2009, **4**, 467.
26. http://www.nanotox.com/about-us/good-laboratory-practices-glp.html.
27. D. R. Johnson, M. N. Methner, A. J. Kennedy and J. A. Steevens, *Env. Health Perspect.*, 2010, **118**, 49.
28. D. M. Zalk, S. Y. Paik and P. Swuste, *J. Nanopart. Res.*, 2009, **11**, 1685.
29. D. M. Zalk and S. Y. Paik, in *Assessing Nanoparticle Risks to Human Health*, ed. G. Ramachandran, Elsevier, Waltham MA 2011, p. 139.
30. http://goodnanoguide.org/Nanoparticle+Emission+Assessment+Technique+-+NEAT.
31. M. Methner, *J. Occup. Env. Hyg.*, 2008, **5**, D63.
32. A. L. Miller, P. L. Drake, P. J. Hintz and M. C. Habjan, *Ann. Occup. Hyg.*, 2010, **54**, 504.

Subject Index